ION EXCHANGE
AND SOLVENT EXTRACTION

ION EXCHANGE
AND SOLVENT EXTRACTION

A SERIES OF ADVANCES

Volume 12

EDITED BY

Jacob A. Marinsky
Department of Chemistry
State University of
New York at Buffalo
Buffalo, New York

Yizhak Marcus
Department of Inorganic Chemistry
The Hebrew University
Jerusalem, Israel

CRC Press
Taylor & Francis Group
Boca Raton London New York

CRC Press is an imprint of the
Taylor & Francis Group, an **informa** business

CRC Press
Taylor & Francis Group
6000 Broken Sound Parkway NW, Suite 300
Boca Raton, FL 33487-2742

First issued in paperback 2019

© 1995 by Taylor & Francis Group, LLC
CRC Press is an imprint of Taylor & Francis Group, an Informa business

No claim to original U.S. Government works

ISBN-13: 978-0-8247-9382-1 (hbk)
ISBN-13: 978-0-367-40184-9 (pbk)

Visit the Taylor & Francis Web site at
http://www.taylorandfrancis.com

and the CRC Press Web site at
http://www.crcpress.com

Preface

This twelfth volume of the *Ion Exchange and Solvent Extraction* series provides the reader with timely considerations of important aspects of the ion-exchange phenomenon.

In-depth examinations and expansions of important topics introduced in earlier volumes are presented. Significant applications of the ion-exchange phenomenon continue to be emphasized, while new and important directions of research in various aspects of ion exchange are introduced to the reader.

The authors of the chapters in this volume have sought to extend the boundaries of the topics under consideration. As a consequence, the presentations continue to be broader and more informative than one ordinarily encounters in technical and review papers.

An important feature of Volume 12 is that it makes current studies of the ion-exchange phenomenon by Russian and Chinese scientists easily accessible to the reader. The fact that much of their research remains inaccessible to many researchers has been especially frustrating to the authors of these studies.

In Chapter 1, there is a scholarly presentation of the development of an important application of the ion-exchange phenomenon. Liquan Chen and colleagues detail the development of their high-pressure ion-exchange separation of the rare earths. They describe how fundamentally based studies improve the efficiency of separation by a factor of hundreds while reducing cost and the time of separations. This chapter also provides an example of the kind of research and development that is an essential part of any successful ion-exchange process.

In Chapter 2, Vladimir Gorshkov considers the advantages that accrue from the countercurrent movement of resin and solution when applying ion-exchange technology to large-scale industrial tasks. Countercurrent columns are easily automated. The quantity of exchanger needed is less, as are the expenditures for reagents to regenerate exchangers. The dimensions of equipment are also reduced. In defining the optimal application of this technique to a spectrum of separation problems, Gorshkov shows how to obtain maximum efficiency through the selective use of countercurrent operations. The nature of the separation problem defines the particular countercurrent steps to be taken.

In Chapter 3, Ruslan Khamizov and coworkers describe the economic recovery of minerals from seawater and brines formed in desalination plants and solar evaporation operations. The employment of ion-exchange technology for this purpose is critically reviewed. They carefully examine the economic aspects of optimal ion-exchange recovery of minerals from these sources by combining various techniques of the operation discussed in Chapter 2. The great rapid depletion of land-based mineral resources of the world make this a problem of great interest.

A. I. Kalinitchev, in Chapter 4, deals with the kinetics of ion exchange in selective systems. In particular he uses the kinetic model developed in his research to show that phenomenological regularities and criteria describing intraparticle diffusion kinetics for conventional ion exchange are not applicable to selective ion-exchange systems.

In Chapter 5, Hirohiko Waki describes the employment of spectroscopic methods to compare the spectra of complexes formed in the resin phase with the spectra of complexes formed concurrently in the solution phase at equilibrium. By bringing the concentration levels of electrolyte, ligand, and metal ion in the two phases as close as possible, differences in the spectra observed are attributable to differences in the physical environment provided by the two phases. Waki attributes the observed differences to the lowering of the dielectric in the aqueous medium of the exchanger phase by the organic matrix of the exchanger.

In Chapter 6, K. Bunzl examines the kinetics of ion exchanger in heterogeneous systems in order to provide a better understanding of the behavior of natural ion exchangers in the soil (clay minerals, humic substances, sesquioxides). With film diffusion as the rate-controlling factor in these mixtures, several important characteristics become predictable.

Tohru Miyajima, in Chapter 7, presents an in-depth examination of the nature of ion-binding equilibria in charged polymer, simple electrolyte systems. He clearly shows that Gibbs-Donnan–based concepts, applicable to systems in which the boundary between the electrolyte solution and the charged polyion gel is well defined, can be employed with equal success to linear charged polyion, simple salt, and even oligomer salt systems in which no distinct boundary is observable between them. In such systems, the counterion concentrating region projected to form around the polyion skeleton is separated from the bulk solution by the invisible boundary. The range of applicability of the Gibbs-Donnan–based concepts has been studied as a function of polyion dimension, shape, and functionality.

In Chapter 8, Zuyi Tao, in order to provide a better understanding of the ion-exchange behavior of amino acids, has compiled their particular acid–base properties, their solubility in water, their partial molal volumes, and their molal activity coefficients in water at 25°C. This information has been used in Gibbs-Donnan–based equations to facilitate a better understanding of the mechanism of amino acid uptake by ion exchangers at low and high solution concentration levels. Measurement of distribution coefficients and separation factors are also described. The eventual resolution of thermodynamic ion-exchange functions (ΔG, ΔH, and ΔS) is provided for the reader.

M. Abe has in Chapter 9 provided sizable insights with respect to the selectivity of inorganic ion-exchange materials. Various factors influence the selective characteristics of a particular exchanger, e.g., method and temperature of preparation, aging, drying, and solution media. Knowledge of such background material is essential for its evaluation as a separation tool. It is pointed out that the absence of such information can mask the potential of an inorganic exchanger for particular separations.

Jacob A. Marinsky
Yizhak Marcus

Contributors to Volume 12

Mitsuo Abe   Tsuruoka National College of Technology and Tokyo Institute of Technology, Tsuruoka, Yamagata, Japan

K. Bunzl   GSF-Forschungszentrum für Umwelt und Gesundheit, Institut für Strahlenschutz, Neuherberg, Germany

Liquan Chen   Department of Modern Physics, Lanzhou University, Lanzhou, People's Republic of China

Changfa Dong   Department of Modern Physics, Lanzhou University, Lanzhou, People's Republic of China

Vladimir I. Gorshkov   Department of Chemistry, Moscow State University, Moscow, Russia

A. I. Kalinitchev   Institute of Physical Chemistry, Russian Academy of Science, Moscow, Russia

Ruslan Khamizov   Vernadsky Institute of Geochemistry and Analytical Chemistry, Russian Academy of Science, Moscow, Russia

Tohru Miyajima   Department of Chemistry, Faculty of Science, Kyushu University, Hakozaki, Higashi-ku, Fukuoka, Japan

Dmitri N. Muraviev   Department of Organic Chemistry, The Weizmann Institute of Science, Rehovot, Israel

Zuyi Tao   Department of Modern Physics, Lanzhou University, Lanzhou, People's Republic of China

Hirohiko Waki   Professor Emeritus of Chemistry, Kyushu University, Hakozaki, Higashi-ku, Fukuoka, Japan

Abraham Warshawsky   Department of Organic Chemistry, The Weizmann Institute of Science, Rehovot, Israel

Wangsuo Wu   Department of Modern Physics, Lanzhou University, Lanzhou, People's Republic of China

Wenda Xin   Department of Modern Physics, Lanzhou University, Lanzhou, People's Republic of China

Sujun Yue   Department of Modern Physics, Lanzhou University, Lanzhou, People's Republic of China

Contents

Contents of Other Volumes

1

High-Pressure Ion-Exchange Separation of Rare Earths

Liquan Chen, Wenda Xin, Changfa Dong, Wangsuo Wu, and Sujun Yue

Lanzhou University, Lanzhou, People's Republic of China

I. INTRODUCTION

A. A Review of the Ion-Exchange Separation of the Rare Earths

The elements with atomic numbers from 57 (lanthanum) to 71 (lutetium) are referred to as the lanthanide elements. These elements and two others, scandium and yttrium, exhibit chemical and physical properties very similar to lanthanum. They are known as the rare earth elements or rare earths (RE). Such similarity of the RE elements is due to the configuration of their outer electron shells. It is well known that the chemical and physical properties of an element depend primarily on the structure of its outermost electron shells. For RE elements with increasing atomic number, the first electron orbit beyond the closed [Xe] shell ($6s^2$) remains essentially in place while electrons are added to the inner 4f orbital.* Such disposition of electrons about the nucleus of the rare earth atoms is responsible for the small effect an atomic number increase from 57 to 71 has on the physical and chemical properties of the rare earths. Their assignment to the 4f orbital leads to slow contraction of rare earth size with increasing atomic number. The 4f orbitals of both europium and gadolinium are half occupied [Xe] ($4f^76s^2$) and [Xe] ($4f^75d^16s^2$), so that there

---

*Lanthanum [Xe] ($5d\ 6s^2$), gadolinium [Xe] ($4f^75d\ 6s^2$), and lutetium [XE]($4f^{14}5d\ 4s^2$) are exceptions.

is even greater similarity between them than between the other neighboring pairs of rare earths.

The above phenomenon is referred to as the "lanthanide contraction." It is responsible for the much smaller differences in the RE properties, such as the acidity of the elements, the solubility of their salts, the stability of their complexes, etc., than are normally associated with neighboring elements.

With the development of special uses for each RE element, the need for their separation from each other became greater. Many methods, such as selective oxidation and reduction, decomposition of fused salts, fractional crystallization, ion exchange and solvent extraction, and so on, were developed for this purpose. The first method with which the separation of the individual pure rare earths from each other became feasible was the ion-exchange method. It was initially developed to separate the RE elements during the 1940s [2–6]. It became widely used because of its simplicity, the high purity of products and the controllable operating environments. The eluents used at that time were citric acid, NTA, and especially EDTA. Although there were many advantages inherent in the ion-exchange process, there were disadvantages as well. The process was not continuous; it was time consuming and production efficiency was low. When compared with solvent extraction methods available in the 1970s it suffered in the comparison and was almost displaced [7] in industrial processes used for their separation.

High-pressure ion-exchange chromatography (HPIEC) developed from Scott's experiments in 1968 [8]. By using a microsize resin particle with diameters from 5–10 μm as the stable phase, and transporting fluids under a pressure of 150 kg/cm^2, Scott succeeded in separating various ionic species in urine; he resolved as many as 140 chromatographic peaks. This new technique provided a significant breakthrough in classical ion-exchange chromatography (CIEC). HPIEC has been used successfully for separating transplutonium elements, RE elements, amino acids, and nucleic acids since Scott's experiment. It has been widely used in many fields, such as nuclear chemistry, radiochemistry, analytical chemistry, biochemistry, pharmaceutical chemistry, and environmental chemistry. We have been studying the basic theory of HPIEC and HPIEC separation of RE elements for many years [9]. In this chapter, we detail these studies.

B. A Comparison of Classical Ion-Exchange Chromatography with High-Pressure Ion-Exchange Chromatography

The differences between CIEC and HPIEC are that microsize resin beads are used as the stable phase; fluids are forced to flow through the ion-exchange column at a high flow rate by using a high-pressure pump; the separation process proceeds at a high temperature; and the process is controlled by using an on-line monitor system. Because of these innovations, HPIEC provides the following advantages.

1. *High Speed and High Efficiency*

HPIEC is a high-speed process characterized by rapid migration of the effluent band. It is also a high-efficiency process, measured by "the bed efficiency," usually defined by "the height equivalent of a theoretical plate" (HETP). The lower the HETP value, the higher is the "bed efficiency." For an ion-exchange displacement process this means that as HETP decreases the boundary between two neighboring zones becomes more precipitous; the overlapping area between them is smaller as a result.

To speed up a separation process one must accelerate the migration rate of the effluent band. The migration rate of a band depends on two factors: One factor is the linear flow rate, which is directly proportional to the migration rate of the band. The other factor is the distribution ratio, the ratio between the equilibrium concentration of an ion in the resin and aqueous phases, which is inversely proportional to the migration rate of the band. In a CIEC process, the linear flow rate cannot be increased because it may increase the magnitude of the HETP, and thus lower "bed efficiency."

In industrial CIEC processes, when the linear flow rate is less than 1.0 cm/min, the band can only migrate at a speed of 0.02 cm/min, but the HETP reaches a value as high as 0.86 cm. In HPIEC processes, the linear flow rate can be greatly increased because of the use of microsize resin beads. When the concentration of the chelate displacer (e.g., EDTA) is 0.075 mol/L, and the linear flow rate is 10 cm/min, the band can migrate 0.86 cm/min, while the HETP is only 0.40 cm. This observation illustrates the advantages of HPIEC, namely in high speed and high efficiency.

The influence of distribution ratio on the migration rate of the band is primarily determined by the size of the resin particle used. The magnitude of the HETP depends on the characteristics of the mass-transfer process encountered in the distribution of species between the resin and aqueous phases. The higher the speed of mass transfer, the lower is the HETP. In CIEC, larger resin particles are used and the resultant low mass-transfer rate of an ion in the resin phase contributes sizably to the rise in the HETP value. This gain in the HETP value depends on the distribution ratio. The smaller the distribution ratio, the slower is the mass transfer in the resin phase and the larger is its contribution to the HETP. Thus distribution ratios larger than 15 should usually be used in CIEC. For HPIEC, mass transfer in the resin phase can be neglected in the whole mass-transfer process because microsized resin is used. Thus distribution ratios barely influence the HETP over a large range of magnitudes. So very small distribution ratios can be used in HPIEC. For instance, ratios of 3–8 are usually used in the HPIEC displacement process. This also demonstrates the advantages of HPIEC, namely high speed and high efficiency.

The use of microsized resin particles is the principal factor that leads to the high speed and high efficiency of HPIEC. With the improvement of ion-exchange resin manufacturing techniques the resins used in industrial processes can be conveniently provided at the size (micro) and uniformity desired. Therefore advantage can easily be taken of at the high mass-transfer speed provided by microsize resin to improve greatly separation speed and efficiency. Other technical aspects of HPIEC, such as column filling, fluid transporting, and even the column design facilitate high speed and high efficiency as well.

2. Enhancement of Eluate Concentration

For the ion-exchange separation of the rare earths the chelate eluate most preferred is EDTA. The reason for this is that the stability constants of the complexes formed by the RE elements with EDTA increase more rapidly with RE atomic number than with other complexing agents (e.g., citric acid). Only the stability constants of the complexes of Gd ($4f^7 5d 6s^2$) and Eu ($4f^7 6s^2$) remain essentially unchanged. In CIEC, the concentration of EDTA used is usually less than 0.015 mol/L. It is produced at low concentration levels (only about 2 g/L) and the volume of solution produced is necessarily large. The consequence of this is lower efficiency and lower recovery of EDTA. The largest obstacle to increasing the concentration of EDTA in CIEC is the solubility factor, EDTA crystallizing at low concentration levels. Even though the pH of the EDTA solutions is elevated, higher concentrations of EDTA form crystals of EDTA easily in the separation process at room temperature and lower pressures because H^+ dissociated in the process may combine with salts of EDTA and form H_4Y, which is difficult to dissolve. Formation of H_4Y crystal in a column may block the flow of solution through the column, stopping the operation.

In HPIEC, higher concentrations of EDTA can be used as eluate. Some salt added to the EDTA increases the concentration of displacing ion, making the formation of H_4Y crystals more difficult; counteracting this somewhat, weak acid is added to enhance the buffer capacity of the EDTA. The fact that it becomes possible to increase the concentration of EDTA is a consequence, primarily, of the high pressure and high temperature employed. As mentioned above, the linear flow rate of the band can be increased by a factor of 10 as well. In our research, the concentration of EDTA has been increased to 0.10 mol/L. This is six times larger than the concentration of EDTA in CIEC. The migration rate of the separation band is 40 times larger than in CIEC, and the concentration of RE in the product solution is more than 15 g/L. Increasing the concentration of eluate reduces process complexity and makes the prognosis for the development of a successful ion exchange process for the highly effective separation of rare earths most promising.

3. Increasing Temperature and Removing Gases

The mass-transfer rate is slow in CIEC. By increasing temperature this parameter is sizably enhanced and the rate of ion exchange becomes more rapid. The increase in temperature, however, may cause the evolution of gas bubbles from the solution and result in their entry into column systems. The gases are formed in chemical reactions and are due to the air dissolved in the solutions employed. They can influence the stability of fluid flow, distort the bandshape, and even promote the formation of cavities in the exchange bed, thereby disturbing the separation process. Because the bubbles rise while the fluid flow is down in a column, it is difficult to remove these bubbles in CIEC.

Increasing the temperature in HPIEC lowers the pressure buildup in a column. Table 1 indicates that at 70°C it is half that experienced at 30°C. Because higher pressure is employed in HPIEC, the gas cannot

Table 1 Relationship between Temperature and Pressure Reduction (Z_0 = 84 cm, r_0 = 20–30 µm, $\bar{V}$ = 16 cm/min)

Temperature (°C)	Viscosity coefficient of water	ΔP (meas.) (kg/cm^2)	ΔP (calc) (kg/cm^2)
30	0.802	50.5	51.8
35	0.721	46.5	46.5
40	0.653	42.5	42.2
45	0.596	38.5	35.5
50	0.550	35.5	35.5
55	0.507	32.5	32.7
60	0.470	30.2	30.3
65	0.436	28.3	28.2
70	0.406	26.0	26.2

Pressure reduction (Δp) was calculated with the Ergun equation (12):

$$\frac{\Delta P}{\Delta Z_0} = \frac{150\eta\bar{V}(1-\alpha_0)^2}{4r_0^2\alpha_0^2}$$

where ΔP = pressure reduction, ΔZ_0 = height of resin bed, r_0 = radius of resin particle, α_0 = porosity coefficient of resin bed, η = viscosity coefficient of water.

emerge from the solution. It is removed at the bottom of the column with the rapidly flowing fluid, even at higher temperatures. As a result gas bubbles rarely appear in HPIEC.

4. *Column Stability, Simple Operation, and Process Control*

In CIEC, the resin bed swells and shrinks as ion-exchange proceeds. Since gas bubbles evolve in the process, it is difficult to control the flow rate exposed only to the pressure due to a flowing fluid. In HPIEC, the flow rate of a fluid is kept constant by transporting the fluid with a high-pressure pump. As a result, the migration rate of the separation is constant. Thus HPIEC processes are easily controlled. For example, when the compositions of separated species are known and the concentration of eluate and its flow rate are given, the times when breakthrough occurs in a column and when fluid flow needs to be transferred from one column to another as well as the production yield can be calculated. Moreover, on-line monitoring techniques are easily carried out in HPIEC.

5. *Separation Advantages Provided by HPIEC*

The advantages of using HPIEC for the separation of RE elements are that it shortens the separation time, increases the recovery of products, minimizes the need for complicated equipment, and makes the process amenable to automatic control. To illustrate this the separations of Tm-Yb and Lu-Yb are presented as examples. The results obtained with high pressure and classical chromatography are compared in Tables 2 and 3, respectively, for this purpose.

From the results obtained we can see that when the concentration of the eluate is identical and the flow rate in HPIEC is 5–10 times greater than that employed in CIEC for the separation of Tm-Yb the recovery of Yb in HPIEC is considerably higher than in CIEC. The Tm can be obtained as product in HPIEC but not in CIEC as well to demonstrate the separation advantage afforded by HPIEC.

Table 2 Results of Separating Tm and Yb with CIEC and HPIEC

Mode	Linear flow rate (cm/min)	Percent recovery of Yb_2O_3			Percent recovery of Tm_2O_3	
		>99.9%	>99%	98%	>98%	>97%
CIEC	1	25.27	60.35	68.82	—	—
HPIEC	5	39.90	67.13	79.98	50.23	62.15

Table 3 Separation Yb and Lu in CIEC and HPIEC

Mode	Linear flow rate (cm/min)	Composition of the sample (%)		Percent recovery of Yb$_2$O$_3$		Percent recovery of Lu$_2$O$_3$
		Yb$_2$O$_3$	Lu$_2$O$_3$	>99%	>98%	>98% (purity)
HPIEC	5	75.59	24.40	43.77	65.33	37.10
CIEC	0.5	88.00	—	—	—	—

II. EQUIPMENT AND EXPERIMENTAL METHODS IN HPIEC

The equipment used in HPIEC is different from that used in CIEC. The columns, valves, and pipes are exposed to high pressures. This requires them to be not only corrosion resistant to acids, bases, and other chemical agents, but they also need to be pressure resistant. Because of this, the materials used in their construction are required to be of higher quality than the material used to construct the CIEC devices. The devices, e.g., the column, valves, pipes, sieves, pressure vessels, and pressure gauges used by the authors, were primarily made of special stainless steel. Some junctions were made of polytetrafluoroethylene plastic. Usually the last column of a system of columns was made of glass so as to permit observation of the separation affected.

A. Three-Column HPIEC System

The three-column system is primarily used in basic studies and simple separations. Important parameters and functions are listed in Table 4 while its configuration is shown in Fig. 1.

Table 4 Column Parameters and Functions of HPIEC Three-Column System

Column number	Material	Function	Cross section (cm^2)	Height of column (cm)	Height of resin bed (cm)	Diameter of resin (mm)
1	Stainless steel	Adsorption and separation	2.90	110	100	40–60
2	Stainless steel	Separation	1.13	110	100	40–60
3	Glass	Separation	0.38	100	80	30–40

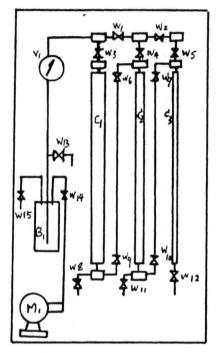

M_1: high pressure pump
V_1: pressure gauge
B_1: charing hooper
C_1: adsorption columns
C_2, C_3: separation columns
W_1-W_{15}: valves

Figure 1   HPIEC three-column system.

The first column provides the two essential functions of adsorption and separation. If the mixed species are easily separated lower column ratios are required and the loading capacity can be increased or reduced. For a basic study only the third column, a glass one, is used. The change of bandshape can, as a result, be directly observed while the migration rate of the separation band can be easily monitored by watching its movement in this glass column.

B. Five-Column HPIEC System

The five-column HPIEC system, because of the sizable scope of its functions, was used by the authors to study those rare earth separations that were too difficult to examine with the three-column system. The largest rare earth oxide sample employed with it was 50 g.

Every column in the five-column system could be used separately as well as together, the various column combinations depending on the particular sample under investigation. The maximum discharge rate of the pump, available to all columns, was 10 l/hr, while the minimum discharge rate was 0.15 l/hr.

Figure 2 and Fig. 3 show its appearance and the details of its structure, respectively. Table 5 lists its parameters and functions.

C. Apparatus and Chemicals

1. *Apparatus*
a. 730 UV-visible spectrophotometer (made in Shanghai, China)
b. N664A scintillation counter (made in Great Britain)
c. N530 scaler/timer (made in Great Britain)
d. pH meter (made in Shanghai, China)

Figure 2 HPIEC five-column system.

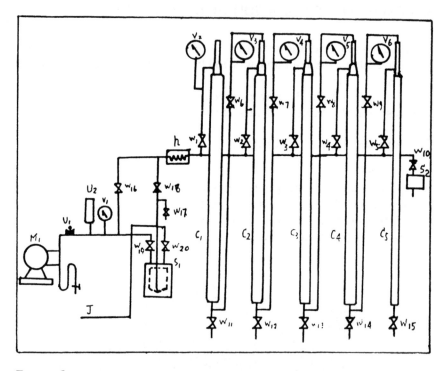

Figure 3 HPIEC five-column system: C_1, adsorption column; C_2–C_5, separation column; M_1, high-pressure metrical pump; U_1, countercurrent-stopping valve, U_2, safety valve; S_1, displacer hopper; S_2, volume extending device; H, preheating device; V_1–V_6, pressure gauges; W_1–W_5, loading valves; W_6–W_9, connecting valves; W_{11}–W_{15}, discharging valves; W_{17}, exhaust valve; W_{10}, valve of volume extending device; J, junction to vacuum pump.

Table 5 Column Parameters and Functions of HPIEC Five-Column System

Column number	Material	Function	Cross section (cm²)	Height of column (cm)	Height of resin bed (cm)	Diameter of resin (mm)
1	Stainless steel	Adsorption	4.45	100	75.0	60–70
2	Stainless steel	Separation	4.49	100	75.0	40–60
3	Stainless steel	Separation	2.18	100	76.0	40–60
4	Stainless steel	Separation	0.80	100	78.7	30–40
5	Glass	Separation	0.36	100	85.0	30–40

2. Chemicals

a. Different RE_2O_3 with purities of 99.5% to 99.99%.

b. Tracer elements: Tm-170; Yb-(169, 175); Tb-166; Eu-(152, 154); Gd-153; Y-90.

c. Other chemicals are analytical or chemical reagent grade.

3. Analytical Methods

Spectrophotometric and tracer-labeled methods were primarily used to define the migration boundaries of more than 10 REs. The first approach was used to assess Pr, Nd, Sm, Dy, Ho, and Er concentrations. The second method was used for monitoring Tb, Gd, Eu, Yb, Tm, and Y. The other RE elements, inaccessible with either of these methods, were determined by difference. The total concentration of RE in a particular solution was evaluated by titrating the free EDTA in solution with a standard solution of $Y(NO_3)_3$ or $Zn(NO_3)_2$. The respective concentrations of two overlapping RE elements were determined in the following manner: after determining the concentration of the one rare earth accessible with the labeled or spectrophotometric method the concentration of the other one became determinable by subtracting the concentration of the measured RE from the total RE concentration made accessible by titration. The concentration of a plateau (see Fig. 9) was verified in the following way: a given volume of solution from the plateau was oven dried and then ignited to obtain RE_2O_3, the RE_2O_3 was then dissolved in HNO_3, and the concentration of RE was determined by titration with a standard EDTA solution.

4. Experiment

a. Filling columns with resin Microsize resin particles of uniform dimensions have become synthesizable with the advances made in synthetic technique. Before their use, however, they had to be examined under the microscope to assure the size distribution sought. If size differences were too large, the resins had to be classified by flotation or other size-determining methods. The largest quantity of resin was used to fill the loading column, smaller quantities of resin were used in the separation column while the smallest quantity was added to the last column, which was the one from which the product was obtained. Before using the resin it had to be allowed to swell. It was first treated with acid and base; then it was brought to constant weight and its exchange capacity was determined; finally resin particles of the desired size were transferred to different columns at quantities defined by the function of the column. When the columns were filled the resins were suspended in salt solutions, such as $Zn(NO_3)_2$, and allowed to settle uniformally. The resin was finally converted into the $Cu^{2+}-H^+$ form and rinsed with distilled water until the pH of the effluent solution was about 3.

b. Loading The mixed RE solution scheduled for separation was prepared at a predetermined concentration level. Usually the concentration was less than 0.3 mol/L. It was then passed through the column at a high flow rate during the RE adsorption step. The resin, with the RE elements adsorbed, was then rinsed with water at a pH of 3.

c. Displacement separation The mixed RE was then separated by using the selected eluate. The elution pattern of the RE elements was monitored. When breakthrough of the RE from the loading column was detected the transfer of effluent to a separation column was made, the flow rate being adjusted to duplicate the rate in the loading column.

d. Product yield When the RE ratio in the separation process reached the value fixed in advance, the migration rate of the band reached a stable value and its shape no longer changed. The yield of different RE products from different columns was dependent on their initial magnitude in the RE mixture. If the magnitude of one RE was larger than another RE it could be removed with higher yield from a column with larger cross section. The reverse was true when the magnitude of one RE was smaller than another. Such considerations were not only useful for improving the yield of a RE, but also saved time. Collection of the product solution was defined by the purity of the RE element in it.

e. Treatment of the product solution The EDTA present in the product solution was allowed to settle prior to filtration. Then the RE was precipitated by adding oxalic acid to the filtrate. The RE oxalate precipitate, recovered by filtration, was rinsed with dilute oxalic acid and oven dried before conversion to RE oxide by ignition in the air. The metal ion containing EDTA solution, e.g., Cu-EDTA left after separation of the RE, remained for recovery of Cu^{2+} and EDTA.

III. BASIC STUDIES

A. Study of the Action of the Displacing Ion

In separating natural RE elements by chelate-enhanced ion-exchange displacement chromatography, the chelate acids (e.g., EDTA) were usually neutralized by adding base (such as $NH_3 \cdot H_2O$, NaOH) to adjust the pH of the solution. In this way rare earth displacing ions (such as NH_4^+, Na^+) were introduced to the solution. In addition, neutral salts (such as $NaNO_3$, $NaClO_4$, etc.) could also be added to the solution to increase the concentration of the RE displacing ions. Ions added as neutral salts are identified as extra displacing ions. The action of displacing ions has been studied by Hale and Love [10], who used the chelates, HEDTA or DTPA, as the eluates. The results obtained showed that the migration rate of the band formed increased with the increasing concentration of the displacing ions; this reduced the separation time. They did not, however, study seriously other factors, such as the quantitative relationship between the band migration rate and band length, the influence of the concentration of the displacing ion on the zone of the ion being displaced, the length of over-lap between neighboring zones, and the relationship between the composition of the two phases in the steady-state zone and the concentration of the displacing ion. These problems have been seriously studied by the authors, using EDTA as the chelate eluate.

1. *The Relationship Between the Concentration of the Displacing Ions and the Migration Rate and Length of the Effluent Band*

The migration rate of the band and its length can be enhanced by increasing the concentration of the displacing ion. This band behavior has been noted by many researchers. The quantitative relationships between the band migration rate, its extension in the column and the concentration of displacing ions, however, was not determined. As a consequence, it was mistakenly presumed that the time needed to yield product could be shortened by increasing the concentration of displacing ion. In the authors' studies of these relationships EDTA was the chelate eluate and NH_4^+ ion functioned as the displacing ion. The concentration of EDTA was fixed at 0.015 mol/L while the concentration of NH_4^+ ion was varied from 0.048 to 0.50 mol/L. The variation of the migration rate of the rare earth band as well as its length in the column was determined. The results that were obtained are shown in Figs. 4 and 5. The figures indicate that both the migration rate and the length of the band increase linearly with the increasing concentration of the displacing ion. The slopes of the two lines are 0.585, and 0.590, respectively. They are nearly equal to each other, i.e., the rate of increase in the band migration rate is nearly equal to the rate of increase in band length. From these observations it was learned that the time saved by increasing the migration rate of the band is offset by the increase in time needed to obtain product by removal of the extra length of band. Therefore, even though the time of the complete separation process is shortened by increasing the concentration of displacing ion, the time needed to obtain the product is not.

2. *Relationship Between the Variability in the Shape and Length of the Ion-Retaining Zone and the Concentration of the Displacing Ion*

Since increasing the concentration of a displacing ion can cause the RE band to lengthen without influencing the time needed to obtain product

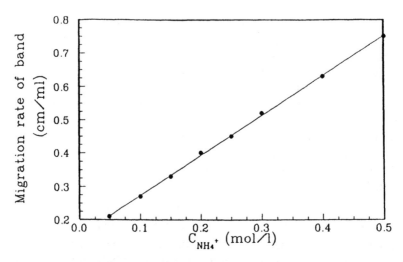

Figure 4 Relationship between migration rate of band and the concentration of the displacing ion: C_{EDTA}, 0.015 mol/L; $C_{NH_4}+$, 0.048–0.50 mol/L; linear flow rate, 10 cm/min; pH, 7.5; temperature, 75°C.

its effect on the shape of the ion-retaining zone was sought to determine its influence on the time needed to obtain product.

To initiate examination of this aspect the shape of the ion-retaining zone was determined first in the absence of extra displacing (NH_4^+) ion. For this purpose, 0.06 mol/L EDTA solution was neutralized to pH = 7.50

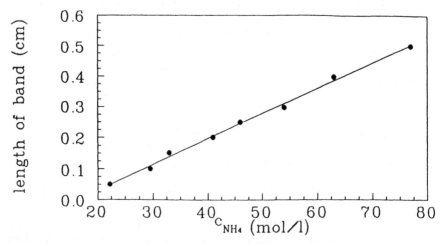

Figure 5 Relationship between length of the band and concentration of the displacing ion.

with $NH_3 \cdot H_2O$. At this pH value, 3.2 of the four hydrogen atoms in EDTA were displaced by NH_4^+ and the concentration of NH_4^+ was 0.048 mol/L.

The RE elements were separated by using the above solution as eluate together with Cu^{2+} as the retaining ion. The concentrations of Cu^{2+}, NH_4^+, and H^+ in the ion-retaining zone were determined. The results that were obtained are shown in Fig. 6. From curve 1 we can see that the concentration of Cu^{2+} increased with increase in the volume of the product solution. When the volume increased to 600 mL, the concentration of Cu^{2+} increased to 22.5 mol/L, and maintained this value until RE breakthrough. At this point the Cu^{2+} ion concentration rapidly fell to zero in the RE zone. Curve 3 indicates that free Cu^{2+} ion did not appear until the volume reached 200 mL. From 200 to 500 mL, the concentration of free Cu^{2+} ion increased gradually. When the volume of the solution was 500 mL, it reached a value of 7.3 mol/L, and kept this value until RE breakthrough. Curve 2 shows the variability of pH. Curve 4 indicates that the displacing ion NH_4^+, did not enter the ion-retaining zone. From the above the following mechanism for the variation of composition in the ion-retaining zone has been deduced.

When the chelate eluent (displacer) at pH = 7.5 entered the RE zone, the RE ions in the resin were exchanged for the displacing ion, NH_4^+, and complexed by the EDTA as shown in the following reaction:

$$Ln^{3+} + HY^{3-} = LnY^- + H^+ \tag{1}$$

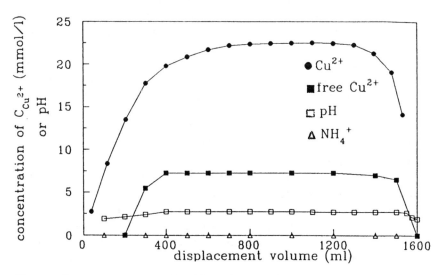

Figure 6 Concentrations of Cu^{+2}, NH_4^+, and H^+ in the ion-retaining zone.

Release of H^+ ion to the eluent lowered the pH value in the RE zone to 2.

When the complex of Ln^{3+}, LnY^-, entered the ion-retaining zone, it reacted with the retaining ion, Cu^{2+}, to form its EDTA complex, CuY^{2-}, freeing Ln^{3+}, for recapture by the resin:

$$3\overline{Cu^{2+}} + 2LnY^- = 2CuY^{2-} + 2\overline{Ln^{3+}} + Cu^{2+} \qquad (2)$$

The bar placed above the metal ions to show their association with the resin phase in Eq. (2) provides a demonstration of the Ln^{3+} ion-retaining function of the Cu^{2+} ion.

The stoichiometry of Eq. (2), where 3 moles of Cu^{2+} are exchanged for 2 moles of Ln^{3+} is written as shown because of the formation of CuY^{2-} with EDTA. This leads to 1 mole of Cu^{2+} existing as the free ion. Support of this estimate of the situation is provided by Fig. 6. The total concentration of Cu^{2+} in Fig. 6 is 22.5 mmol/L, and that of the free Cu^{2+} ion is 7.3 mmol/L. The ratio of the two concentrations is about 3:1. The fact that there is no free Cu^{2+} ion and that the pH is lower before the volume of the eluate reaches 200 mL is attributable to the presence of a certain percentage of H^+ in the retaining bed. When the retaining ion, Cu^{2+}, is removed from the resin phase to the solution it releases H^+ ion when it enters the resin phase once again. This lowers the pH of the solution and keeps free Cu^{2+} ion from appearing in the product solution initially.

a. The Variation of the Ion-Retaining Zone with Extra Displacing Ion Added Different quantities of neutral salt, NH_4NO_3, were added to the chelate displacer to study the effect of changing the concentration of displacing ion on the shape of the ion-retaining zone. The displacement curves of Cu^{2+} in different concentrations of displacing ion that were determined for this purpose are shown in Fig. 7.

In the figure, curve 1 represents the displacement of Cu^{2+} without extra added displacing ion; curves 2–7 correspond to the displacement affected at different concentrations of the displacing ion. All the curves exhibit maxima, which increase with the increasing concentration of the displacing ion. When the concentration of NH_4^+ ion is 0.50 mol/L (curve 7), the maximum is nearly 10 times as large as it is without extra displacing ion (curve 1). With increase of the concentration of the displacing ion the maximum appears earlier and earlier. The total output volume of the retaining ion is smaller and smaller. It can also be seen from Fig. 7 that, when the concentration of NH_4^+ ion is less than 0.20 mol/L, the displacement curves of Cu^{2+} (curve 2 and 3) overlap the curve obtained without extra added NH_4^+ once the peaks appear; when the concentration of NH_4^+ is larger than 0.25 mol/L the curves (curves 4–7) do not overlap curve 1.

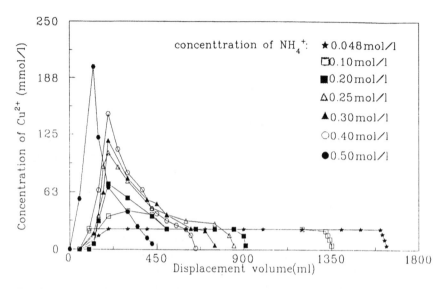

Figure 7 Displacement curves of Cu^{2+} at different concentration levels of displacing ion: C_{EDTA}, 0.015 mol; pH, 7.50; temperature, 75°C; linear flow rate, 10 cm/min; concentrations of NH_4^+ (mol/l): curve 1, 0.048; curve 2, 0.10; curve 3, 0.20; curve 4, 0.25; curve 5, 0.30; curve 6, 0.40; curve 7, 0.50.

Such behavior is due to the extra added NH_4^+ entering the RE and Cu^{2+} zones. Only a part of the extra added NH_4^+ ion exchanges with RE ions in the trailing boundary of its band; the rest enters into the RE and Cu^{2+} zones to exchange for the RE^{+3} and Cu^{2+} ions, their withdrawal from the resin phase being facilitated by the presence of EDTA. As a consequence, the displacement curves of Cu^{2+} have maxima which increase in magnitude as the concentration of displacing ion increases. Because NH_4^+ ion is not complexed by EDTA, there is no tendency for it to withdraw NH_4^+ ions from the resin. This may lead to lengthening of the RE bands as observed and conversion of the ion-retaining band from the mixed $Cu^{2+}–H^+$ form to the mixed $Cu^{2+}–NH_4^+$ form. After NH_4^+ distribution in the RE zone reaches equilibrium, the concentration of Cu^{2+} no longer changes, and the plateau in the displacement curve of Cu^{2+} appears. When the concentration of displacing ion is less than 0.20 mol/L and the column ratio is 1:3.5, the retaining zone eventually reaches equilibrium and the displacement curves overlap with curve 1. When the concentration is over 0.25 mol/L, and the column ratio is the same the band lengthens and the retaining zone cannot reach equilibrium when RE ion breakthrough occurs. As a result the curves (curves 4–7) do not overlap curve 1. In addition, because NH_4^+ ion enters the ion-retaining zone and replaces Cu^{2+} in the ion-retaining bed, both the loading capacity of Cu^{2+} in the bed and its ion-retaining ability decrease. Therefore, not only the migration rate of the RE band can be increased, but the displacement volume of the Cu^{2+}–EDTA solution can be decreased by increasing the concentration of the displacing ion. This led to the shorter time for product recovery that was being sought.

b. The Relationships Between the Length of Overlapping Area of Neighboring RE Zones, HETP, and the Concentration of the Displacing Ion   Since the length of the RE band at steady-state increases with increasing concentration of the displacing ion, it is important to know how the overlapping area of neighboring RE zones will vary at steady state. It has been shown experimentally that, without extra added displacing ion, the migration rate of the RE band is slower and that the displacement volume of the overlapping area for two ions is larger with increasing concentration of the displacing ion, the displacement volume of the overlapping area decreasing gradually. By plotting the mole fraction (x) of an element as ordinate, the displacement volume of the element as abscissa and by letting the focal point of the boundary curves define the origin, a group of boundary curves can be obtained. The displacement time of the overlapping area can be calculated by dividing the flow rate of the displacer ion by the displacement volume of the overlapping area. From the determined migration rate of the band, the length of the overlapping area and the HETP can be easily calculated. The results obtained in this manner are listed in Table 6.

Table 6   Relationship Between the Concentration of Displacing Ion and Both the Length of the Overlapping Area and the HETP

Concentration of NH_4^+ mol/L	Migration rate of band (cm/min)	Length of overlapping area (cm)	HETP (cm)
0.048	0.19	6.3	0.56
0.10	0.25	5.5	0.41
0.15	0.30	5.4	0.42
0.20	0.36	5.3	0.38
0.25	0.42	5.3	0.41
0.30	0.50	5.2	0.41
0.40	0.62	5.8	0.58
0.50	0.75	5.8	0.60

C_{EDTA} = 0.015 mol/L; pH = 7.50; linear flow rate = 10 cm/min; temperature = 75°C.

From Table 6 we can see that, when the concentration of EDTA is 0.015 mol/L, and the concentration of the displacing ion is changed from 0.11 to 0.30 mol/L, the length of the overlapping area remains constant; when the concentration of the displacing ion is greater than 0.40 mol/L, the length of the overlapping area increases slightly. When the concentration of NH_4^+ exceeds 0.40 mol/L, a larger portion of the Ln^{3+} ions accessible to the solution by exchange with NH_4^+ ion is in equilibrium with the EDTA complexed LnY^-. Such disturbance of the separation process leads to a more diffuse boundary between the rare earth ions being separated in the course of their displacement by NH_4^+ ion. However, the length of the overlapping area increases only about 5% as a result of such disturbance of the separation. The fact that such increase of displacing ion produces only this minor disturbance in the rare earth separation affected is of significance with respect to its potential application as an industrial process. For example, by increasing the concentration of a displacing ion for the separation of natural or synthesized RE samples the band will be lengthened while the overlapping area is increased so slightly that the percent recovery of each RE will be increased.

B. The Influence of Increasing Concentration of EDTA on the Migration Rate of the RE Effluent Band and the Concentration of Displacement Fluids

Because microsize resin particles are used in HPIEC, the approach to equilibrium is rapid. Through their use the migration rate of RE band is speeded up as a consequence. When the band reaches stability the length of the zone associated with each species in the resin no longer changes. If the linear flow rate is constant, conservation of mass considerations tells us that the total millequivalents of an ion removed from the resin phase to the aqueous phase must be equal to the total equivalents of the ion leaving the system along with the displacer over a particular time period. Thus, the migration rate of the band will increase linearly with the concentration of EDTA. In a CIEC separation of RE, the concentration of EDTA is usually 0.015 mol/L, and the concentrations of RE elements in the displacement solution are only about 3 g/L, so the displacement volume is large. This causes the separation efficiency to be low. In the HPIEC process, the concentration of EDTA is usually raised to 0.075 mol/L to provide good separation efficiency. From Table 7 we see that when the linear flow rate is constant in HPIEC both the migration rate of the RE band and the RE concentration in product solution are increased by increasing the concentration of EDTA. HPIEC is thus characterized by rapid production and high separation efficiency.

Table 7 Influence of C_{ECTA} on the Migration Rate of Band and the RE Concentration in Effluent Solution

C_{EDTA} (mol/L)	Linear flow rate (cm/min)	Migration rate of band (cm/min)	RE concentration in effluent solution (mol/L)
0.015	7.5	0.14	0.014
0.025	7.5	0.24	0.024
0.050	7.5	0.49	0.047
0.075	7.5	0.75	0.075
0.100	7.5	1.01	0.092

Temperature = 75°C.

C. Selecting the pH of the Displacer

In the ion-exchange process for the separation of RE elements, EDTA is often used as displacer. It is important to select suitable pH values for these displacer solutions.

EDTA, the abbreviation for ethyldiaminotetraethyl acid, usually is identified as H_4Y. When dissolved at different pH values it exists in different forms. In strong acid solutions (pH < 1), it exists primarily as H_6Y^{2+}. In the pH range of 2.67–6.16 it exists mainly in the H_2Y^{2-} form. At a pH > 10.26, it exists mainly in the Y^{4-} form. The pH value employed for the separation of RE elements with EDTA as the chelate displacer is the key to a successful operation. The RE elements can be separated by using EDTA because the stability constants of the complexes they form are different. These stability constants increase with the increasing atomic number of the RE elements. In the steady-state zones obtained for the RE elements in the course of their separation, the pH value in each element separation zone is sizably different. The more stable the complex, the lower is the pH. Conversely, the less stable the complex, the higher is the pH. The pH values associated with some of the RE elements in their respective steady-state zones are listed in Table 8.

Table 8 lists the pH values of the solution associated with some RE elements in their respective steady-state zones for a separation process

Table 8 pH Values of Different RE Elements in Steady-State Zone

Element	Yb	Tm	Er	Ho	Dy	Tb	Gd	Sm	Nd
pH	1.65	1.70	1.75	1.84	1.94	2.04	2.20	2.34	2.44

employing EDTA solution as displacer at a particular pH value. When the pH value of the EDTA solution is changed from 7.00 to 8.80, the pH value of the solution associated with each RE element zone may also change but the pH pattern does not change.

Considering the effect of the pH value of the displacer alone on the separation one notes that use of a lower pH leads to improvement. When the pH value of the displacer is lower, however, the free EDTA in the displacement solution may increase and crystallize in the column. Three methods are available to avoid this. In the first the temperature is increased. This increases the solubility of the EDTA, and the likelihood of crystal formation is reduced. The second is facilitated by adding salts of weak acids. The increase in buffering capacity of the EDTA solution keeps the complex of RE-RDTA from decomposing and forming free EDTA. The third is facilitated by increasing the pH value of the EDTA solution. Experiments show that with this approach the pH value of the EDTA solution must not exceed a value of 9.20 while the pH value in the boundary between the RE band and the ion-retaining zone must be smaller than 3. If such pH control is not exercised, the cation of the base added to increase the pH of the EDTA, e.g., NH_4^+ ion, will break through the RE zone and enter the ion-retaining bed, thereby reducing its ion-retaining capability.

D. The Effect of Increasing the Concentration of EDTA on the Separation

As mentioned in Sec. I.B. the efficiency of separation is measured by the HETP, so it is important to study the influence of increasing concentrations of EDTA on the HETP. The effect of this variable on HETP was monitored by using the separation of Tm-Er as a suitably representative system. The results obtained are listed in Table 9.

Table 9 Influence of C_{EDTA} on HETP

C_{EDTA} (mol/L)	Temperature (°C)	Linear flow rate (cm/min)	Migration rate of band (cm/min)	HETP (cm)
0.015	75	7.5	0.14	0.41±0.01
0.050	75	7.5	0.49	0.38±0.01
0.075	75	7.5	0.75	0.38±0.02
0.100	75	7.5	0.99	0.53±0.02

From Table 9 it is observed that, under the conditions of constant temperature and constant linear flow rate, HETP is essentially unchanged over a wide EDTA concentration range. This illustrates that advantage gained from increasing the concentration level of EDTA is not canceled by loss of separation efficiency. It is also apparent from Table 9 that this insensitivity of HETP to EDTA concentration is lost when the concentration of EDTA exceeds 0.075 mol/L and reaches 0.100 mol/L. The concentration of EDTA obviously needs to be confined to a predefined range to assure reproducibility in the separation process.

E. The Effect of the Linear Flow Rate of Ion Displacer on the Migration Rate of the RE Band

One method that increases the migration rate of the RE band and shortens operating time increases the linear flow rate of ion displacer. With CIEC, however, large-size resin particles are used, and ion-exchange equilibrium is reached slowly. As a consequence, the effect of increasing the flow rate of the ion displacer on the separation may be very detrimental. For example, when the concentration of EDTA is only 0.015 mol/L and the linear flow rate is 0.5–1.0 cm/min, the migration rate of the rare earth band is only 0.02 cm/min, and the HETP reaches the undesirably high value of 0.86 cm. In HPIEC equilibrium is reached rapidly because microsize resin particles are used. Because of this the linear flow rate can be greatly increased. Experiment shows that the migration rate of the RE band increases linearly with increasing linear flow rate of the ion displacer.

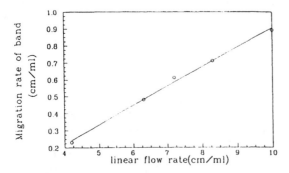

Figure 8 Relationship between migration rate of the band and linear flow rate of the displacer.

IV. SEPARATION OF NATURALLY OCCURRING RE ELEMENTS

Most reserves of RE that are distributed in China occur in the south of China. The minerals contain all the RE elements except Pm. Their abundances vary in the different mineral sources. The RE composition of the sample used in the separation example summarized in Table 10 is presented in this table.

A. Experiment

The sample described in Table 10 has been separated by employing the following conditions dictated by the results of earlier fundamental studies. The five-column system was employed for the separation. Neutral salt added to 0.075 mol/L EDTA at a pH value of 7.5 was used as the ion displacer. The column ratio was 1:2. The temperature of the process was fixed at 75°C. All products exited from the last column. Yb-(169, 175) was used to define the boundary curves of Lu-Yb and Tm-Yb, while Tb-160 was used to define the boundary curves of Tb-DY and Tb-Gd. In addition, Tm-170 and Gd-(153, 159) were used to determine the displacement curves of Tm and Gd. Other element boundaries were defined by using spectrophotometry. The results obtained are shown in Fig. 9.

It is immediately apparent that the elements can be separated in a single operation of the five-column system. Even though Lu, Tm, and Pm, elements present at the lower percentages, were not completely separated they were directed to their respective zones. It can be predicted on the basis of this result that, if large amounts of sample are separated, and the cross section of the last column is properly defined, these pure elements will emerge as products.

B. Purity of Products

With one sequence of operations the separation of elements such as Yb, Er, Ho, Dy, Tb, Gd, Sm, and Nb with the five-column system have yielded

Table 10   Contents of Components in the Mixed RE Oxide Samples

Re oxide	Y_2O_3	La_2O_3	CeO_2	Pr_6O_{11}	Nd_2O_3	Sm_2O_3	Eu_2O_3
content (%)	0.3	0.41	0.66	2.83	12.0	9.0	0.13
Re oxide	Tb_4O_7	Dy_2O_3	Ho_2O_3	Er_2O_3	Tm_2O_3	Yb_2O_3	Lu_2O_3
content (%)	3.7	21.5	4.8	10.2	1.82	12.0	2.11

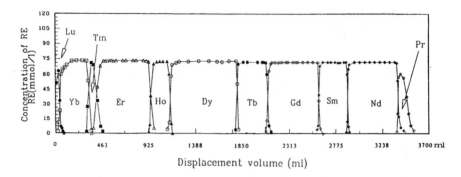

Figure 9 Displacement curve of RE elements: C_{EDTA}, 0.075 mol/L; column ratio, 1:2; linear flow rate, 7.5 cm/min; pH, 7.5; temperature, 75°C.

each of them in a highly purified state. For example the purities reached were for Yb_2O_3: 99.96%; Er_2O_3: 99.96%; Dy_2O_3: 99.94%; Ho_2O_3: 99.95%; Tb_4O_7: >99.0%; and Nb_2O_3: 99.96%, respectively. The impurities encountered were Fe: 0.0005%; Cr: 0.0003%; Ca: 0.0003%; Ni: < 0.0003%; Mg: < 0.0003%; Mn: <0.00003%; Cu: 0.001%;Al: 0.003%. The impurity percentages were lower than those obtained with the solvent extraction method. High purity is one important advantage of the HPIEC method. Because the percentages of La, Ce, Y, Eu in the sample were too low to separate the zones they form from all of the other separated zones (their zones are covered in overlapped areas), they were not analyzed.

C. Percent Recovery of Products

Because displacement chromatography can be used to treat large quantities of mixed samples, it is always used for preparation purposes. The various species cannot be separated completely with this method, however, overlapping areas of the zone boundaries of the two neighboring species always being present. The larger their separation factor the smaller is the overlapping area of the two species. If the reverse is true, it will be larger. Thus the percentage recovery of a species depends primarily on the content of the species in the sample, the cross section of the output column and the separation factor characterizing the species and its neighbors once the band reaches stability and the overlappihng areas in the column remain constant as a result. The larger the content of a species, the bigger is the separation factor for the species and its neighbors, the smaller is the overlapping area and the higher is the percentage recovery of the species. As shown in Fig. 9 the percentage recoveries of Er, Dy are greater than 90%, for Ho, Gd they are higher than 70%, while for Tm, Tb they are less than 50%. This result is for one special experimental condition. The zone of a species can be lengthened by reducing the cross section of the output column, thereby providing higher percentage recovery for lower content elements when the cross section of the output column is reduced.

V. SEPARATION OF Gd AND Eu

Because of the similarity of the outer electron configuration of Gd and Eu their separation is more difficult than the separation of the other RE elements. Neither the ion-exchange or solvent extraction methods employed for this purpose earlier had enjoyed much success. On this basis it was thought that separation of Gd and Eu through use of the ion-exchange method would not be industrially practicable. Many researchers have therefore resorted to methods based on the prior reduction of Eu^{3+} to Eu^{2+}. They then separate Eu^{2+} from Gd^{3+}.

When use of the ion-exchange separation method developed in this study was considered it was determined from the separation factor calculated from the stability constants [4] of their complexes with EDTA that a column ratio greater than 40 would be needed to separate them. Experiments showed that a column ratio nearly 10 times larger would be needed to affect their separation with CIEC. Theoretically based studies [7–9] led to success in separation Eu and Gd by the use of HPIEC and a binary displacer. This technique has made the separation of Gd and Eu simple and useful enough to warrant its use as an industrial process.

A. Technological Process

Use of the three-column system shown in Fig. 2 has been examined to judge its potential for rare earth separations. The purities of Sm, Eu, and Gd that were obtained from the mixed Sm-Eu-Gd sample used for this basic study were: Sm_2O_3: 99.5%. Eu_2O_3: 99.99%, Gd_2O3: 99.5%. The composition of the natural Sm-Eu-Gd sample separated in this study was: Nd: 13%. Tb: 4.5%, Sm: 25%, Eu: 4.8%, Gd: 27.0%, and the other RE elements: 25%. Gd and Eu alternatively labeled with Gd-(153, 159) and Eu-(152, 154) were used to determine the displacement curves of these two elements while the displacement curve of Sm was determined by using a spectrophotometric method.

Based on these theoretical studies, Gd and Eu can be separated by employing the following conditions: displacer: 0.030–0.075 mol/L of EDTA with suitable neutral salt added to it; pH of the displacer: 8.0–8.5; column ratio: 2.5–3.0; linear flow rate of the displacer: 4.0–7.5 cm/min.

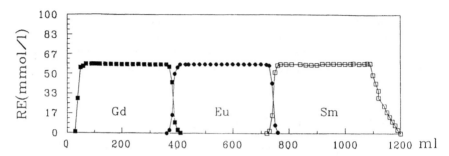

Figure 10   Separation of synthesized sample of Sm, Eu, and Gd at the same molal concentration.

B.   Results of Experiment

1.   *Separation of Synthesized Sample*

Sm, Eu, and Gd at the same molal concentration were combined in solution. The Gd was labeled with its radioisotope, Gd-(153, 159), and the conditions specified above were employed to separate them. The results are presented in Fig. 10.

Figure 10 shows that the mixture of Sm-Eu-Gd was separated effectively. The fact that the overlapping area of Gd-Eu is longer than that of Sm-Eu is consistent with theoretically based estimates.

2.   *Separation of an Enriched Sm-Eu-Gd Sample*

Additional Eu was mixed with Sm-Eu-Gd from natural RE ores to increase its Eu content to 8.5%, a commercially based value. The separation results are shown in Fig. 11.

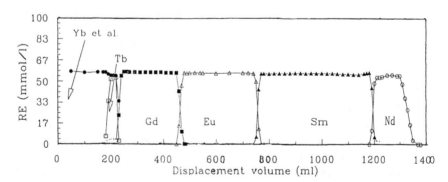

Figure 11   Displacement curve of enriched Sm-Eu-Gd mixture: C_{EDTA}, 0.60 mol/L; column ratio, 1:3; linear flow rate, 6–7 cm/min; pH, 8.0; temperature, 71°C.

Table 11 Analysis of RE Impurities in Eu_2O_3 Product (ppm)

La_2O_3	CeO_2	Pr_6O_{11}	Nd_2O_3	Sm_2O_3	Gd_2O_3	Tb_2O_3
<3	<1	<5	<6	<3	<3	<10
Dy_2O_3	Ho_2O_3	Er_2O_3	Tm_2O_3	Yb_2O_3	Lu_2O_3	Y_2O_3
<10	<10	<10	<3	<3	<3	<3

3. Purity of the Products

The purity of one product, Eu_2O_3, is listed in Table 11. Its value is greater than 99.99%. This excellent result means that the separation of Gd-Eu with this method is feasible. This result is not only of scientifically significant importance, but shows that such separation of Eu and Gd is practicable industrially.

VI. CONCLUSIONS

Through this fundamentally based study a key problem of how to improve the separation efficiency of the ion-exchange method has been solved by increasing the concentration of EDTA from 0.015 to 0.075 mol/L.

The introduction of HPIEC has overcome many disadvantages such as longer period of production, lower efficiency, and higher cost, that are encountered with CIEC. Under the conditions studied in this chapter, production efficiency can be increased by a factor of hundreds. HPIEC has provided new and more useful ways of separating the RE elements.

Once separated, the purities of RE products can reach 99.9–99.99%.

The problem of separating Gd-Eu without prior reduction of Eu has been solved. This development is not only of fundamental significance, but is also important to industrial practice.

REFERENCES

1. E. A. Cotton and G. Wilkinson, *Advanced Inorganic Chemistry*, Interscience, New York, 1962, pp. 870–890.
2. F. H. Spedding et al., *J. Am. Chem. Soc.*, 69: 2777–2781 (1947).
3. F. H. Spedding et al., *J. Am. Chem. Soc.*, 69: 2812–2818 (1947).
4. J. A. Marinsky, L. E. Glendenin, and C. D. Coryell, *J. Am. Chem. Soc., 69*: 2781 (1947).
5. W.E. Cohn, E. R. Tompkins, and J. X. Khym, *J. Am. Chem. Soc., 69*: 2769 (1947).
6. B. Ketelle and J. E. Boyd, *J. Am. Chem. Soc., 69*: 2800 (1947).
7. L. Qiu, High-Pressure Ion-Exchange Separation Chromatography, Pi, (1982), Beijing.

8. D. C. Scott, *Clin. Chem.*, 14: 521 (1968).
9. L. Q. Chen et al., *A New Technical Process of Separating Heavy RE Elements by HPIEC*, A monograph of Journal of Lanzhou University (III), 1984.
10. W. H. Hale and J. T. Love, (*S. R. L.*). *Inorg. Nucl. Chem. Lett.* 5(5): 363 (1969).
11. J. T. Lowe et al., *Ind. Eng. Chem., Process Ces. Dev., 10*: 131 (1971).
12. J. T. Lowe et al., *Ind. Eng. Chem., Process Ces. Dev., 10*: 131 (1971).

2

Ion Exchange in Countercurrent Columns

Vladimir I. Gorshkov

Moscow State University Moscow, Russia

I. INTRODUCTION

The application of ion-exchange resins in scientific research and industry is growing. Fixed-bed columns of granulated resins are commonly used in the laboratory as a research tool. In the case of larger-scale industrial tasks, countercurrent movement of resin and solution is employed to provide the increase in convenience and efficiency that such operation insures. Countercurrent columns may be easily automated. They need smaller amounts of ion exchanger. The expenditure for reagents needed for regeneration of ion exchangers is less than for columns with fixed beds. Their application allows one to reduce the dimensions of the equipment used.

Ion-exchange processes can be differentiated by assignment to one of two groups: Group I: replacement of ions of one kind by ions of another, for example, purification of nonelectrolytes or weak electrolytes from strong electrolytes, softening and demineralization of water, recovery of substances from solutions, etc.; Group II: separation of mixtures of dissolved electrolytes.

The problems related to the first group are most easily solved. In the sample cases listed, passing of solution through a resin bed results in the replacement of the electrolyte solution ions by resin ions. Depending on the nature of the exchange equilibrium isotherm and the kinetic peculiarities of the process, which are characterized by the height equivalent of a theoretical plate (HETP), the different lengths of ion-exchanger bed required for the complete replacement of ions of one kind by ions of another can be estimated. The higher the affinity of the ion exchange resin for the ion removed from solution, the smaller is the number of theoretical plates and the shorter is the ion-exchange resin bed needed for the process.

As a rule in these processes that can be classified as ion replacement or recovery operations, the conditions are selected to assure a highly convex sorption isotherm. With the development of such a sorption isotherm, the full recovery of the ions of interest does not require sizable multiplication of the separation reached with one theoretical plate. In view of this, the application of countercurrent columns has been initiated to facilitate further the resolution of these relatively simple problems.

A more complex field of ion-exchange application is associated with the separation of mixtures of substances that react similarly with ion-exchange resins. In this case the separation is based on differences in resin selectivity toward the mixture's components. These differences are characterized by the magnitude of the equilibrium separation coefficient which can be represented for a pair of ions as follows:

$$\alpha = \frac{y(1-x)}{(1-y)x}$$

where y and x are equivalent fractions of one of the mixture components in the resin and solution phases, respectively. Usually this component is the more strongly sorbed one.

The particular problem of separating mixtures of substances with similar properties when the equilibrium separation coefficient, α, does not differ greatly from unity can be solved only by using effective columns. The application of countercurrent operations in these processes is still confined to laboratory or semi-industrial experimental installations.

In order to select conditions for the separation of any mixture of substances it is necessary to consider three separate groups of questions: (i) What type of ion exchanger? What solvent? What should the concentration of the treated solution be to obtain an equilibrium separation coefficient, α, as different from unity as possible? (ii) What process scheme will minimize expenditures for auxiliary operations such as the regeneration of resin and auxiliary substances? (iii) What equipment will be most effective and efficient?

Since the literature, original and review, is devoted to theoretical and experimental studies of ion-exchange equilibria, these aspects are not considered in this presentation. Instead it is concerned primarily with the essentials of carrying out continuous ion-exchange processes. The fact that the design and construction of countercurrent ion-exchange ensembles are the subject of a number of reviews in the literature as well (see, for example, [1–4]), consideration of this topic has been confined to a few general problems. Particular attention is focused on the design of effective contactors, that is, devices which provide the greatest number of equilibrium separation stages per unit length.

In designing countercurrent processes, particularly those for separating mixtures of substances with similar properties, the choice of a suitable process scheme is of prime importance.

Along with selecting conditions which ensure obtaining products with the required degree of purity at a maximum yield, the choice of a process scheme in which expenditures for auxiliary operations are reduced and wastes are minimized determines, in many respects, the efficiency of the process and the expediency of its use.

Countercurrent process schemes for replacing ions of one kind by ions of another are analogous to those carried out in fixed ion-exchange resin beds. The use of countercurrent columns enables one, in these cases, to employ ion-exchanger capacity more effectively, to reduce expenditures for resin regeneration, and to decrease the amount of waste. It also permits automated control of the process.

At the same time, in the case of separating mixtures of dissolved substances, use of a countercurrent mode opens up the possibility of developing processes which are inaccessible to an ion-exchanger fixed bed.

Also to be considered are processes with closed cycling of an ion exchanger. In these processes a certain amount of ion exchanger repeatedly participates in the same operations.

To attain a sufficiently high degree of separation, multiplication of the single stage separation factor is needed. This can only be accomplished by providing circulation of the separated substances. With this approach most of the enriched mixture is returned to the column.

Two fundamental principles applicable to the design of continuous countercurrent separation processes are known. They are identifiable with (i) flow reversal* of the separated mixture at the edge of the contact sys-

---

*By "flow reversal" of separated ions it is meant that the enriched flow is returned to the separation process (this term is analogous to "reflux" in distillation). Applied to the ion-exchange method, as well as to other methods of separation such as extraction and chemical exchange, this term is more exact than the commonly used term "phase reversal" since not all of the phase, but only the mixture separated, is returned to the column (as distinguished from distillation).

tem and (ii) change in the direction of the transfer between contacting phases at different edges of the system, at the expense, for example, of a sharp temperature change (two-temperature method).

The first principle is predominantly used in ion-exchange practice.

In analyzing schemes for separation processes the separation of two-component mixtures or the purification of components with sorbability extremes (lowest or highest) from multicomponent systems will be considered. When separating multicomponent mixtures the problem of providing flow circulation is solved in the same way as in separating two-component mixtures. Some peculiarities in the recovery of components with intermediate sorbability will be dealt with in Sec. II.

II. SEPARATION PROCESSES WITH FLOW REVERSAL

A general scheme for the countercurrent ion-exchange separation process with flow reversal is analogous to the schemes for such two-phase separation processes as rectification, chemical exchange, etc. (Fig.1).

The process in question is performed in two countercurrent columns. In both columns, ion-exchange resin moves down, i.e., from top to bottom and solution moves in the opposite direction, i.e., from bottom to top.

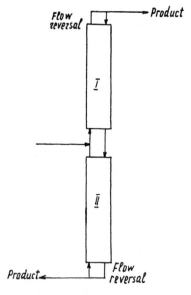

Figure 1 A scheme for an ion-exchange separation process.

A solution of the mixture of ions to be separated is fed into the bottom of column I. At the top end of the column flow reversal is affected, i.e., ions to be separated are transferred from the solution to the ion-exchange resin and directed back into the column. Leaving the column the ion-exchange resin, saturated with the mixture of ions to be separated, is moved into the top part of column II. At the bottom part of column II flow reversal is performed, i.e., ions to be separated are transferred from the ion-exchange resin to the solution and directed back into the column. Leaving the top part of column II the solution is moved at once, or after a particular treatment, for example after a concentration change, to the bottom part of column I.

As a result of the fact that the selectivity of the ion exchanger toward mixture components is different the more weakly sorbed component is gradually concentrated in the top of column I while the more strongly sorbed one is concentrated at the bottom of column II. Once the required degree of enrichment is achieved it is possible to remove the product while adding the original mixture of components at the bottom of column I.

Flow reversal, i.e., the transfer of the mixture of ions from solution to ion exchanger in the top portion of the column (sorption) and from ion exchanger into solution in the bottom portion (desorption) is effected with the help of specially selected auxiliary ions which differ in their affinity for the ion exchanger from the ions separated.

For example, in purifying cesium salts from other alkali metal impurities using sulfophenolic ion exchanger, barium ion [5] may be employed as auxiliary ion for flow reversal in column II to facilitate separation of pure Cs^+ ion from the initial mixture. Ba^{2+} ion is more strongly sorbed than the cesium ion under certain conditions. Put into one end of the column is the ion exchanger loaded with the ion mixture to be separated; barium chloride solution is in the other end. Barium ions displace the ions separated from the ion exchanger transferring them into the solution phase while the ion exchanger is being transformed into the Ba^{2+} ion form.

Problems connected with accessory operations accompanying the ion-exchange process arise. They include such operations as ion-exchanger regeneration, i.e., its transformation from the Ba form into another form (hydrogen) suitable for the sorption of ions undergoing separation and the regeneration of an auxiliary electrolyte through the transfer of Ba^{2+} from ion exchanger back into the solution as $BaCl_2$ which can be repeatedly used in column II.

The proper choice of an auxiliary ion allows the number of accessory operations to be minimized. With this goal approached, economic efficiency of the separation process is more likely to be achieved.

Three means for flow reversal in separations employing countercurrent ion-exchange can be distinguished. They are (i) flow reversal in the separation column itself, (ii) flow reversal outside the column, (iii) partial flow reversal in the column and the completion of the process outside the column. The choice of one of the three is determined by the likelihood of a convenient auxiliary ion being selected. It is connected, as well, with solving the problems of automated column operation. In the region of the column where replacement of ions under separation by auxiliary ones takes place, properties of the ion-exchanger – solution system such as electrical conductivity, refractive index, color, etc., change sufficiently enabling one to accomplish automatic control at the flow reversal zone and to regulate the operating conditions of the column.

On this basis, the automation of countercurrent column operation is easier to accomplish by flow reversal in the separation column than outside the column. However, it is not always possible to select an auxiliary ion which permits the flow reversal operation to occur solely in the column. This necessitates conduct of this operation outside the apparatus as well. If ions with similar properties are to be separated or the desired substances are to be purified by separating them from a small amount of impurity, difficulty in automating the column operation arises. Under these conditions it is expedient to carry out partial flow reversal inside the column and to complete the process outside the column.

The choice of the most suitable path to flow reversal must therefore consider whether the selection of an auxiliary ion is helpful and how the problem of column automation is best solved. Then the means for flow reversal, mentioned above, and the auxiliary operations accompanying them must be examined. To demonstrate the exercise of these judgments processes accomplished with the aim of obtaining both components of the mixture will be considered first. Some details of the process for purifying a substance, where the impurities are of no value, will then be examined.

A. Flow Reversal inside the Separating Column

To effect flow reversal of the ions to be separated inside the separating column, while at the same time avoiding product pollution by an auxiliary ion, it is necessary to form a sharp stationary front between the ions to be separated, and the auxiliary ion C in column 1 (Fig. 2) and the second auxiliary ion D and the separated ions in column II (indicated first

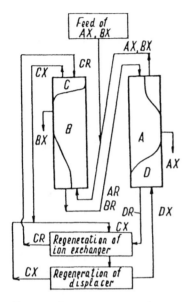

Figure 2 A scheme for process with flow reversal inside separation columns upon applying two auxiliary ions flow reversal.

are the ions fed to the column in the solution). The stationary front can be formed, with sufficient length of the column, if ion C, the auxiliary ion entering the column with the ion exchanger, has a lower affinity for the resin and ion D, the second auxiliary ion displacer added with the solution phase, has a higher affinity than the ions separated at any concentration ratio between the auxiliary ions and the ions separated. The stationary front should remain immobilized in the column provided the flow of ions entering the front zone with the solution and resin phases are equal:

$$\chi VC_0 = (1 - \chi)Wm_0$$

Here χ is the volume fraction of the column available to the solution phase, V and W are the linear rates of the solution and resin motion, and C_0 and m_0 are total concentrations of ions in the solution and ion exchanger, respectively.

In the general case [6] different auxiliary ions are used in columns I and II. In order to affect a closed process the regeneration of ion exchanger, that is, its conversion from the DR-form into the CR-form is required. Regeneration can be effected either in the column with a fixed bed of resin or in the countercurrent column. The quantity of C^+ ion needed to effect such regeneration can be determined from the mass-action-based description of the equilibrium distribution of C^+ and D^+ ions between resin and solution that is presented next. In the equation

$$K_C^D = \frac{y(1-x)}{(1-x)y}$$

y and x are equivalent fractions of D (the more strongly sorbed ions) in ion exchanger and solution, respectively, and K is the equilibrium constant. For complete regeneration to the C^+ ion form quantities of CX salt at least equal to K_C^D multiples of the ion-exchanger capacity is required [7]. The minimum regenerant that is needed to reduce y to a value closely approaching zero is accessible, as well, from the slope at the angle tangent to the equilibrium isotherm if x_D is approaching zero.

As a result of such regeneration of the ion exchanger (CR form) a solution mixture of DX and CX salts is obtained. In the optimum case DX and CX are separated and reapplied to the process, DX being used for displacement in column II with CX being used for resin regeneration.

With the discussed process scheme a certain amount of resin and auxiliary ions will be circulating in the system. Supplementary operations include resin regeneration and separation of CX from DX. Commonly with cation separation the H^+ ion is one of the auxiliary ions and the latter operation offers no difficulty, for example, if X is the volatile acid anion.

The output solutions contain only enriched products without auxiliary ion impurities. At the same time that the ion separation takes place concentration of the mixture components occurs. The product concentration in column I equals the total concentration of the ion mixture put into the column and the product concentration in column II, that is, the ion-displacer concentration. This is why, other conditions being equal, the use of more concentrated solutions is beneficial. However, it should be taken into account that increase in the total concentration of the mixture of ions separated may result in a decrease in the sorbability differences between the separated and auxiliary ions. Moreover, the resultant decrease in the single-stage separation coefficient reduces the efficiency of the column operation.

It is to be noted that with complete flow reversal the separation process in column I is the analogue of the frontal separation process or the frontal analysis in a fixed bed of the sorbent; the separation process in column II is the analogue of the reverse frontal analysis; and the process in both columns, with conditions, phase composition, and temperature unchanged in passing from one column to the other is the analogue of displacement chromatography. The analogy is related only to the modes of flow reversal rather than to the process efficiency, degree of product recovery, reagent expenditures, and so on.

If product dilution by auxiliary ions is permissible, then it sometimes appears to be suitable to carry out incomplete flow reversal of ions separated inside the column selecting as product some part of the enriched mixture flow exiting the column with the auxiliary ions. In this instance the total flow in column I of substances injected with solution must exceed the total flow of substances injected with the ion exchanger:

$$\chi VC_0 > (1 - \chi)Wm_0$$

In column II the total flow of substances in solution should be less than the flow of substances in the ion exchanger:*

$$\chi VC_0 < (1 - \chi)Wm_0$$

As a result a part of the separated ion mixtures in the two phases, which is enriched, respectively, with the component either less or more effectively sorbed by the ion exchanger, exits column I with the solution and column II with the ion exchanger. Similar schemes have been used to separate sulfuric and sulfonic acids [8], to concentrate cesium [9], to purify and concentrate uranium [10] and plutonium [11].

The advantage of the processes mentioned above lies in the fact that no special devices are needed in the column to control output of the product. However, in case auxiliary electrolytes are needed in the process again, it becomes necessary to separate them from the solution leaving column I and from the resin leaving column II and this presents certain difficulties.

In some cases a special stage of ion-exchanger regeneration may be avoided. By using only one auxiliary ion for this purpose the conditions are created by altering the solution concentration, by changing temperature, or by introducing a complex-forming agent that insure that ions separated are sorbed more strongly than the auxiliary ion in column I and more weakly than the auxiliary ion in column II. This can be most simply accomplished by using auxiliary ions with charge different from that of the ions to be separated and by regulating differences in sorbability of auxiliary and separated ions through changes in solution concentration.

Suppose, for example, the A and B ions to be separated are divalent and C, the auxiliary ion, is univalent. It is known [12,13] that with ex-

---

*The conditions for breakthrough of a portion of the mixture flow through the column are indicated here for the case when an auxiliary ion predecessor C (the initial ionic form of resin is in the C-form) is sorbed less effectively than the ions being separated and the ion displacer, D, is sorbed more effectively than ions of the mixture undergoing separation.

changing ions of different charge an increase in solution concentration results in an increase in the sorbability of the ion with smaller charge. Hence, by using dilute solutions in column I upon frontal separation and concentrated solutions upon displacement in column II the process can be performed with the same auxiliary ion. The supplementary operations will consist of diluting the solution of the separated AX_2 and BX_2 mixture exiting column II and concentrating the CX solution coming out of column I.

Another method of avoiding the ion-exchanger regeneration stage uses complex-forming reagents. This method is based on creating the conditions where complexes can be formed only in one of the columns. As a result the stationary front is maintained due to the difference in sorbability of separated and auxiliary ions in one column and the difference in the stability of complexes in the other.

As an example let us consider a likely scheme for the recovery of Ra^{2+} from its mixture with Ba^{2+} using the operating conditions recommended in [14]. In column I the ion mixture to be separated is sorbed from the solution at pH = 5–6 on cation exchanger introduced in the Na^+-ion form; in column II Ra^{2+} is concentrated by the elution of sorbed ions with the sodium salt of ethylenediaminetetraacetic acid (EDTA) at a pH of 8–9. The very stable complexes formed by divalent cations with EDTA ensure their complete displacement from ion exchanger by Na^+ ion. In this process, even though there is no regeneration stage for the ion exchanger, there are supplementary operations connected with change in solution pH that require reagent expenditures.

A similar scheme has been used to remove Zr and Nb while concentrating plutonium [15]. In this case use is made of the strong dependence of the stability of nitrate complexes of plutonium on the concentration of HNO_3. Nitrate complexes are sorbed from concentrated (6 mol/L) solution of HNO_3 on anion exchanger introduced in the NO_3^- form. At lower concentrations of HNO_3 the complexes dissociate and the sorbed ions are displaced by nitrate ion in the diluted solution (1 mol/l of HNO_3) employed for flow reversal in column II.

A convenient scheme for separating nitrogen isotopes in the course of their exchange between aqueous ammonia in solution and the NH_4^+-ion form of cation exchanger is described in Ref. 16. Also employed is dimethylaminoethanol (DME) which is sorbed more weakly by sulfonate cation exchanger cross-linked 5% by weight with divinylbenzene than ammonia at 10°C and more strongly than ammonia at 40°C. The separation is carried out in a fixed ion-exchanger bed. Performance of the process in two countercurrent columns is much more suitable than in a fixed bed. The frontal separation on ion exchanger in the DMAE form is carried out at 10°C and the displacement of ammonium ions is effected at 40°C by the solution of DMAE exiting column I. Leaving column II, ion exchanger, saturated with DMAE, is moved to column I upon cooling.

Separations may also be carried out without a special ion-exchanger regeneration stage in case it is impossible to select the conditions under which the stationary front forms [17] in one of the columns. To consider the scheme for processes such as these take as examples the purification of substances from impurities when impurities are sorbed more strongly than the substance purified and when impurities are sorbed more weakly than the substance purified. Assume also in these examples that in the column where the purified product is obtained flow reversal takes place by forming a stationary front.

1. *Purification from Impurities Sorbed More Strongly than the Substance to Be Purified*

The product BX, with decreased content of impurities, is obtained in column I. Carried out in column II is the combined process of separation and recovery of the AX and BX mixture from the ion exchanger and resin regeneration. The process scheme is shown in Fig. 3. For complete displacement of the ions separated by the auxiliary C ions in column II it is necessary [7] that the ratio of solution and ion exchanger flow inputs should satisfy the condition

$$\frac{\chi V C_0}{(1-\chi)W m_0} \geq K_C^A$$

where $K_C^A = y(1-x)/(1-y)x$, and y and x are equivalent fractions of A in the ion exchanger and solution, respectively, at equilibrium. In the solution leaving column II there will be auxiliary ions in addition to the separated ones. Therefore, before feeding the solution to column I, CX should be removed from it. The impurity AX concentrated in column II is removed from time to time or continuously.

A similar scheme has been used for example, for the recovery on anion exchanger of uranium as a nitrate complex from its mixture with thorium [17, p. 317], or for the purification of nickel from calcium on cation exchanger [18].

Similarly the process may also be performed when the purpose is to obtain A as AX. The recovery of Rb^+ from its mixture with K^+, using NH_4^+ as an auxiliary ion, can be used as an example [19] of this.

In the scheme in question the processes of recovery of a given component from ion exchanger and resin regeneration are combined. However, some difficulties may arise when removing CX from the solution leaving column II. With ion exchanger having a high affinity for the ions

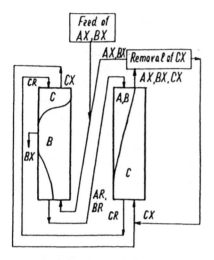

Figure 3 A scheme for purification of product from more strongly sorbed impurities.

separated, such difficulties can be avoided by omitting this operation. Such a process is described in Sec. III.

2. Purification from Impurities Sorbed Less Strongly than the Purified Substance

The product AX, containing a small amount of impurities, is obtained in column II. Ion exchanger is loaded with ions to be separated in column I. Placed at the top of the column is ion exchanger in the auxiliary D-ion form while the AX, BX solution flows into the bottom of the column. The ratio between resin and solution flow is selected to insure that D ions are fully displaced from the ion exchanger. To that end it is necessary that the relation

$$\frac{\chi V C_0}{(1-\chi) W m_0} \geq K_B^D$$

be satisfied. The solution leaving column I contains AX, BX, and DX, the BX content being larger than the sum of AX and BX in the original solution. After removing DX the mixture of AX and BX is returned to column I again and the B impurity is gradually concentrated in column I. Depending on the impurity content of the substance purified, the mixture circulating in column I must be renewed from time to time.

If, in the solution leaving column I, the content of BX is determined, through combination of α, the separation coefficient for A and B ions, the content of B ions in the initial mixture, and the column efficiency, to be high enough, the scheme described may be used to obtain both components of the mixture. In this case the solution leaving column I, after removing the DX compound, is collected as product while DX is moved to column II for displacement. Such a scheme has been applied [20] to separate zirconium and hafnium.

B. Flow Reversal Outside the Column

When for some reason it is not possible to effect flow reversal in one of the columns, e.g., column II, ion-exchange resin leaving the column in the pure A form or as a mixture more enriched in A than in the feed solution would then be completely regenerated in a separate device with the auxiliary salt (CX) solution used for flow reversal in column I. Recovered from solution of the salts, CX and AX, obtained from the regeneration, would be AX. A portion of salt AX would be collected as product and the main portion would be directed to column II; CX would then be utilized upon subsequent regeneration of the ion exchanger. In selecting the CX salt the ease of its separation from AX is an essential requirement. Thus a certain amount of ions A and C are constantly involved in supplementary operations, e.g., ion exchanger regeneration and separation of AX from CX.

In the general case, when it is impossible to carry out flow reversal in both the columns, the number of supplementary operations increases. The scheme for such a process is shown in Fig. 4. The ion exchanger AR, leaving column II, is directed to regeneration which is carried out with electrolyte, CX, solution that should be easily separable from both AX and BX. A part of AX that is separated from the solution of AX and CX obtained upon regeneration is collected as product and the remainder is directed to column II. The regenerated ion exchanger, CR, is treated with solution BX leaving column I. Then it is transformed into BR and sent to column I. The BX and CX components which are obtained during this operation are separated, a fraction of BX being collected as product. The remainder is joined with the solution leaving column I and is used to obtain BR. The solution of CX remaining after the separation of AX and BX is employed in regenerating the ion exchanger. Thus the advantage of the process is that only one auxiliary ion is used. Its drawback arises from the need to perform the operations of ion-exchanger regeneration and separation of auxiliary ions from the product ions.

Flow reversal outside the column has been employed in separating Rb^+ and Cs^+ on sulfophenolic cation exchanger KU-I [21] and in recovering Rb^+ from other alkaline metals with type A zeolite [22]. Hydrogen ion served as auxiliary ion in the first case and NH_4^+ ion served as auxiliary ion in the second.

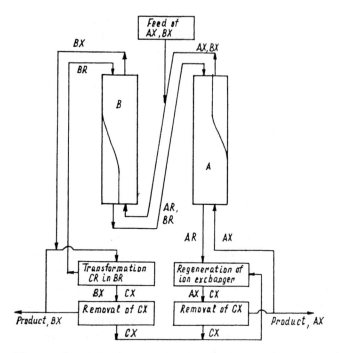

Figure 4   A scheme for a process with flow reversal outside separation columns.

C.   Partial Flow Reversal in the Column

The process scheme with partial flow reversal in column II is shown in Fig. 5. The solution of the mixture of salts, AX and DX, is used for displacement, the auxiliary ion having to be sorbed better than the A ion over the range where the equivalent fraction of the D ion changes from 0 to x_D, the equivalent fraction of the D ion in the mixture used for displacement. The ion exchanger, loaded with ions A and D, leaves column II for regeneration to CR and transfer to column I. The CX is removed from solution upon regeneration and the resultant solution of salts AX and DX is directed to column II.

We see from the above that with flow reversal a certain amount of the mixture of ions A and D is circulating in the apparatus. As a consequence the D ion can play the role of indicator. The presence of the transition zone, from the mixture of ions A and D to the A ions alone at the bottom of the columns, is characterized by appreciable change in the properties of the ion exchanger–solution system. This enables the establishment of automatic control for column operation. A similar scheme of flow reversal may be utilized in column I.

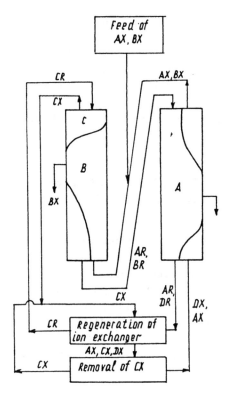

Figure 5 A scheme for a process with partial flow reversal inside the separation columns.

Such a method for flow reversal proved rather convenient in the purification of cesium salt with cation exchanger KU-I [5]. The Ba^{2+} ion was used as the D ion and H^+ ion as the C ion.

Comparison of the different schemes employable in various ion-exchange processes shows that the most suitable and simplest are those with flow reversal carried out directly in the column using one auxiliary ion. In the processes employed with this scheme the number of supplementary operations is kept to a minimum since the ion exchanger leaving column II can be used at once to sorb separated ions in column I.

III. SEPARATION OF MULTICOMPONENT MIXTURES

Either a more strongly or a more weakly sorbed component may be recovered from mixtures in the separating column with flow reversal at one end if the column is long enough. The recovery and purification of mixture components having intermediate sorbability present a more complicated problem.

This problem is generally solved in a two-stage approach. During the first stage the substance to be purified is separated from either more strongly or less strongly sorbed impurities. During the second stage the remaining impurities are removed.

Let us consider first the likely ways to solve the problem dealing with recovery of B ion from a three-ion mixture, in which the order of sorbability is C<B<A [23].

A. Preliminary Separation from More Strongly Sorbed Component Mixtures and Subsequent Resolution of Pure Product

The first step of the above process involves the frontal mixture separation in countercurrent column I on ion-exchange resin where auxiliary D ions are sorbed more weakly than the ions to be separated.

If the column is sufficiently long, the concentration distribution presented by the scheme in Fig. 6a will be reached in the prescribed time interval. The less strongly sorbed C component is concentrated below the auxiliary ion zone. Somewhat lower than the C zone are the C and B zones. Contained in the zone near the entry of the original solution are the three ions.

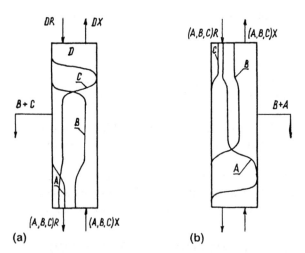

(a)   (b)

Figure 6   Distribution of component concentrations as a function of column height during separation of a three-component mixture: (a) frontal separation; (b) displacement separation.

It is possible to collect either solution or ion-exchange resin suspension from the zone which contains C and B ions and an A content which is reduced to the level required for the subsequent production of B.

For a more complete utilization of the original mixture the recovery section, in scheme 6b, is needed. In it, ion-exchange resin leaving column I is treated with solution containing the auxiliary ion that is sorbed more strongly than the separated ions. Upon displacement, A ions are concentrated near the auxiliary ion zone.

The C and B mixture collected is directed to the second countercurrent column where purification from C ions is affected. These process schemes are shown in Fig. 7.

If ion-exchange resin suspension has been removed from column I, the release of B is affected by displacing the mixture from the ion-exchange resin with the auxiliary E ion. The solution leaving column II is directed to column I (Fig. 7a).

Provided that the mixed solution of CX and BX is removed from column I, column II, with flow reversal at both ends should be used for the ion-exchange recovery of B. The C^+ ion is obtained as a by-product on column II's upper section which serves for the mixture transfer into the ion-exchange resin. The B substance is concentrated in the lower section upon displacement (Fig. 7b).

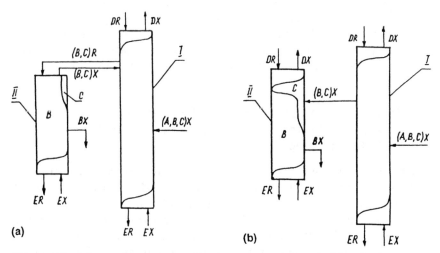

Figure 7 Schemes for recovery of a component with sorbability both stronger and weaker than other components after removing the better sorbed ones: (a) collection of resin suspension (B, C)R; (b) collection of solution (B,C)X.

In case impurities are of no value and the only task is to recover the intermediately sorbable B component, the ion-exchange resin on which frontal separation is carried out must not contain impurities that sorb better than B and the solution of E displacer must not contain impurities that sorb less effectively than B. On this basis, A and C ions can be applied as auxiliary ones. This is sometimes done when separating mixtures of rare earth elements.

A scheme for the process of recovering the component with intermediate sorbability is simplified if it can be separated from the less sorbable C component in some simple way, such as by evaporation, crystallization, or by transformation into a poorly soluble compound.

It is suitable to perform frontal separation on the ion-exchange resin loaded with the more weakly sorbed C mixture component and to carry out the recovery of the desired B component from the ion-exchange resin obtained upon frontal separation either with the same C ion or with A and C mixtures at such a ratio of the phase flows which provides complete displacement of B from the ion exchanger.

To complete the cycle the ion-exchanger regeneration stage for replacement of the A ions remaining in it with C ions is required.

An example of a process such as this is one where strontium is recovered and concentrated from the mixtures in which it is present together with a large excess of sodium and a considerable amount of calcium, natural brines, and concentrates of seawater treatment.

The scheme for such a process is shown in Fig. 8 where $C=Na^+$, $B=Sr^{2+}$, and $A=Ca^{2+}$, and concentrations in the starting solution of this example are 2.5, 0.01, and 0.07 equiv/dm^3, respectively. The separation has been performed on carboxylic cation exchanger KB-4. Strontium carbonate has been precipitated from the product solutions of $SrCl_2$ and NaCl [24].

B. Preliminary Separation from More Poorly Sorbed Impurities Followed by Production of the Pure Substance Sought

Purification from more weakly sorbed C ions is affected in the countercurrent column with flow reversal at both its ends and by providing additional feed solution to the inlet at the middle of the columns.

Unlike the scheme considered in the previous section output is provided for below the zone accessible to the original solution.

If the column is sufficiently long, the more weakly sorbed C ion will be removed from the bottom portion of the separated ion zone in the time interval selected. Ion distribution in the bottom part of the column is as represented by the diagram in Fig. 6b.

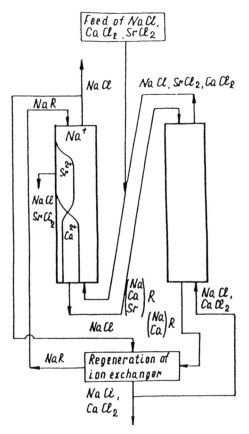

Figure 8 A scheme for strontium recovery from mixtures of sodium and calcium chloride.

Solution can be collected from the zone containing B and A while the BX can be obtained from it upon frontal separation (Fig. 9). The ion-exchange resin leaving column II returns to the zone of column I from which solution is collected. Concentrated A and C impurities are removed from time to time at the ends of column I near flow reversal zones. It is also possible to use an auxiliary section in column II to displace separated ions from ion exchanger by the better sorbed E ions.

C. Separation of Mixtures with Any Number of Components

The schemes described above can also be employed to separate one substance from mixtures containing more than three components. In this

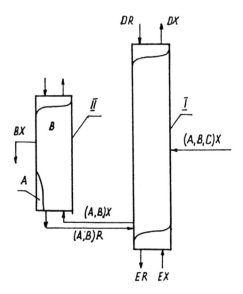

Figure 9 A scheme for recovery of a component with sorbability both stronger and weaker than other components after removing the weaker sorbed components.

case C is used to represent the sum of all ions which are sorbed less effectively than the one recovered while A represents the sum of all ions which are more effectively sorbed than the one recovered.

Now consider a scheme for the resolution of all components of the above mixture (Fig. 10) which is fed into the middle inlet to countercurrent column I, and in which flow reversal is effected at both ends. Suppose the sorbability of the separated substances bear the following relationship: $A > B > C > D$. Some time after the beginning of column operation the most strongly sorbed A component would be removed from the upper part of the column while the most weakly sorbed D component would be removed from the bottom part.

Two directions for proceeding with the separation seem most suitable. In the first [25], a solution containing mixtures B, C, and D is selected for subsequent separation from the upper part of the column. This solution is fed into column II. The component B is concentrated in the bottom part of column II while C, D mixtures are withdrawn from the upper part and separated in column III. In the second approach, A, B, and C solution is selected from the bottom part of column I. Recovered from this mixture in column II is either A or C while the binary mixture recovered from another section is moved to column III.

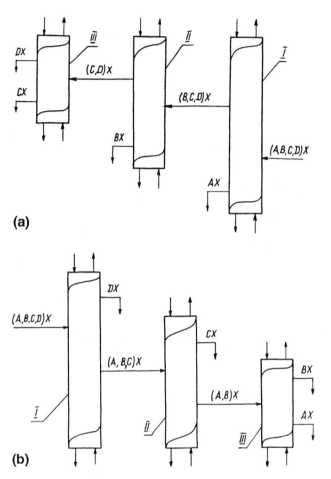

Figure 10 Schemes for complete separation of four-component mixtures after removal of (a) more strongly and (b) less strongly sorbed components.

IV. SEPARATION OF MIXTURES BY CHANGING THE DIRECTION OF INTERPHASE TRANSFER

In ion exchange practice, the means to the design of continuous separation is flow reversal with the help of auxiliary ions. However, this approach has a number of disadvantages. The application of auxiliary ions results in reagent expenditure and supplementary ion-exchange regeneration operations. In effecting the high degree of purification sought for a particular component auxiliary ions may be a source of impurities. As a result a large number of ion exchangers highly selective toward one or another ion mixture are not used for separation. The infrequent use of synthetic zeolites for such separations is supportive of the above statement.

The drawbacks mentioned above can be avoided or minimized if another approach to the design of a continuous process, namely, through the change of interphase transfer direction at the column ends, is used. Such a path is employed in the technique of isotope separation by chemical exchange in a two-temperature scheme [26,27]. Auxiliary substances are not needed with this method. The concentrated component is recovered from the phase leaving the column by the original composition mixture which is in the phase entering the column. For separation by this method it is necessary that equilibrium separation coefficients at the column edges should be different. For example, the displacement of the separated mixture from the ion-exchange resin must be carried out with solution of the same mixture having the initial composition under conditions providing decreased selectivity of the ion exchanger for the recovered component.

Further clarification is provided by Fig. 11. Given in Fig. 11a is the scheme for conventional ion-exchange separation in a countercurrent column to obtain a more effectively sorbed component. To displace the separated ions from the ion-exchange resin auxiliary ions are applied. The distribution of the concentrated component along the column height is

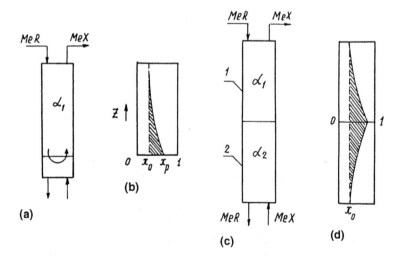

Figure 11 Concentration of a more strongly sorbed component: (a,b) the process with flow reversal; (c,d) the process with changes of interphase transfer direction; (a,c) schemes for the process; (b,d) distribution of the concentrated component along the column height.

shown in scheme 11b. Represented in Fig. 11c is the scheme for changing the interphase transfer direction. The countercurrent column consists of two sections. The upper one is analogous to the separation section of the column in scheme 11a while the bottom section replaces the section of the column where flow reversal has occurred. Fed into it from the top is ion exchanger from section I and fed from the bottom is the original solution mixture. The conditions are fixed to insure that the equilibrium separation coefficient α_2 is less than α_1. In this case excess amount of concentrated component proportional to the difference between $\alpha_1 - \alpha_2$ is eluted from the ion-exchange resin by the solution and returns to section I. If $\alpha_2 = 1$ section I (scheme 11c) will operate in the same way as the column with flow reversal. Nearly the same process will take place if $\alpha_2 \neq 1$ as long as $\alpha_2 < \alpha_1$. In other words, inequality of α_1 and α_2 is essential.

A scheme like this is employed when separating hydrogen isotopes in the course of isotopic exchange between its various compounds. The change in the separation coefficient, α, achieved by changing temperature, gives the method its name.

In the two-temperature method [26] separation efficiency depends on differences in the values of α_1 and α_2. In addition, it is characterized by the specific dependence of enrichment on the ratio of flows of the separated mixture with the phases injected into the column providing a sharp optimum.

The application of a two-temperature scheme for separating Li^+-NH_4^+ and Cs^+-Na^+ mixtures in a fixed bed of sulfonate cation exchanger is described in the literature [28,29]. However, such an approach has not been developed further because of the complexity of moving temperature zones and the need to shift points of solution input and output. Besides temperature change in simple cases of ion exchange with sulfonated cation exchanger does not influence changes in selectivity sufficiently.

This was confirmed by the results of research [30] in which the two temperature method was employed for separation of mixtures of Ca^{2+}-K^+ and Fe^{3+}-Cu^{2+}-H^+ with sulfonate cation exchanger while employing countercurrent motion of phases.

By using countercurrent columns the two-temperature concept becomes more acceptable since the selectivity of ion-exchange resins may be changed by varying such process conditions as solution concentration, solution pH, the composition of solvent, and the presence of complexing reagents.

Presented in Fig. 12 is a separation scheme in which use is made of the change in ion-exchanger selectivity affected by altering the solution concentration [31,32]. Selectivity changes such as these are particularly large for exchanging ions of different charge.

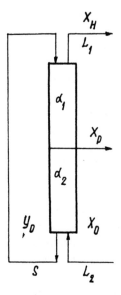

Figure 12 Separation effected by solution concentration change at the sections' boundary.

The separation is carried out in a two-section countercurrent column. Ion exchanger loaded with the ions to be separated is moved from top to bottom. After its removal from the bottom of the column, the resin is redirected to its upper section without any treatment. Solution of the same ions is fed into the bottom of the column and is removed at the top. Dilution or concentration of the solution is accomplished at the section boundary.

For simplicity consider the case where the content of one of the components is small ($x \ll 1$ and $y \ll 1$).

At the beginning of the column operation the composition of ion exchanger leaving the column, y_0, is equilibrated with entering solution of composition x_0

$$y_0 = \alpha_2 x_0$$

The composition of ion exchanger fed into the top part of the column and that of the solution leaving it are also relatable by the equilibrium condition

$$y_0 = \alpha_1 x_H$$

It follows from these relationships that

$$x_H = \frac{\alpha_2}{\alpha_1} x_0$$

In the solution being diluted the selectivity toward ions of greater charge is increased. In this case, then, $\alpha_2 < \alpha_1$ and ions of greater charge are concentrated in the middle portion of the column.

When the solution increases in concentration the opposite is true and $\alpha_2 > \alpha_1$ so that ions of smaller charge are concentrated. The range of ion ratios associated with the solution, $L = XVC_{03}$, and ion exchanger, $S = (1 - X)WM_0$, where separation is occurring, is estimated as follows.

For a particular component to be concentrated it is necessary that at each edge of the column the amount entering with one of the phases should be larger than the amount leaving the column with the other, i.e., the following inequalities should prevail:

$$L_2 x_0 > S y_0 \quad \text{and} \quad L_1 x_H < S y_0$$

Operating without collection of product $L_2 = L_1$ (i) it follows from the above that

$$\frac{y_0}{x_0} < \frac{L}{S} < \frac{y_0}{x_H}$$

the difference between the composition of phases will reach a maximum when interphase equilibrium prevails at the edges of the column. For linear equilibrium relations

$$\alpha_2 < \frac{L}{S} < \alpha_1$$

Similarly for concentration of the second component the following inequalities should apply:

$$\frac{1}{\alpha_2} < \frac{L}{S} < \frac{1}{\alpha_1}$$

A method such as this has been used [31,32] for the separation of Ca^{2+} and K^+. The sulfonate cation exchanger, KU-2×9, was saturated with these ions by employing 0.7 M mixtures of calcium and potassium chlorides, containing either 5% or 50% KCl. Initiation of the procedure consisted of introducing the original solution into the column from below. The ion exchanger, having been previously equilibrated with the original solution, was then injected from above. Water was fed into the middle part of the column until the concentration in the upper section decreased to 0.17 M. The equilibrium separation coefficient, α_K^{Ca}, varied from 2.0 to 4.7 (for the mixture containing 5% KCl). The distribution of separated ions along the column at different times is shown in Fig. 13. Solution containing less than 0.1% KCl was collected from the middle part of the column at the conditions selected.

Dual-temperature based theory for linear equilibrium isotherms has been developed [26]. From the results of this analysis the optimal value of flow ratio in the simplest case of nonselective conditions for concentrating the least strongly sorbed component can be found with the following expression:

$$\left(\frac{L}{S}\right)_{opt} = \frac{\alpha_1 N_1 + \alpha_2 N_2}{N_1 + N_2}$$

Here N_1 and N_2 are numbers of transfer units in the first and second column sections, respectively. For concentration of the more easily sorbed

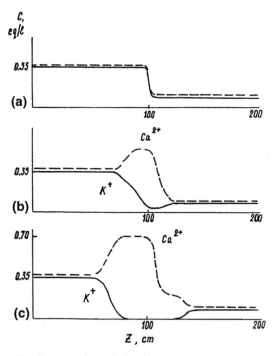

Figure 13 Distribution of ion concentrations along the column (sulfonate ion exchanger) KU-2 × 9; C_2 = 0.7 g-equiv/dm³; $\overline{X}$ = 0.5; V_2 = 0.38 cm/min; W = 0.1 cm/min; C_1 = 0.17 g-equiv/dm³: (a) initial distribution; (b) distribution after 22 hr; (c) distribution after 70 hr.

component a similar relation can be obtained where α_1 and α_2 are replaced by $1/\alpha_1$ and $1/\alpha_2$. These relationships were confirmed by the results of research [32] carried out in studies of the separation of Ca^{+2} from K^+ in their chloride mixtures. Kinetic data obtained for the process when the ratio of flows, L/S, was close to optimal are presented in Fig. 14.

With L/S > $(L/S)_{opt}$ concentrated calcium is distributed mainly in the upper section of the column; in the lower section the mixture retains a composition close to the original one. With L/S > $(L/S)_{opt}$ the zone enriched by calcium is shifted into the lower section. The greater the deviation from optimum conditions, the lower are the curves $q = f(z)$. This way maximum enrichment is identifiable near the water inlet. The differences noted above can, on this basis, serve as a criterion for determining the direction of deviation of the chosen flow ratio from the optimum value.

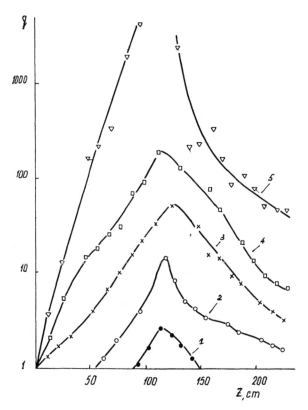

Figure 14 Enrichment coefficient (q) along the column (z) for different periods of time after starting column operation (resin KU-2×9; C_2 = 0.7 g-equiv/dm³; $\overline{\overline{X}}$ = 0.95; V = 0.62 cm/min; W = 0.24 cm/min; water injection 1.95 cm/min): (1) in 4 hr; (2) 11 hr; (3) 28 hr; (4) 49 hr; (5) 97 hr.

The combination of two countercurrent columns, one where dilution of solution takes place and the other where concentration occurs, enables one to design the process for obtaining both of the mixture components. In this process the solution leaving one of the columns is used to feed the other one. No auxiliary ions are needed to effect separation nor is the ion-exchanger regeneration stage required. The only auxiliary operations consist of diluting and concentrating the solution.

Another example in which change in the interphase transfer direction is made use of is provided by separation processes based on the dependence of selectivity and sorbability of polyfunctional ion exchangers upon the solution pH [33].

Ion exchange with different exchange groups of polyfunctional ion exchangers takes place at various pH values, selectivity of the different exchange groups toward the same ions being different as a rule.

For example, the equilibrium distribution of cesium and rubidium between sulfophenolic cation exchanger KU-I and neutral solution is characterized by the value $\alpha_{Rb}^{Cs} = 1.5–1.6$ while in alkaline solution $\alpha_{Rb}^{Cs} = 1.8–2.0$. In the latter case sorption capacity is 2.5–3.0 times as great as it is in neutral solution [34].

Let us consider process schemes by using the bifunctional sulfophenolic cation exchanger in a sample separation. In scheme "a" (Fig. 15) the ion exchanger, converted to a mixed ion form with ions sorbed only by its sulfonic groups, is fed into the column from above and alkaline solution of the same ion mixture is introduced to the column from below. Phenolic groups of the ion exchanger react with the alkali, neutralizing it. A rather sharp stationary boundary dividing the alkaline and neutral solution zones is established in the column. When the ion flows of alkali solution and ion exchanger (phenolic groups only) are equivalent, the boundary will remain stationary and the equilibrium separation coefficients, α_1 and α_2, in the top and bottom sections will differ, creating the conditions necessary for "two-temperature" separation. Since the enrichment attainable depends on the ratio of the separate mixture flows, a mixture of salts is added to the solution to obtain the desired magnitude of flow ratios.

At the border between the different pH zones the transfer of a portion of the ions from the solution to the ion exchanger proceeds by sorption of phenolic groups, thereby combining "two-temperature" separation with partial flow reversal*. Upon separation with flow reversal in the upper part of zone II the weaker sorbed component should be concen-

---

*The idea of providing a combined scheme of separation in general form, irrespective of the particular technological process, has been described in Ref. 35.

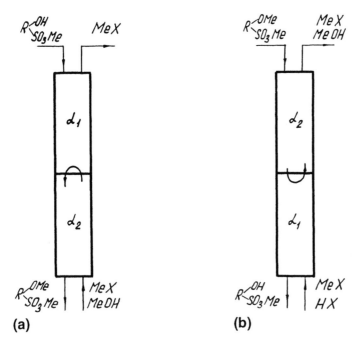

Figure 15 Separation affected by change of selectivity and sorbability of bifunctional ion exchanger: (a) sorption of mixture by weakly dissociated groups; (b) partial desorption of mixture.

trated. The separation direction in a "two-temperature" scheme is determined by the ratio of the equilibrium separation α_1 and α_2. For example, in exchanging Cs^+ and Rb^+ the selectivity of sulfophenolic cation exchanger toward cesium ions in alkaline medium is higher than in a neutral one [34], i.e., $\alpha_2 > \alpha_1$. In this case the separation with a "two-temperature" scheme follows the same direction of Rb^+ enrichment as it does with flow reversal.

Partial displacement of the separated ions from ion exchanger completely saturated with them is provided (Fig. 15b) to concentrate the second component of the mixture. To accomplish this, acidified solution of salts of the separate ions is used. Formed in the column is the stationary front between the acidic and alkaline solution zones. Separation coefficients in the two zones are different. Thus the two principles of operation are combined here as well.

Shown in Fig. 16 is a scheme for the continuous production of both components of a mixture. Polyfunctional ion exchanger containing the ions to be separated is passed successively through columns I and II and

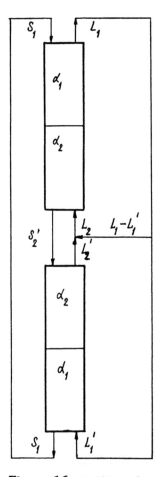

Figure 16 A scheme for recovery of two components from their mixture.

is returned to column I without any treatment. Solution of the separated mixture circulates in the opposite direction. Its pH value, changed before feeding it to the column, provides both additional sorption in column I and displacement of separated ions from a part of the exchanging groups in column II. Concentrated components are withdrawn from the middle part of the columns and an additional quantity of the original solution mixture is added to the earlier column I input.

Since the same amount of ion exchanger moves through the columns, i.e., S_2 and S_1 flows are fixed, the solution flow leaving one of the columns will not correspond to the flow needed to perform optimum separation in the other column. For this reason only part of the solution flow leaving column I is utilized to feed column II while the remainder is combined with the solution feeding column I.

The separation theory applicable to the "combined" scheme has been considered in the references listed [36, pp. 70–82, 37–40]. Here only the results of this analysis are presented.

The rate of concentrating components and the efficiency of their separation depend on (1) the change in ion-exchanger capacity upon the transfer from neutral to alkaline solution (or from alkaline to acidic solution); (2) the difference in equilibrium separation coefficients on completely saturated ion exchangers and on ion-exchanger saturation limited to its sulfonic groups; (3) the ratio of mixture flows injected into the column with solution and ion exchanger.

The third item, as shown in [33], depends on the ratio of the concentration of ions in the feed solution to the concentration of alkali or acid in the same solution.

The increase in capacity and differences between α_1 and α_2 result in the growth of separation efficiency. The dependence of the enrichment coefficient upon the ratio of flows (or concentrations) is represented by a curve with a maximum.

One important feature characteristic of the processes under consideration should be noted. Movement by the sorption front between alkaline and neutral zones or acidic and alkaline ones relative to the column walls is matched by the movement of the section 1 and 2 border. In particular it becomes possible to carry out separation in a fixed bed of ion exchanger. In both cases, namely using a fixed bed or employing a countercurrent column ion exchanger, regeneration is not required.

Typical experimental curves obtained with this new chromatographic method are presented in Fig. 17. In these experiments the solution of cesium and rubidium ($x_{Cs} = 0.1$), salt mixture (0.45M), and alkali (0.3M) was passed through a fixed-bed column filled with sulfophenolic cation exchanger KU-I (the bed height was 28 cm., its diameter was 8 mm) previously equilibrated with 0.45 M solution of cesium and rubidium nitrate (1:9).

With the help of experiments such as these optimum flow ratios for continuous separation in countercurrent columns can be found. By carrying out a series of experiments in the column with different ratios of salt and alkali concentration the dependence of maximum degree of separation upon flow ratio may be established (Fig. 18) and then used to provide the optimum composition for separation in the countercurrent column.

The procedure for finding optimum flow ratio is considered in detail in [41].

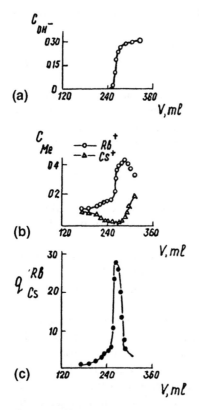

Figure 17 Plots of (a) OH⁻-ion concentration versus (b) separated ion concentrations; (c) enrichment coefficient versus volume of eluted solution.

Kinetic curves for cesium purification in a column with sequential phase motion are given in Fig. 19 to illustrate this.

V. PROCESSES FOR REPLACING IONS OF ONE KIND BY IONS OF ANOTHER KIND

Countercurrent ion exchange, besides being used for separating mixtures, may also be used to solve other practical and important problems, e.g., for recovery of valuable and/or toxic substances from solution and pulps, for concentrating solutions, for softening and demineralizing water, for separating nonelectrolytes from electrolytes, and for synthesis of different compounds.

The particular scheme for processes such as these is simpler than those applicable to separation processes. Two countercurrent columns

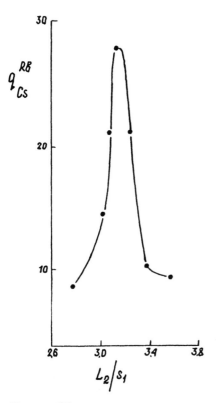

Figure 18 Plots of maximum enrichment coefficient values versus flow ratios.

are employed. In one of them ions are sorbed from the solution by ion exchanger, and in the other column ions are recovered from the ion exchanger which at the same time is prepared for subsequent use in the first column.

Shown in Fig. 20 is a scheme for countercurrent softening of water [9,42]. Divalent cations, Me^{2+}, are sorbed from strongly acidic cation exchanger in the Na^+ form. Ion exchanger regeneration is then affected with the concentrated solution of NaCl softened earlier by the replacement of all multicharged cations with sodium.

A similar scheme is used [15, p. 312] for recovering uranium from sulfonic acid solutions. The solution containing $5 \times 10^{-4} - 10^{-3}$ moles per liter of uranyl sulfate at pH = 2 is treated with anion exchanger (Fig. 21). Uranium is almost completely recovered from solution by the resin. Solution containing 20 g/L of uranyl sulfate is then eluted with H_2SO_4. Anion exchanger in the sulfate form is transferred to column I.

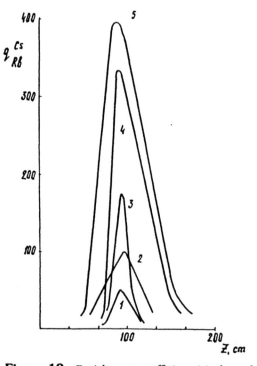

Figure 19 Enrichment coefficient (q) along the column during countercurrent concentration of cesium for different periods of time after starting column operation: (1) 4 hr; (2) 13 hr; (3) 19 hr; (4) 20 hr; (5) 21 hr; (0.207 mol/dm³ solution of $RbNO_3$ and $CsNO_3$ mixture with molar ratio of 1:1 in 2.25 M acetic acid; V = 87 cm/min; W = 27 cm/min).

The recovery of uranium from hydrochloric acid solutions and its concentration are described in [43]. In this operation, 10 M HCl solution containing uranium traces is treated with strongly basic anion exchanger. Upon washing the ion exchanger with water the effluent of uranyl chloride at a concentration level as large as 200 g/L uranium is obtained.

Countercurrent ion exchange is rather widely used for the demineralization of water.

A scheme for the demineralization of bicarbonate waters [9] using carboxylic acid exchanger is presented in Fig. 22. In column I divalent cations react with the ion exchanger as shown: $Me(HCO_3)_2 + 2\ HR = 2H_2O + CO_2 + MeR_2$. Ion exchanger is then regenerated in column II with an equivalent quantity of acid.

In general, multivalent and univalent cations, as well as anions such as chloride, sulfate, and silicate, are removed from water sources in the course of their demineralization. In these instances water must be treated

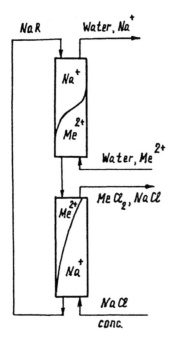

Figure 20 A scheme for countercurrent water softening (cation exchanger Dowex-50 × 8).

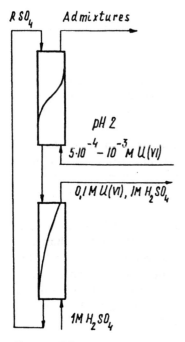

Figure 21 A scheme for uranium recovery from sulfonic acid solutions.

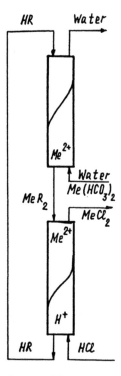

Figure 22 A scheme for bicarbonate water demineralization (cation exchanger Amberlite IRC-50).

with cation exchangers in the H^+ form and with anion exchangers in the OH^- form. Exhausted ion exchangers are regenerated and reused in the process. As a consequence, each operation is carried out in countercurrent apparatus consisting of two columns (or in a two-sectional column). The operational schemes are analogous to those shown in Fig. 20. Complete schemes of the processes can be rather complicated. The difference between them lies both in the order of water treatment by cation or anion exchangers and in their chemical nature (strongly dissociated or weakly dissociated).

Successive water treatment with strong acid and strong base exchangers is described in [44,45]. Water is passed through exchanger in the H^+ ion form. The H^+ ion released to the water is neutralized by OH^- ion released from the anion exchanger. The net result is transfer of solute ions from the water to the ion exchangers. The application of fully dissociated ion exchangers makes it possible to remove both strong and weak electrolytes from water. However, considerable excess of acid and alkali is required for regeneration of the strong-acid and -base ion exchangers. As a consequence, the schemes using weak (acid, base) ion exchangers are frequently employed [44, 46–49]. Their full regeneration is provided by quantities of acid and base no greater than their capacity.

The following demineralization scheme has been used [50]. In the first column water is treated with a fully dissociated ion exchanger in the H^+ ion form. Then after removing CO_2, treatment with a weak-base anion exchanger is carried out. At this stage anions of strong acids are removed from the water. To eliminate silica from the water it is passed through a fixed bed of strong-acid ion exchanger.

In one paper [51] examining cation-exchanger treatment of water a mixture of equal amounts of strong- and weak-acid ion exchangers was employed.

Schemes using only weak-acid and weak-base ion exchangers have been considered as well [52,53]. In the first column illustrating this (Fig. 23) water is treated with a weak-base anion exchanger in the bicarbonate form. All anions are replaced by bicarbonate ion. In the second column the bicarbonate solution obtained is treated with weak-acid ion exchanger in the H form. Because carbon dioxide is formed all the reactions of the type

$$MeHCO_3 + RH \rightarrow H_2O + CO_2 + RMe$$

are shifted to the right and the water obtained contains only carbon dioxide. Regeneration of cation exchanger is accomplished with the equivalent amount of sulfuric acid, whereas regeneration of anion exchanger is with solutions of ammonia or lime slurry. When passed through regenerated weak-base anion exchanger in the OH form in column II, water is freed from carbonic acid and the ion exchanger is partly transformed into the bicarbonate form. After additional treatment of the anion exchanger with CO_2 it is directed into column I.

The scheme in question has been somewhat modified [53]. Only two columns are used. The original water is saturated with CO_2 and is treated with weak-base exchanger in the OH^- form. The bicarbonate mixture obtained is treated with weak-acid exchanger in the H^+-ion form to remove cations. Demineralized water is finally obtained as the product after the removal of carbonic acid.

Countercurrent demineralization of water in a mixed bed is described in [44]. The main difficulty in using a mixed bed is encountered in the separation of cation and anion exchangers before regeneration. Use is made of the difference in ion-exchanger densities to achieve this end. Ion exchangers regenerated in countercurrent columns are put into a mixer.

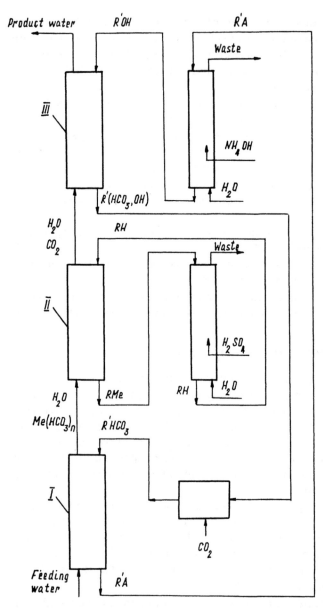

Figure 23 A scheme for water demineralization by treatment with weak-acid and weak-base exchangers.

An industrial unit [46] consisting of several countercurrent columns has been described. In the first stage water is treated with a weak carboxylic acid exchanger in the H^+ form (see Ref. 9). Next, after removing CO_2, the water, demineralized to a considerable extent, is treated with a mixed bed. Demineralization processes with a mixed bed are described in Refs. 48 and 54.

Ion-exchange-based production of a number of different compounds has been described in some detail [43,55]. The conversion of NaCl to NaOH was carried out in the countercurrent column using strong-base exchangers [43]. The ion exchanger was regenerated with 1 mol/L $Ca(OH)_2$ suspension.

To obtain potassium nitrate from chloride the strong sulfonic acid cation exchanger was converted to the K^+-ion form and treated with 4 M HNO_3 [43] or with 3–5 M NH_4NO_3 [56].

To produce substituted potassium phosphate, phosphoric acid was passed through a weak-base exchanger. The ion exchanger, eluted with KCl solution, then yielded the solution of K_2HPO_4 being sought. The chloride form of ion exchanger produced was regenerated with $Ca(OH)_2$ slurry.

The production of KOH from KCl has been described [55]. In this paper the use of countercurrent columns for demineralization of sugar is recommended.

The expediency, in practice, of a number of ion-exchange processes such as synthesis has been demonstrable, for example, by the ready accessibility, with this approach of sufficiently pure products as concentrated solutions with the volume of waste water kept to a minimum.

Such success was encountered when producing KNO_3 from NH_4NO_3 and KCl [56]. The process scheme is presented in Fig. 24.

Production of KNO_3 and regeneration of cation exchange with KCl solution are provided in countercurrent columns 1 and 2. In performing both processes stationary sorption fronts form.* In units 3 and 4 the solution in contact with ion exchanger is replaced. For this purpose ion exchanger is first washed with water. The solutions of NH_4NO_3 and KCl are used to prepare 3–4M solutions. Then any residual water is removed and the ion exchanger is filled with the product solution for transfer to the proper countercurrent column.

---

*It was discovered in Ref. 57 that the selectivity of sulfonate cation exchangers toward K^+ and NH_4^+ ions was reversed once the concentration of their nitrates exceeded 2 M. At > 2 M NH_4^+ ions are better sorbed.

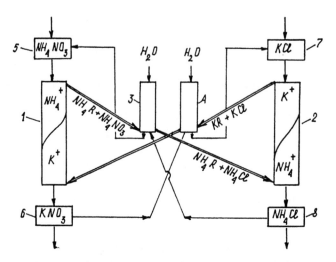

Figure 24 A scheme for KNO_3 production (sulfonate cation exchanger KU-2 × 8; 3.0–4.0 mol/dm^3 solutions): (1) column for KNO_3 synthesis; (2) resin regeneration column; (3,4) systems for electrolyte displacement; (5–8) tanks.

VI. UNITS USED FOR COUNTERCURRENT ION-EXCHANGE PROCESSES

For more than a half century of countercurrent ion-exchange history a great number of different contactors have been recommended for the countercurrent movement of solutions and granulated ion exchangers.*
Several literature reviews [1–4,59–61], where the design of units and their operation principles are discussed, have appeared. For this reason discussion of this subject is restricted to an outline of the principles of operation of contractors suitable for carrying out the processes considered above. Particular attention is focused on efficient countercurrent units, that is, columns characterized by rather low values of the height equivalent of a theoretical plate (the height of a transfer unit).

Contactors in which countercurrent motion is used can be classified as follows: (1) contactors with continuous motion of an ion-exchanger compact bed; (2) contactors with a suspended or fluidized bed of ion-

---

*It should be noted that long before the advent of the first countercurrent ion-exchange column (Nordel, [58]) research using a countercurrent solution and a granulated sorbent for the purification of sugar solution by activated coal had been performed. However, poor mechanical stability of the sorbent [59] discouraged wide applicability of the approach.

exchange resin; and (3) contactors with movement, in turn, of solution through a fixed compact bed and of the bed, after solution movement is stopped.

Units with vertical columns in which the motion of granulated sorbent as a dense bed is gravity-inspired counter to the upward flow of solution are the simplest countercurrent systems. They are characterized by rather low HETP values. Their efficiency, however, is not high since a sufficiently dense bed of ion exchanger is maintainable only at low rates of solution flow. The solution flow rate is not more than 1–3 m/hr.

In columns with a fluidized bed, where solution moves up and ion exchanger moves down, advantage can be taken of these solution velocities to widen the ion-exchanger bed by a factor of 1.5 to 2.0. Under these conditions ion exchanger and solution are vigorously mixed and the HETP values reached are much higher than can be reached with a fixed dense bed. Cylindrical or conical columns are employed. To enhance efficiency (lower HETP values) they are divided into sections or different packings are used. A fluidized bed is created by mechanical (vibrating) or pneumatic (pulsating) mixing. Units with a fluidized bed are considered in detail [4,61]; the operation of particular types of columns is dealt with in [2–4,59,61].

Recovery of substances from muddy solutions or pulps can be effected in contactors of this group. They are comparatively simple in design and are easy to maintain. Their specific efficiency is considerably higher than with dense-bed columns in continuous motion (10–20 m/hr). But high values of HETP affected in such operation do not allow them to be employed for separating mixtures of components with rather similar properties.

Units with phases alternately in motion (up to 150 m/hr) are characterized by high specific efficiency and satisfyingly low HETP values. As a consequence, they have found wide application in large-scale production in spite of their more complex design.

A. Columns with Gravity-Induced Ion-Exchanger Motion

Columns whose flow of ion exchanger is gravity controlled have been described in the literature [62–67]. The scheme for such column operation is presented in Fig. 25 where the combination of a screw for the removal of ion exchanger from the column at the desired speed and a system consisting of two tanks for the output of ion exchanger is pictured. Continuous motion of ion exchanger is facilitated with such columns.

Ion exchanger, exposed to gravity, moves down and is transferred to the bottom of tanks 5 and 6 via screw action. The solution, moved at a constant rate, is transferred to tank 5 with a hose pump. After filling tank

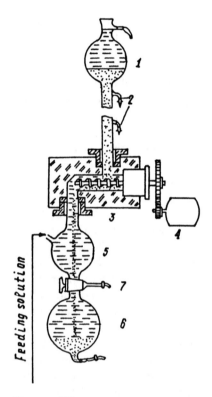

Figure 25 A scheme for countercurrent column operation in the laboratory: (1) top resin tank; (2) glass column; (3) screw; (4) electric motor; (5,6) bottom resin tanks; (7) resin valve.

6 with ion exchanger communicating tap 7 is turned off and ion exchanger is removed. At the same time the tank is filled with solution feeding the column through an inlet controlled by valve 7. Along the height of the column there are capillary taps with filters for collecting samples for examination.

Automated units with columns of the same type have been described in [68]. The operations of loading, unloading, and transferring ion exchanger (hydrotransport) are mechanized and controls for positioning the sorption fronts and regulating the operating conditions are automated. The scheme of this unit is represented in Fig. 26.

Constancy of the position of flow reversal regions with respect to the column walls is achieved by automated regulation of solution fed into the column.

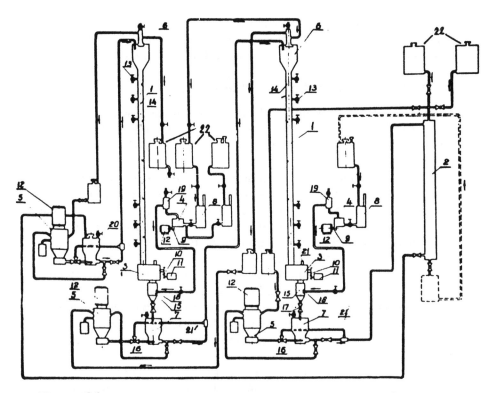

Figure 26 A scheme for automation unit: (1) countercurrent columns; (2) regeneration column; (3) screws; (4) feed pumps; (5) pumps for moving resin; (6,7) top and bottom resin tanks; (8) solution feeders; (9) electric motors; (10) valves-collectors; (11) platinum contacts; (12) screw tanks; (13) jets; (14) resin valves; (15) solution distributions; (16) a delivery cap; (17) tanks for regenerated resin; (18) bypass pipe; (19) solution tanks.

Two shoulders that are neighboring parts of the column provide support for a bridge sensor that is employed to measure conductivity. The application of a scheme such as this allows one to compensate, to a large extent, for interference arising from the motion of phases.

Change in the position of the sorption front, which is in the part of the column positioned between electrodes, results in bridge imbalance that is registered by a potentiometer. Divergence of the needle, as a result, breaks the contact and a relay reacts. Resistance is increased if the sorption front has risen. If the front has lowered, resistance is reduced. Response of the electric motor of the proportional pump controlling solution flow in response to such signaling with respect to sorption front position decreases or increases the rate of solution input and the front begins to return to its original position. The imbalance, so reduced, returns the potentiometer needle contact and relay (and resistance) cut off.

B. Columns with Alternate Phase Motion

The results of investigating the dynamics of ion exchange (see Sec. VI) show that the lowest value of HETP or HTU is obtained by exchange of ions in a dense bed. When the bed is less densely packed column efficiency is decreased.

Since continuous transfer of ion exchangers as dense beds in the face of solution counterflows at a high rate in cylindrical columns is not practicable, the problem of exchanger transport was solved by alternating the movement of solution through the fixed dense bed with the movement of the sorbent for the transfer of its exhausted portion and for the addition of its replacement.

One of the most remarkable units of such a type is the Higgins contactor. The apparatus, in the form of a U, has sections with different diameters and heights. Depending on the process performed the number of sections and valves isolating them can be varied. Both branches of the loop are connected to a piston capable of producing extremes of pressure.

The operation scheme for one of the simplest contactors, where water softening is carried out, is shown in Fig. 27 [9]. In the operating

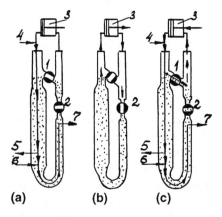

Figure 27 Scheme for Higgins contactor: (a) operating cycle; (b) resin movement; (c) repeat of operating cycle; (1,2) valves; (3) piston; (4) original solution inlet; (5) treated solution output; (6) regeneration solution inlet; (7) waste output.

section solution moves from top to bottom, the ion-exchanger bed moving from bottom to top.

During the operating cycle (Fig. 27a) valve 1 is opened and valve 2 is shut with the piston on the left. The initial solution passes from top to bottom through the ion-exchanger bed in the left branch of the unit and is drained. Regeneration is accomplished in the right branch. The part of the column located between the solution input and the regenerant output serves for washing excess regenerant from the ion exchanger.

After exhausting a part of the bed in the left branch the input and output of solution is stopped, valve 1 is shut and valve 2 is opened. Upon moving the piston to the right the ion exchanger, regenerated, is pumped into the bottom part of the left branch while the exhausted top portion of the bed moves to the top portion of the left branch (Fig. 27b). This operation lasts for 15-60 s. The pressure developed is as high as 8 atm.

Then valve 2 is shut, valve 1 is opened, the flow of solution is resumed and the piston moves to the left position (Fig. 27c), the exhausted ion exchanger moving to the left branch through a connecting tube and being collected over valve 2. The operations are then repeated. Various designs of Higgins contactors are described in the literature [9,10,43,47, 50,69,70].

Commercial units are produced by the Chemical Separation Company. They have been employed for solving numerous problems particularly in the radiochemical industry and for water treatment in the processing of brines and pulps.

The scheme for the ion-exchange apparatus developed by Permutit (Great Britain) [44,71] is shown in Fig. 28. There are some tanks for the ion exchanger between the sorption and regeneration sections. During the operating cycle the solution being treated is moved from bottom to top through the ion-exchanger bed when valves 5 and 6 are shut. In the regeneration section regenerant solution and the rinsing water move from top to bottom and leave the column through valve 17. After exhausting a part of the ion-exchanger bed in the sorption column valve 8 is closed with valves 13 and 6 being opened when valve 7 is turned off. The ion-exchanger bed, except for the exhausted portion which is transferred into the lower tank, remains. Then valves 13 and 6 are closed, 10 and 5 are opened, and the regenerated ion exchanger is transferred to the sorption column. Next valve 5 is turned off, 8 is turned on and the process is repeated.

During the sorption cycle the supply of solution to the regeneration section is stopped several times every 3–10 min, valves 12 and 7 are opened to develop excess pressure, and the exhausted ion exchanger is

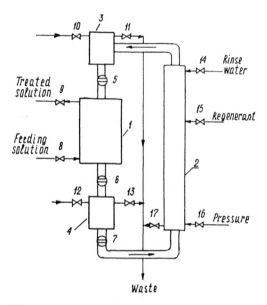

Figure 28   The Permutit contactor: (1) sorption column; (2) regeneration column; (3) tank for regenerated resin; (4) tank for exhausted resin; (5–7) valves; (8–17) taps.

transferred to the bottom part of the regeneration column, the regenerated ion exchanger being moved into the top tank.

The Japanese firm Asahi has developed the ion-exchange apparatus [44,48,71–74] that is sketched in Fig. 29. The liquid to be treated is injected from below, passes through the ion-exchanger bed and is filtered. The rate of solution flow is high enough to compact the ion exchanger effectively next to the filter. A ball-like check valve at the top of the column, isolating the upper tank with ion exchanger, is closed to the pressure of water from below.

After exhausting a part of the ion-exchanger bed, valve 5 is closed and solution input is stopped. At the same time valve 7 is opened, the ion-exchanger suspension settles and fills the bottom part of the column, ball-check valve 3 being opened and the ion exchanger from the upper tank being transferred into the column. Valve 7 is then closed, valve 5 is opened, and the operation is repeated. The ion exchanger being at a level lower than the level of the solution distributor is moved into the upper tank of the regeneration column.

The regeneration column operates similarly except for the addition of rinse water below the solution inlet.

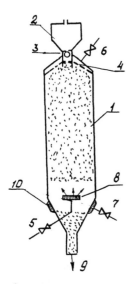

Figure 29 The scheme for an Asahi contactor: (1) operating section; (2) resin hopper; (3) ball-check valve; (4) filter; (5–7) valves; (8) solution distributor; (9) exhausted resin outlet; (10) drain.

In the Asahi contactor there are no valves to direct transfer of ion exchanger. A successful solution to this problem has been affected through the development of a ball-check valve system for the loading of ion exchanger.

In Asahi contactors each operation (sorption, regeneration) is performed in a separate apparatus allowing the particular contactor to be applied to processes with a mixed bed of ion exchangers. A scheme of the unit used for demineralization in a mixed bed [44,48] is shown in Fig. 30.

The exhausted mixture of ion exchangers is put into a separating-rinsing column into which rinse water is injected from below at a fast enough rate to suspend all resin particles.

Anion exchanger beads of lower density move up and higher density beads of cation exchanger move down. The separated ion exchangers are transferred into regeneration columns with water. The ion exchangers, regenerated and rinsed, are directed into the top tank of main column I, mixed, and enter the operating cycle.

To demineralize water of low salt content an apparatus [75] has been developed in which several simultaneously operating units with mixed beds are used with one apparatus for separating and regenerating

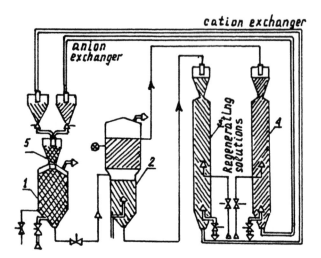

Figure 30 The scheme for a mixed-bed unit: (1) mixed-bed sorption column; (2) separating-rinsing column; (3) regeneration column for cation exchanger; (4) regeneration column for anion exchanger; (5) mix-hopper.

ion exchangers. The scheme and detailed description of units such as these are dealt with in Ref. 54.

The columns whose schemes for operation are shown in Fig. 31 are suitable for developing individual stages of a process in the laboratory. In the first of them solution moves from top to bottom through a dense bed of ion exchanger in section 1 during the operating cycle. Upon exhausting a part of the bed valves 7 and 8 are shut, valves 5 and 9 are opened and a portion of ion exchanger prepared for use is transferred to the operating section by the application of excess pressure emanating from supplementary section 2. The exhausted part of the ion-exchanger bed is removed through valve 5. After that the process is repeated. At a pressure excess of the order of 2 atmospheres movement of ion exchanger together with its solution-filled free space is piston-like.

In the column of the second type the solution under treatment moves from bottom to top and is removed through the drain along the perimeter of the column. When the solution motion is interrupted ion exchanger is moved from the tank by the action of excess pressure.

In a larger column of this type (Fig. 32) hydraulic thickening of the bed in a supplementary section has been used [77] in place of valve 5 in the top part of the column.

After transferring the ion-exchanger bed simultaneously with the beginning of the operating cycle the ion exchanger, prepared for use, is

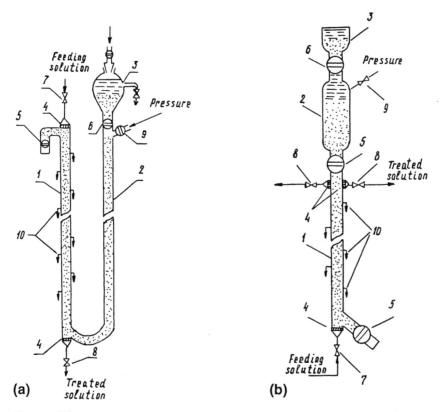

Figure 31 Schemes for laboratory columns with alternate movement of phases of the (a) first and (b) second type: (1) an operating section; (2) a supplementary section; (3) resin tank; (4) filters; (5,6) resin valves; (7–9) solution valves; (10) solution sample collectors.

loaded into the upper tank from tank 14 with the help of pump 5. Valve 11 is shut and valve 10 is opened for this purpose. The transfer is accomplished with a part of the solution leaving the operating section circulating along the following path: tank 6–pump 5–tank 14–supplementary section 2–filter 3–tank 6. During this operation the bed of ion exchanger in the supplementary section thickens. The "plug" of ion exchanger formed on the filter then prevents removal of the ion-exchanger bed from the operating section once the loading of ion exchanger and the circulation of solution through the supplementary section has stopped.

The apparatus in which it has become possible to realize continuous motion of solution at a high rate through a dense bed of ion exchanger moving in the opposite direction of the solution counterflow has

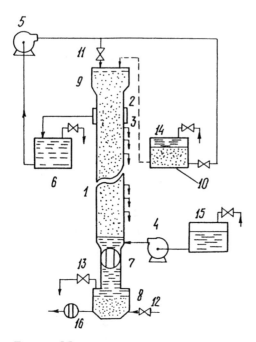

Figure 32 The scheme for a scaled-up unit: (1) a column with capillary collectors for solution samples; (2) a supplementary section; (3) ceramic filter; (4,5) pumps; (6) tank for treated solution; (7) a resin valve; (8,9) top and bottom resin tanks; (10–13) solution valves; (14) tank for regenerated resin; (15) tank for solution under treatment; (16) resin tank.

been described in the literature [78]. The countercurrent column is in the form of a truncated cone with a 2°–3° angle at the apex. The sorbent is moved in the direction of the increase in the column diameter considerably reducing friction against the column walls. This enables one to transfer the ion-exchanger bed without interrupting input of the solution under treatment.

The scheme of the column is shown in Fig. 33. The solution to be treated is injected either through a drain along the column perimeter or through vertical jets. The solution is removed through filter 4.

Transfer of ion exchanger into the operating section is effected with solution periodically injected through valve 10. The empty space is filled with a suspension of ion exchanger injected through jet 14. Excess solution is removed through filter 4.

Columns of the type described offer advantages over cylindrical columns with alternate movement of solution and resin when (a) ion-

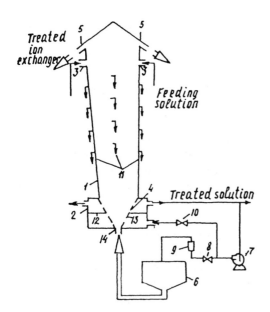

Figure 33   The scheme for a unit with a conical column: (1) operating section; (2) resin feeding section; (3) solution distributor; (4) filter; (5) inlet resin pipes; (6) feeding resin hopper; (7) pump; (8) valve; (9) rotameter; (10) valve; (11) probe collectors; (12) partition; (13) solution feeding filter; (14) resin feeding pipe.

exchanger swelling is increasing with exchange, when (b) ion-exchanger stripping by solution in the course of resin bed transfer (i.e., the exchange when ion concentrations in solution and ion exchanger are close in value) should be minimized or even eliminated and when (c) it is necessary for the sorption front or the zone of the substance purified to be kept stationary with respect to the column walls.

C. Comparison of Countercurrent Apparatus and Columns with a Fixed Bed of Ion Exchanger

There are no universally accepted criteria for comparing ion-exchange apparatus. Parameters to be taken into account in choosing the contactor for performing a certain process are considered in detail in Ref. 4; however, a number of features are pointed out in another publication of interest [3].

The parameters considered in Ref. 4 are classified with respect to suitability of equipment, technological efficiency, and economic efficiency.

Suitability of apparatus is determined by physicochemical characteristics of the solutions under treatment (density, viscosity, concentration, the presence of insoluble substances, toxicity, radioactivity, etc.) and the scale of production.

Technological efficiency involves the following features: (a) efficiency characterized by the height of a transfer unit (HTU) or the height equivalent of the theoretical plate (HETP); (b) specific productivity and the possibility of operating over a wide range of phase rates; (c) ease of putting into and out of operation, reliability.

Economical efficiency is characterized by the magnitude of capital and maintenance expenditures for ensuring desired levels of productivity and recovery. The former include the cost of equipment, the cost of building, indirect costs, the volume of apparatus and area occupied, and the quantity of sorbent.

Maintenance expenditures include costs for electrical energy, labor and service, reagent expenditures, regeneration of ion exchanger, and waste handling. Unfortunately, there are insufficient data in the literature for quantitative comparison of the various types of apparatus. In [4] tables compiled on the basis of the limited data available in the literature some industrially applied equipment and their most promising designs are compared according to the features mentioned above.

Here we restrict ourselves to qualitative consideration of the process of replacing ions of one kind by those of another, an operation that is well understood.

In columns with fixed bed above and below the sorption front, i.e., the zone with varying concentrations of exchanged ions, there are zones with ion exchanger in its initial form before involvement and with exhausted ion exchanger already in equilibrium with the solution injected. There is far more ion exchanger in these zones than in the operating one. In countercurrent units the width of such nonworking zones is minimized. Therefore in the case where HETP does not differ greatly the necessary amount of ion exchanger needed in the countercurrent column is considerably reduced.

In columns with a fixed bed, ion-exchanger capacity is not completely used since the process is performed up to "breakthrough," that is until the appearance of the substance injected into the column in the outlet of the column. So a considerable part of ion-exchange capacity which is within the zone of the sorption front is not used. The countercurrent process is carried out so that completely exhausted ion exchanger is in equilibrium with the injected solution when it leaves the column. In this case the capacity is fully used.

Upon regeneration in a fixed bed conducted until the complete removal of the sorbed ions from the bulk of ion exchanger in the column a considerable part of the reagent is expended for treating partly regenerated ion exchanger. The reagent expenditure, as a consequence, is much higher than with countercurrent regeneration where the process removes the sorbed ions from those ion-exchanger beds right at the point of their exiting the column.

Thus in the countercurrent process less ion exchanger is required, its capacity being more efficiently used. Accordingly, apparatus volume and area of use are decreased. Expenditures for regeneration and volumes of solutions obtained upon regeneration are reduced. The operation is simplified.

If the product recovered from solution is valuable, there is advantage gained from obtaining it in a more concentrated form.

There are additional advantages connected with decreasing the time of contact between ion exchanger and solution and shortening the interval between sorption and regeneration. For example, there is danger of contaminating ion exchanger with rather quickly polymerized substances such as are encountered with silicic acid during water treatment. Such contamination of ion exchanger has also arisen from its degradation upon treatment of radioactive solutions or solutions containing strong oxidizing agents.

Another important advantage of countercurrent exchange is introduced by removing complexities associated with the change of ion-exchange swelling when transformation from one ion form into another occurs and when solution concentration varies. In columns with a fixed bed the increase in swelling results in raising the hydraulic resistance of the bed and in particular cases even in the rupture of columns. When swelling decreases, channels are formed and uniformity is lost. In addition, with a fixed bed all the operations are accomplished in one apparatus without unloading ion exchanger. As a consequence, the array of pipes needed to supply the different solutions required complicates process automation.

Moreover, if the continuous production of product is needed it is necessary to carry out the process in duplicate units. During the regeneration of ion exchanger in one, another should be in operation.

Countercurrent processes are carried out in multicolumn or in multisectional units. Every section (or column) is used for accomplishing one of the operations constituting the whole process. Units are fed the various corresponding solutions while ion exchanger is circulating between them. As a consequence, countercurrent processes are much more easily automated.

Sometimes sizable attrition of ion exchanger has been considered a disadvantage of countercurrent exchange (see, for example, Ref. 79). In arriving at such a conclusion, however, account has not been taken of the fact that much smaller quantities of ion exchanger are required for countercurrent operation than in a fixed bead. Hence, expenditures for ion-exchanger recovery, when the same amount of solution is treated, is roughly the same [3].

It should be noted here that the attrition of ion exchanger is largely independent of external factors, being due, instead, to volume change in transforming from one ion form into another and in changing solution concentrations. Since a smaller amount of ion exchanger is involved in countercurrent operations, each bead undergoes ion transfers far more frequently than in the fixed bead.

In a number of papers [43,44,46–48] the economic advantage provided by replacing fixed-bead operation with sequential movement of solution and exchanger in the course of water treatment has been examined. It is concluded that capital expenditures are reduced by 20–50% and maintenance expenditures are reduced by 20–40%, the area used being decreased by three to four times.

The data base available in the literature for comparing the efficiency of different ion-exchange units varies having been obtained under conditions that were not comparable. For example, the nature of ion exchanger and ions exchanged, the bead size, dimensions of apparatus, and solution concentrations used could differ sizably. Consequently, their comparison is rather complicated. Nevertheless, the qualitative conclusions (see, for example, [2]) reached with respect to the advantage of countercurrent units are not subject to uncertainty. HTU proves twice as large as in columns with a fixed bead and approximately one order of magnitude smaller than in columns with a fluidized bead. Data such as these are not available in the literature for Asahi units. These results of a more detailed study of the dependence of HTU on the rate of phase motion are presented next.

D. Dependence of HTU on the Rate of Phase Motion

Studies have been performed on the same ion-exchange system in order to provide a more quantitative comparison of the effectiveness of different ion-exchange units. In these studies, it was convenient to compare the dependence of the HETP or the HTU on the rate of phase motion with certain columns exposed to the same conditions. The methods for calculating the HETP for stationary distributions in the countercurrent column are described in detail in [27, p. 197].

A number of experimental methods have been developed to evaluate the HTU, or the effective HTU, of the columns. These methods are based on the distribution of ions exchanged along the stationary sorption front [17, pp. 192; 80, 81] by a dispersing pulse [82–84] or step [85,86] signals, on the initial period of separation column operation [87,88], and on the stationary distribution in the separation or exchange column [17, Chap. 8; 27, p. 197]. Used in these methods are the solutions to the problems based on describing mass transfer in exchange systems with the help of simplified, one-dimensional mathematical models such as a non-equilibrium model of plug flow (PF) or another of equilibrium diffusion (ED).

The problems to be solved and their solutions are presented in Ref. 36, Chap. 2.

The application of ion-exchange dynamics for the determination of HETP or HTU values should be studied in ion-exchange systems characterized by invariable static and kinetic parameters such as ion-exchanger swelling, electrolyte sorption, separation coefficient, and interdiffusion coefficients in solution and resin phases over experimentally investigated ranges of component variation.

These requirements may be satisfied either upon the exchange of ions with similar properties* or upon slight alteration in the relative content of ions exchanged.

One of the systems convenient for studying dynamic characteristics of the ion-exchange column involves exchange of the ion pair, K^+-NH_4^+, with sulfonic cation exchangers.

The results [86] of studying the dynamics of the K^+-NH_4^+ exchange in a fixed bed of Dowex-50 × 10 with 0.5–0.7 mm beads and a column diameter of 17 mm are presented in Fig. 34. They were obtained by studying the dispersion of a step signal.

These results yield information with respect to HETP values at various solution flow rates and show how HETP is affected by the change in solution concentration. Similar observations for Ni^{2+}-Co^{2+} are presented in Ref. 17, chap. 5.

Estimate of the HETP intradiffusion coefficients, when based on HETP dependence on concentration, has been shown to be close in value to the quantity calculated with the theoretical formula obtained by using the model of centrosymmetrical diffusion in an ion-exchanger bead [89,90]. The values in this formula of r^2/D, where r is the effective particle radius, were determined in thin-layer kinetic experiments with the same ion-exchanger fraction.

The contribution of the intradiffusion coefficient to the HETP is increased by raising the solution flow rate and the concentration. For example, by raising flow velocity of a 0.4-equiv/dm³ solution from 0.40 to

---

*In this case, as shown for example in [24, p. 22], HTU is equal to HETP.

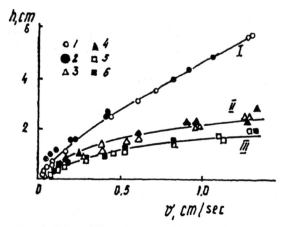

Figure 34 HETP values as a function of solution flow rate at different solution concentrations (g–equiv/dm$_3$): (i) 0.4; (ii) 0.2; (iii) 0.04; (1,3,5) K-1 < 0 (a concave isotherm); (2,4,6) K-1 > 0 (a convex isotherm).

1.0 cm/s the contribution is raised from 41% to 68%. Other contributions to the HETP are strongly related to outer diffusion resistance and longitudinal dispersion.

The dynamics of K^+-NH_4^+ exchange from 0.2g-equiv/dm^3 solution has been studied on cation exchanger KU-2×9 using spherical 0.25–0.50 mm beads. The experiments were performed in countercurrent columns with diameters of 26 and 79 mm and heights of 2–5 m and in columns with a fixed bed 100–200 cm high provided at the bottom with a filter or net to hold the ion-exchanger bed; capillary collectors were placed strategically at different column heights for collecting samples of solution (Figs. 25, 26, 31b, 32). It has been found from preliminary experiments that the transfer of fronts parallel to each other is achievable. By comparing concentration profiles obtained in different periods of time during ion exchange and solution counterflow, it was determined that stationary distribution of the exchanging ions was controlled. These results are presented in Fig. 35.

Curve 1 illustrates the effect of the solution rate of flow on HETP values in the dense fixed beds of ion exchanger when the solution moves down the column from top to bottom. The data have been obtained in columns with inner diameters of 18, 26, and 79 mm. The fact that the points lie along one curve indicates that column diameter does not influence HETP values.

The rise in HETP with increasing solution rates of flow is largely due to the increase in the HETP intradiffusion component. In this case the

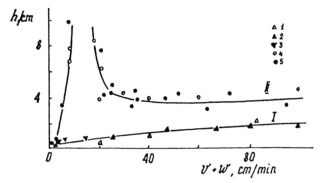

Figure 35 HETP values versus average solution flow rate relative to the particular resin phase examined: (I) ion exchange in fixed beds with columns of different inner diameters (mm); (1) 79; (2) 26; (3) 17; (II) ion exchange exposed to phase counterflow in columns of different inner diameter (mm); (4) 79; (5) 26.

longitudinal dispersion constituent of HETP is not high and has little dependence on the flow rate.

Upper curve 2, was obtained when the solution moved from bottom to top through the ion-exchanger bed. The right branch shows the effect of sequential motion of the solution and the ion exchanger. Curve 2, on the left, shows the strong dependence of HETP values on the rate of continuous motion by the ion exchanger under gravitational influence. The differences in the HET values for the columns with diameters of 26 and 79 mm are well within the range of experimental error in the HETP values. This duplicates the absence of column diameter influence on HETP values observed earlier in curve 1.

The observed differences in the HETP in moving the solution through the ion-exchanger bed from bottom to top and from top to bottom are connected to the contribution of longitudinal dispersion to the magnitude of HETP. At low flow rates, $V - W < 1$ cm/min, the differences are not great. Increasing the rate of solution flow from bottom to top results in expansion of the ion-exchanger bed, the HETP increasing considerably because of the increased contribution of longitudinal dispersion.

When $V - W$ is equal to 6 –10 cm/min, fluidization of the ion-exchanger results. At higher rates of solution flow resin phase compaction begins to occur at the upper filter. HETP values decrease with rates increasing from 15 to 30 cm/min. The fact that HETP changes slightly while flow rates range from 30 to 100 cm/min is accounted for by decrease in the longitudinal dispersion constituent cancelling gain in the intradiffusion constituent. The values of HETP at high rates of flow approach those obtained when the solution moves from top to bottom.

The rather sizable scatter in the HETP values obtained in experiments where the phases move alternately can be explained by using results obtained from studying the characteristics of ion-exchanger moving beds that are reported in [91]. The results show that at excessive pressure, P > 0.5 atm, rather dense packing of the ion-exchanger bed occurs, with the fraction of free bed space, $\chi^3 \le 0.38$; at the densest packing $\chi_0 = 0.36$. The motion of solution barely keeps up with the pressure-induced movement of the ion-exchanger bed. At P = 1 atm the bed moved at the rate of 2.2 cm/s, and at P = 2 atm the bed moved at the rate of 3.8 cm/s. Under these conditions from cycle to cycle, a fraction of bed-free space is well reproduced and changes in the front width during motion of the ion exchanger are not great. For 1–2 min after stopping the movement of ion exchanger and resuming the movement of solution the sorption front is the same width as it was before beginning transfer of the bed.

At excess pressures smaller than 0.5 atm and the less compact packing that ensues, significant scattering between results is observed to occur.

The performance of the operation of transferring ion exchanger at a very low excess pressure on the ion-exchanger bed, 0.03–0.05 atm, results in increases in HETP as large as 12–15 cm when V – W $\cong$ 60–70 cm/m.

The dependence of HETP upon solution flow rate when exchanging the H^+–Na^+ pair of ions between 0.2 M chloride solution and cation exchanger, KU-2×8, with beads 0.25–0.50 mm in diameter, has been compared in the literature for columns of different types [76,84]. The "pulse" method which has been used possesses the advantage that with the pulsed injection of a small amount of substance all physicochemical and hydrodynamic characteristics of the system remain invariably determined by the primary substance, Na^+ ion in this case. Dynamic parameters obtained characterize the process when the concentration range of impurity is low.

The exchange process has been studied in columns 20–23 mm in diameter with a fixed bed, with continuous-gravity-inspired ion-exchanger motion (Fig. 25) and with alternate motion of bed and solution (Fig. 31).

The results are presented in Fig. 36. The dependence of h on V + W in columns with continuous motion of ion exchanger is similar to the dependence shown in Fig. 35. The dependence of h upon the linear rate of solution motion for both types of column with phases moving alternately and for columns with a fixed bed are virtually the same when performing the operations of ion-exchanger bed transfer under expansionless

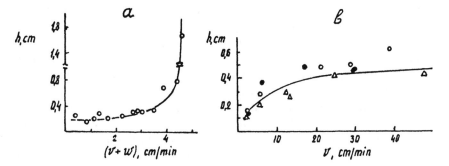

Figure 36   HETP values (H^+-Na^+ exchange) as a function of a) relative phase motion rates under continuous counterflow and b) solution flow rates in columns with fixed bed (●); in columns with alternate motion of phases in the first (Δ) and second (○) type of operation.

piston-like conditions. At any solution flow rate it was essential that for effective change in the column of the second type with solution motion from bottom to top that the column be completely filled with ion exchanger.*

The results obtained indicate that the alternate movement of solution and dense ion-exchanger bed provides high rates of phase motion, the exchange being characterized by HETP values which are from 1.5 to 2.5 times as high as those for the fixed bed. This agrees with the data presented earlier for the Higgins contactor where solution was moved from top to bottom and ion exchanger was moved from bottom to top.

The data obtained are rather general in nature since with the same resin particle sizes and the same solution densities they do not depend on the nature of the exchanging ions. On this basis, then, countercurrent units with alternately moving phases are the most effective and highly productive.

It should be noted that the dependence observed on particular variations of resin-solution system characteristics (swelling, solution, and resin densities) when transferring from one ionic form to another may prove unreliable. The effectiveness of countercurrent columns of different types may differ from the effectiveness of columns with a fixed bed that respond differently.

The influence of the difference in densities of the original solutions and the subsequent solutions that result develops upon exchange from

---

*With the data obtained (Fig. 35, curve II) this requirement was not realized and there was some solution bed over the solution distributor.

rather concentrated solutions. Natural convection that arises may lead either to dispersion or sharpening of the sorption front.

Upon the exchange of KCl from 3.0 to 4.0 mol/dm^3 solutions with sulfonate cation exchanger KU-2×8 in the NH_4^+ –ion form [92,93] in both the fixed bed and in the alternately moving solution and ion exchanger the sorption front is narrower when KCl solution is injected from below than when it is injected from above. This effect is particularly noticeable at low rates of solution flow. Stopping the process in the latter case leads to dispersion of the front increasing with time. In the former case even the interruption in operation for one to two days does not result in appreciable disturbance of the sorption front.

Thus, it is more suitable to perform ion-exchange processes in columns where the direction of phase motion is such that the denser solution is below the sorption front while the less dense one is above it.

The study in 3.0–4.0 mol/dm^3 solutions of KCl and NH_4Cl, of ion-exchange dynamics in countercurrent columns with solution motion from bottom to top (Fig. 37) has shown that in this system the HTU values are the same as in the fixed bed (from 2 to 5 cm over a flow rate ranging from 8 to 50 cm/min).

Changes in ion-exchanger swelling in the course of the exchange process affects the efficiency of countercurrent columns with an alternating mode of phase transfer. If the resin beads increase in size as a result of exchange the efficiency reduces while contraction increases the efficiency of operation.

For example, consider the exchange from 3.0 mol/dm^3 NH_4NO_3 solution of cation exchanger KU-2 in the K^+ form [92,93]. At such con-

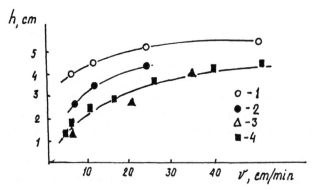

Figure 37 HETP values versus solution flow rate: (1,2) NH_4^+ -K^+ exchange in 3.0 mol/dm^3 nitrate solution; (1) phase counterflow; (2) fixed bed; (3,4) K^+ -NH_4^+ exchange in 3.6 mol/dm^3 chloride solution; (3) fixed bed; (4) phase counterflow.

centrations the exchange of NH_4^+ ion for K^+ ion from nitrate solutions is preferred by the resin phase [57]. Countercurrent operation leads to formation of a stationary front. It has been found that the exchange in columns with sequential phase motion (Fig. 37) lead to HTU values that are 1.3–1.4 times larger than those obtained with the exchange in fixed bed. This observation is explained by the enhancement of cohesive forces between column walls and the swelling but otherwise static resin beds. In countercurrent operation, transfer of the ion-exchanger bed leads to movement that deviates from a "piston-like" pattern to one where interbed solution somewhat outstrips the resin plugs.

A similar phenomenon was observed upon the exchange of Cs^+ and Rb^+ with sulfophenolic cation exchanger in alkaline solution [36, p. 172].

ACKNOWLEDGMENTS

The author wishes to express his gratitude to Professor Erik Högfeldt for his support, help, and encouragement with respect to this publication. Dr. Dmitri N. Muraviev is gratefully acknowledged for editorial work and assisting in the preparation of this manuscript. Dr. Lidia V. Tarasenko is acknowledged with thanks for her tireless work in translating the text into English.

REFERENCES

1. N. Lenborn, *Svensk. Kem. Tidskrift.*, *70*: 255–266 (1958).
2. M. I. Slater, *Brit. Chem. Eng.*, *14*: 41–46 (1969).
3. G. Lermigeaux, and H. Roques, *Chem. Ind. Gen. Chim.*, *105*: 725–748 (1972).
4. E. I. Zakharov, B. E. Ryabchikov, and V. S. Dyakov, *Ion Exchange Equipment for Nuclear Industry*, Energoizdat, Moscow, 1987, (in Russian).
5. V. I. Gorshkov, G. M. Panchenkov, G. M. Gulyaeva, et al., *Rare Alkaline Metals*, Nauka, Novosibirsk, 1967, pp. 287–295. (in Russian)
6. V. I. Gorshkov, I. A. Kuznetsov, and G. M. Panchenkov, *Dokl. Akad. Nauk SSSR*, *143*: 643–645 (1962).
7. V. I. Gorshkov, and M. S. Safonov, *Zh. Fiz. Khim.*, *42*: 1466–1469 (1968).
8. R. C. Glogan, D. O. Halvorson, and W. J. Sloan, *Ind. Eng. Chem.*, *53*: 275–278 (1961).
9. I. R. Higgins, *Ind. Eng. Chem.*, *53*: 635–637 (1961).
10. I. R. Higgins, J. T. Roberts, C. W. Haucher, and J. T. Marinsky, *Ind. Eng. Chem.*, *50*: 285–292 (1958).
11. *Chem. Eng.*, *69*: N7 44, 46 (1962).
12. A. N. Ivanov, and E. N. Gapon, *Zh. Fiz. Khim.*, *15*: 659–664 (1941).
13. F. Helfferich, *Ion Exchange Resins* (Russian translation), Moscow, 1962, p. 154.

14. M. E. Fuentevilla, U.S. Patent 2892679 (1959).

15. B. Trémillon, *Les séparations par les résines échangeuses d'ions* (Russian translation), Mir, Moscow, 1967.

16. D. Carrillo, M. Urgell, and J. Iglesias, *An. Quim. Real. Soc. Esp. Fis. y Quim.,* 66: 461–470 (1970).

17. V. I. Gorshkov, *Teor. Osnovy Khim. Tekhnol.,* 4: 168–180 (1970).

18. V. I. Gorshkov, V. A. Chumakov, and T. V. Rudakova, *Vestnik Moskov. Univ., Khimiya,* N. 6: 20–24 (1969).

19. V. I. Gorshkov, and V. A. Chumakov, *Vestnik Moskov. Univ., Khimiya,* N. 1: 41–43 (1973).

20. B. N. Laskorin, G. E. Kaplan, K. V. Orlov, and A. M. Arzhatkin, *Ion-Exchange Sorbents in Industry,* Akad. Nauk SSSR, Moscow, 1963, pp. 118–123. (in Russian)

21. V. I. Gorshkov, G. M. Panchenkov, N. P. Savenkova, and S. U. Savostyanova, *Zh. Neorg. Khim.,* 8: 2800–2805 (1963).

22. V. I. Gorshkov, V. A. Fedorov, and A. M. Tolmachev, *Zh. Fiz. Khim.,* 40 1436–1439 (1966).

23. V. I. Gorshkov, *Production and Analysis of High-Pure Substances,* Nauka, Moscow, 1978, p. 56. (in Russian)

24. V. A. Ivanov, V. I. Gorshkov, N. F. Nikolaev, V. A. Nikashina, and N. B. Ferapontov, *7th Danube Symp. Chromatography and Analytiktreffen,* Leipzig, Abstracts, V. II, TH 096, 1989. *Reactive Polymers,* 18: 25–33 (1992).

25. L. Pawlowski, *Ann. Univ. Maria Curie-Sklodowska, Lublin: Sect. AA, Phis, Chem.,* N. 26/27: 191–202 (1971/72).

26. K. Bier, *Chem.–Ing.–Techn.,* 28: 625–632 (1956).

27. A. M. Rozen, *Theory of Isotope Separation in Columns,* Atomizdat, Moscow (1960). (in Russian)

28. B. M. Andreev, G. K. Boreskov, and S. G. Katal'nikov, *Khim. Prom.,* N. 6: 389–393 (1961).

29. B. M. Andreev, and G. K. Boreskov, *Zh. Fiz. Khim.,* 38: 115–124 (1964).

30. M. Bailly, and D. Tondeur, *J. Chromatogr.* 201: 343–357 (1980).

31. V. I. Gorshkov, A. M. Kurbanov, and N. V. Apolonnik, *Zh. Fiz. Khim.,* 45 2669–2670 (1971).

32. V. I. Gorshkov, A. M. Kurbanov, and M. V. Ivanova, *Zh. Fiz. Khim.,* 49: 1276–1278 (1975).

33. V. I. Gorshkov, M. V. Ivanova, A. M. Kurbanov, and V. A. Ivanov, *Vestnik Moskov, Univ., Khimiya,* 18: 535–550 (1977).

34. V. I. Gorshkov, M. V. Ivanova, and V. A. Ivanov, *Zh. Fiz. Khim.,* 51: 2084–2086 (1977).

35. M. S. Safonov and R. G. Iksanov, *Teor. Osnovy Khim. Tekhnol.,* 7: 770–773 (1973).

36. V. I. Gorshkov, M. S. Safonov, and N. M. Voskresenskii, *Ion Exchange in Counter-flow Columns,* Nauka, Moscow, 1981. (in Russian)

37. V. A. Ivanov and V. I. Gorshkov, *Teor. Osnovy Khim. Tekhnol.,* 17: 723–729 (1983).

38. M. S. Safonov and S. A. Borisov, *Teor. Osnovy Khim. Tekhnol.*, *15*: 676–682 (1981).
39. M. S. Safonov, N. M. Voskresenskii, and B. M. Andreev, *Teor. Osnovy Khim. Tekhnol.*, *15*: 163–169 (1981).
40. S. A. Borisov, and M. S. Safonov, *Teor. Osnovy Khim. Tekhnol.*, *18*: 159–164 (1984).
41. V. A. Ivanov and V. I. Gorshkov, *Vysokochistye Veshchestva*, N. 5: 100–104 (1987).
42. E. G. Stepanyanz, *Trudy VODGEO* (in Russian), *25*: 3–10 (1970).
43. I. R. Higgins, *Chem. Eng. Progr.*, *60* N11: 60–63 (1964).
44. M. E. Gilwood, *Chem. Eng.*, *74* N26: 83–88 (1967).
45. G. A. Medvedev, V. I. Gorshkov, D. N. Muraviev et al., *Zh. Prikl. Khim.*, *51*: 96–99 (1978).
46. J. Newmann, *Chem. Eng.*, *74* N26: 71–74 (1967).
47. I. R. Higgins and R. C. Chopra, in *Ion Exchange for Pollution Control*, CRC, Boca Raton, FL, 1979, V. II, pp. 75–86.
48. C. H. Dallmann, *Combustion, Jan.*: 17–24 (1969).
49. B. A. Henelry, *Wat. Schi. Technol.*, *14*: 535–552 (1982).
50. I. R. Higgins, *Chem. Eng. Progr.*, *65* N6: 59–62 (1969).
51. W. Zebban, T. Fithian, and D. Maneval, *Coal Age., July*: 107–111 (1972).
52. C. H. Thorborg, US Patent 3607739 (1971).
53. C. H. Thorborg, US Patent 3617554 (1971).
54. M. S. Shkrob, in *Water Treatment, Water Conditions and Chemical Control at Steam-Power Units*, Energiya, Moscow, 1974, Vol. 5, pp. 87–93. (in Russian)
55. H. Kakihana, *Chem. Ind. (Japan)*, *11*: 337–341 (1960).
56. V. I. Gorshkov, O. T. Gavlina, B. F. Fedushkin, et al., *Khim. Prom.*, N. 2: 35–39 (1989).
57. V. I. Gorshkov, A. I. Novoselov, O. T. Gavlina, and M. V. Denisova, *Zh. Fiz. Khim.*, *61*: 1679–1681 (1987).
58. C. H. Nordel, US Patent 1608661 (1926).
59. E. A. Swinton and D. E. Weiss, *Austral. J. Appl. Sci.*, *4*: 316–340 (1953).
60. J. M. Hutcheon, In *Ion Exchange and Its Applications*, London, 1955, pp. 101–111.
61. S. M. Korpacheva, and B. E. Ryabchikov, *Pulsation Equipment in Chemical Technology*, Khimiya, Moscow, 1983. (in Russian)
62. E. P. Cherneva, N. N. Tunizkii, and V. V. Nekrasov, *Rare-Earth Elements*, AN SSSR, Moskva, 1958, pp. 129–139.
63. V. I. Gorshkov and I. A. Kuznetsov, G. M. Panchenkov, L. V. Kustova, *Zh. Neorg. Khim.*, *8*: 2790–2794 (1963).
64. E. Herrmann, *Chem. Ing. Technol.*, *27*: 573–577 (1955).
65. W. A. Selke and H. Bliss, *Chem. Eng. Progr.*, *47*: 529 (1951).
66. T. A. Arehart, I. C. Breesee, C. W. Haucher, and S. H. Jury, *Chem. Eng. Progr.*, *52*: 353–359. (1956).
67. C. W. Haucher, *Eng. Min. J.*, *160* W3: 80–83 (1959).

68. V. I. Gorshkov, S. N. Dimitriev, G. M. Panchenkov, and L. I. Krasil'nikov, *Khim. Prom.,* N.9: 693–696 (1967).

69. *Chem. Eng., 64* W7: 184–188 (1957).

70. *Chem. Eng., 66* W14: 84–86 (1959).

71. G. Apfel, in *Proc. 27th Int. Water Conf.*, Pittsburgh, 1966, pp. 165–170.

72. P. Treille and P. Fort, *La Technique Moderne,* May, *60*: 245–249 (1968).

73. J. N. Gossett, *Chem. Process. (USA), 29* W4: 44–50 (1966).

74. Asahi, Pat. 3244561, (USA) (1966); 987021, (Great Britain). (1965)

75. H. Padu, in *Proc. Int. Water Conf.* Pittsburgh, 1968, Vol. 29, p. 155.

76. Ya. N. Malykh, V. I. Gorshkov, and V. A. Ivanov, *Vysokochistye Veshchestva,* N. 1: 120–125 (1989).

77. V. I. Gorshkov, G. A. Medvedev, and D. N. Murav'ev, *Tsvetn. Metally* N. 1: 53–54 (1974).

78. N. B. Ferapontov and V. I. Gorshkov, *Vysokochistye Veshchestva,* N. 5: 209–211 (1988).

79. N. K. Hiester and R. C. Phillips, *Chem. Eng., 61* W10: 161–178 (1954).

80. V. I. Gorshkov, S. S. Epifanova, and M. S. Safonov, *Zh. Fiz. Khim., 45*: 732–733 (1971).

81. V. I. Gorshkov, G. A. Medvedev, and D. N. Murav'ev, N. B. Ferapontov, *Zh. Fiz. Khim., 51*: 980–981 (1977).

82. M. S. Safonov, V. A. Poteshnov, E. V. Sud'in, and V. I. Gorshkov, *Teor. Osnovy Khim. Tekhnol., 11*: 315–324 (1977).

83. V. K. Belnov, V. V. Brey, N. M. Voskresenskii et al., *Teor. Osnovy Khim. Tekhnol., 13*: 339–346 (1979).

84. M. S. Safonov, Ya. N. Malykh, V. A. Ivanov et al., *J. Chromatogr. 364*: 141–152 (1986).

85. M. S. Safonov, V. K. Shiryaev, and V. I. Gorshkov, *Zh. Fiz. Khim., 44*: 975–980 (1970).

86. V. K. Shiryaev, M. S. Safonov, V. I. Gorshkov, and V. A. Lipasova, *Zh. Fiz. Khim., 45*: 2292–2296 (1971).

87. A. Klemm, *Z. Naturforsch.,* B. 1, S. 252–257 (1946); B. A7, S. 418–421 (1952).

88. M. S. Safonov and V. A. Poteshnov, *Zh. Fiz. Khim., 45*: 687–689 (1971).

89. N. N. Tunizkii, E. P. Cherneva, and V. I. Andreev, *Zh. Fiz. Khim., 28*: 2006–2010 (1954).

90. E. Glueckauf, *Trans. Faraday Soc., 51*: 1540–1551 (1955).

91. V. I. Gorshkov, G. A. Medvedev, D. N. Muraviev, and N. B. Ferapontov, *Zh. Fiz. Khim., 50*: 1345 (1976).

92. V. I. Gorshkov, O. T. Gavlina, A. I. Novoselov et al., in *Theory and Practice of Sorption Processes*, Voronezh, 1989, Vol, 20, pp. 72–77.

93. O. T. Gavlina, Author's abstract of candidate dissertation (in Russian), Moscow University, 1990.

3

Recovery of Valuable Mineral Components from Seawater by Ion-Exchange and Sorption Methods

Ruslan Khamizov

Vernadsky Institute of Geochemistry and Analytical Chemistry, Russian Academy of Science, Moscow, Russia

Dmitri N. Muraviev and Abraham Warshawsky

The Weizmann Institute of Science, Rehovot, Israel

I. INTRODUCTION

The world's oceans hold 1.37×10^{18} m^3 of water (97.2% of the total amount of water of the hydrosphere). They cover 71% of the earth's surface, are actually the biggest reservoir on our planet, and contain many important minerals. The overall content of mineral matter in the oceans is estimated to be about 5×10^{16} tons [1,2]. The seas contain virtually all of the naturally occurring elements and are the only universal source of mineral wealth that is available to most nations. For some of them it is the only source. Yet, most of the elements, the "microelements," are available in very low concentrations, i.e., in parts per billion (ppb). The products being extracted from seawater with economic profit at present are sodium chloride, magnesium compounds, and bromine [2–4]. During the last two decades there has been growing interest in the possibility of commercial recovery of additional minerals from seawater [5] and brines [6].

This interest is strongly influenced by the following considerations:

    1.   Land-based mineral resources of the world are being depleted at a very high rate. Those that are still accessible, e.g., copper, tin, and zinc, are expected [7] to become exhausted by the turn of the century, while the oceans contain vast amounts (at a level of $\sim 10^9$ tons) of these and other minerals yet to be recovered.

2.   Intensive development of desalination plants all over the world (see, e.g., [3, p. 453]) is resulting in the production of enormous quantities of concentrated brine effluent. These highly concentrated effluents from treated seawater provide a natural source of minerals that is an attractive alternative to the cumbersome mining sources.

3.   Special attention is being given to mineral extraction from "bitterns" formed by solar evaporation of seawater in arid waters or by seawater freezing in the Arctic areas. Such bitterns represent approximately 30-fold concentrated seawater with depleted concentrations of sodium, calcium, chloride and sulfate ions.

4.   Countries with very limited natural resources find recovery of minerals from natural seawater and brines an attractive alternative to their import.

5.   Whereas the recovery process for minerals, and especially metals, from land-based sources must be totally adapted to the unique ore composition, the recovery of minerals (including metals) from seawater can become universal thanks to the unique ionic composition of sea water all over the world. This makes the development of a unified technology an attractive prospect for mining the seas.

Two additional factors, one positive and the other negative, that need to be considered are:

6.   If minerals are produced from desalination unit effluents, their recovery can provide some additional credit toward the cost of producing fresh water.

7.   The disposal of highly concentrated effluents (after seawater desalination) back into the ocean leads to increased pollution. This will result in the destruction of local and marine flora and fauna, and can be expected to introduce other ecological hazards. Because of this, it probably will be worthwhile to install a mineral recovery unit at the desalination plants, despite the extra cost, just to reduce the possibility of serious ecological problems.

II.   ECONOMIC ASPECTS OF ION-EXCHANGE RECOVERY OF MINERAL COMPONENTS FROM SEAWATER

Current economic and ecological analyses of the various processes available for recovery of minerals from seawater (evaporation, solvent extraction, sorption, ion exchange, flotation, fractional precipitation, distillation, electrolysis, electrodialysis, and electrocoagulation) favor ion exchange and sorption technology.

Even though ion exchange is favored, one must take into account the fact that a combination of different processes may provide additional advantage. In general, each process associated with ion exchange technology (as well as with any other technique) has a certain "range of applicability," where it is competitive with other methods. This multiparametric approach for the selection of such combinations includes both economical and technological aspects of the process under consideration. As an example of the path to such a combination, the two-parameter diagram of the "range of applicability" for different effluent treatment processes is presented schematically in Fig. 1. Here, the metal ion concentration is plotted against the flow rate of the solution under treatment [8]. From this, it follows that ion exchange, particularly in its countercurrent version, is the most suitable technique for recovering microcomponents from seawater at the high flow conditions required for the treatment of hugh amounts of solutions.

The most important economic factors applicable to ion-exchange technology are

1. Reduction in energy expenditure for seawater supply.
2. Choice of the minerals to be extracted from seawater.
3. Reduction in expenditure for ion exchangers, sorbents, and auxiliary chemical reagents.

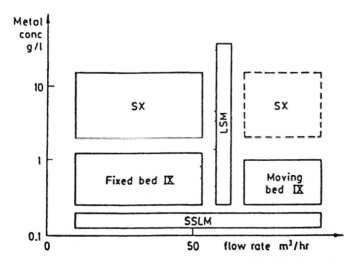

Figure 1 Two-parameter examination of range of applicability of ion exchange and solvent extraction techniques with a semilogarithmic plot of metal concentration in g/L versus solution flowrate in m^3/hr. IX = ion; SX = solvent extraction; LSM = surfactant liquid membrane; SSLM = solid supported liquid membrane [8].

A. Reducing Power Expenditure

The steam power stations in some countries which border the sea (e.g., Russia, United States, Israel, and others) use seawater for cooling the turbine stream. The amount of seawater pumped through the cooling cycle of these stations is approximately 20,000 to 30,000 m^3/hr. If this currently wasted seawater were used for recovering minerals in a continuous mode, the power costs in moving the seawater through the mineral recovery process (that may consist of up to 50% of the overall expenditure for electricity) could be written off. The same conditions apply to installing mineral recovery units at seawater supply facilities of existing desalination plants. At present about 50% of all magnesium consumed in the United States is extracted from the turbine coolant taken from and returned to the sea by electrical power generating stations.

B. Selection of Elements to be Recovered from Seawater

Several comprehensive reviews on the recovery of inorganic materials from seawater have already been published by Tallmadge et al. [9], Hanson and Murthy [10], McInny [11], Seetharam and Srinivasan [12], Senyavin [13], Massie [14], and Senyavin and Khamizov et al. [15]. They all agree that exploitation of the ocean for mineral recovery must be economically competitive with mineral recovery from land-based sources. Hanson and Murthy [10] concluded, on a comparative basis of prices at the beginning of the seventies, that the following elements might be recovered economically from the sea: chlorine, sodium, magnesium, potassium, bromine, boron, and lithium. Waldichuk [16] has added strontium to this list. Gaskel [17] suggested that extraction of molybdenum, uranium, silver, gold, tin, nickel, copper, and cobalt could eventually become economical as well.

According to the latest estimates of Skinner [18], elements potentially recoverable from seawater are sodium, potassium, magnesium, calcium, strontium, chlorine, bromine, boron, and phosphorus because of their practically unlimited presence in the ocean. After improving respective technologies, recovery of the following elements is expected to become profitable as well: lithium, rubidium, uranium, vanadium, and molybdenum. Additional profit can be gained since desalinated water will probably be obtained as a by-product. This could be important for countries with a very limited number of freshwater sources (e.g., Israel, Saudi Arabia).

Khamizov et al. [19,20], Senyavin et al. [15], and Gorshkov et al. [21] have recently studied the recovery of more than 12 elements from seawa-

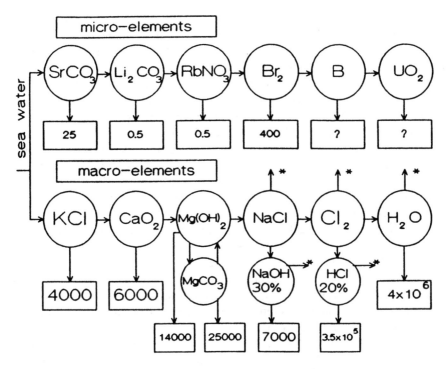

Figure 2 Scheme for recovery of mineral components from seawater. Amounts of components are given in tons per year, assuming the unit capacity is around 1000 m³/hr.* Products which can be partially used as reagents in the technological process.

ter (and desalinated water as a by-product,*) applying a complex scheme of seawater processing. An estimate of the potential productivity of this process is presented in Fig. 2. The component amounts are given in tons per year, assuming plant capacities of 1000 m³ of seawater per hour.

C. Main Approaches to Ion-Exchange Recovery of Valuable Mineral Components from Seawater

Studies related to the recovery of mineral components from seawater have been conducted in several countries, including island countries, such as

*This cycle of investigations was carried out using experimental pilot facilities installed at the seaside power station on Sakhalin Island (Russia). R. K. and D. M. were actively involved in these studies.

Japan and the United Kingdom, and mainland countries, such as the United States of America, Sweden, Russia, and Germany. Until very recently, most of these studies were aimed mainly toward the recovery of single valuable elements. In the majority of cases, these attempts were linked to the extraction of uranium (see below and, e.g., [22]).

Two main approaches are noted: The first deals with sorbents of high selectivity toward target ions [23]. The attractiveness of this approach is probably the increased potential for the removal of certain ions from seawater without any pretreatment. Such highly selective sorbents for a one-stage extraction of valuable mineral components from seawater are still very expensive and in some instances exhibit unsatisfactory chemical stability. Yet, intensive studies in this direction are continuing in many countries (see below).

The other approach deals with the development of multistage technology based on inexpensive adsorbents (e.g., natural zeolites [19,20]) with limited selectivity in the first stage.

Use of a low-cost ion exchanger for the preconcentration stage permits employment of 20–30 m^3 of the adsorbent in this first step of the operation. The adsorbed elements are stripped from the sorbent during the regeneration stage in 100–1000-fold reduced volumes. In the next stage, the regenerated solutions are concentrated using traditional methods and the desired elements are extracted by using either conventional ion exchangers or by using highly selective sorbents. This second approach allows the creation of a more flexible technology with wide perspectives. It can be easily redirected toward the production of desired components in accordance with market demands.

III. RECOVERY OF MACROELEMENTS

A. Magnesium

1. *Traditional Methods*

The overall magnesium content of the world's oceans is estimated to be about 2.3×10^{15} tons. This quantity represents about one sixth of the compounds dissolved in seawater. One thousand cubic meters of seawater [2,3] are required to produce 1 ton of magnesium. At present, around 25% of overall world production of magnesium stems from hydromineral resources (seawater, underground brines, and bitterus of some salty lakes) and is growing. In 1968, 1.18×10^6 tons, in 1973, 1.77×10^6, and in 1986, 2.06×10^6 tons of hydrometallurgical Mg were produced. The annual production of magnesium from hydromineral sources in 1986 is shown in Table 1 for a number of countries.

Table 1 Annual Production in 1986 of Magnesium from
Hydromineral Sources in Different Countries

Country	Mg produced (million tons)	Number of plants
United States	750	8
Japan	513	7
Great Britain	250	1
Italy	217	2
Norway	120	1
Ireland	75	1
Israel	50	1
Mexico	54	1
Canada	30	1

Source: Ref. 21.

The first magnesium-from-seawater plant was erected in England at the beginning of World War II [25]. The simple technology included mixing the raw seawater in special reservoirs with lime milk; $Mg(OH)_2$ slurry was filtered and treated with HCl, followed by evaporation, drying, and electrolysis. All subsequent plants for recovery of Mg from seawater and brines are essentially unchanged. They have been described in detail in numerous reviews (see, e.g., [2, 26–28]). When one considers that natural magnesites contain 300 times more magnesium than seawater, the fact that this metal is processed from hydromineral sources shows how attractive the use of seawater can be for mineral processing.

2. *Ion-Exchange and Sorption Methods for Recovering Magnesium from Seawater*

Traditional methods for producing magnesium by processing hydromineral sources fail to satisfy the newer ecological standards. Consequently, new, alternative technologies, based on sorption methods are being considered despite the profitability of earlier methods.

One of the first and more interesting of these uses strong-acid cation exchangers of the Dowex 50 type (sulfonated copolymers of styrene and DVB) [29]. In this process, the seawater is passed through the Na^+-ion-form resin bed until equilibrium is reached, then magnesium stripping and resin regeneration is carried out in one step with concentrated NaCl solution. The electroselectivity effect [30] promotes complete regeneration and yields a high concentration of magnesium in the eluate (5–10 times higher than in the initial seawater). The equilibrium separation coefficient α for Mg and Na can be expressed as follows:

$$\alpha_{Na}^{Mg} = \frac{\overline{M}_{Mg}}{\overline{M}_{Na}} \cdot \frac{C_{Na}}{C_{Mg}} \tag{1}$$

where $\overline{M}$ and C are the equilibrium concentrations of the components in the resin and solution phases, respectively (α values >1 in seawater and <1 in concentrated NaCl solution). The magnesium-containing eluate is then treated with water-immiscible extractant (diamine), and $MgCl_2$ is selectively extracted by the organic phase. The residual NaCl solution, after being fortified with additional NaCl, is employed in the next sorption-desorption cycle. The organic extractant phase is mixed with water to strip magnesium compounds and the extractant is directed to the next extraction cycle.

This approach has several drawbacks. The extractant gradually contaminates the ion-exchange resin and gradually permeates the seawater. The ion-exchange capacity of the resin is used ineffectively, because of the relatively low selectivity of Dowex 50 cation exchangers for Mg relative to Na in the presence of seawater. The equilibrium capacity of Dowex 50×8 toward Mg by sorption from seawater does not exceed 1 mg-equiv/cm³ of the resin bed.

The Dow Chemical Co. has developed a number of sorbents for selective recovery of magnesium from multicomponent brines containing twice the magnesium concentration of seawater in the presence of high calcium concentrations [31–33]. The composite sorbent is prepared as follows: a macroporous strong-base anion exchanger, Dowex-MWA-1 (containing quaternary ammonium groups), is impregnated with concentrated $AlCl_3$ solution and then treated with aqueous ammonia solution. The dispersion of $Al(OH)_3$ precipitate in the macroporous matrix that results is then treated with MgX_2 solution (where X is a halide). This leads to the formation of a fine $MgX_2 \cdot Al(OH)_3$ crystalline phase dispersed in the resin phase. The composite material can be used for selective recovery of magnesium from brines. The process is carried out at temperatures higher than 50°C and in the pH range of 5.5–9.0. Several inorganic sorbents based on calcium titanyloxalates have been designed to recover magnesium from brines containing alkali earth ions [34].

In recent years, a number of processes have been developed [35–37] for the ion exchange recovery of Mg from seawater by applying weak-acid (carboxylic) ion exchangers. The following monofunctional carboxylic acid cation exchangers (methylmethacrylate-DVB type) were considered for this purpose: KB-4 (Russia), Amberlite IRC-50 (U.S.A.), Zerolite-226, and Zeocarb-226 (Great Britain).

Because of both their high selectivity toward divalent ions and their high capacity, these resins can be used effectively for processing neutral and alkaline solutions containing a high concentration of sodium ions, e.g., seawater with pH = 8.0–8.1. However, effective application of carboxylic acid ion exchangers requires the earlier selective removal of calcium ions.

Such removal of calcium from seawater can be achieved at present by applying two types of sorbents. The first type corresponds to supersulfonated cation exchangers or Activite, which contains 1.5–2 sulfonic acid groups per benzene unit [38]. For strong-acid cation exchangers of this type, the α_{Mg}^{Ca} values (see Eq. (1)) may reach a number as large as several hundred. A much cheaper method of seawater decalcination uses Mg-modified type-A zeolites, which have, after modification, selectivity of α_{Mg}^{Ca} ~30 when combined with natural zeolites of the clinopthylolite type [39].

In the method described in [35], the seawater, after calcium removal by adsorption on zeolites, is passed at a high flow rate through one of the sections of the countercurrent column filled with KB-4 resin. At the same time, the exhausted part of the resin bed, containing Mg ion, is eluted with HCl solution and finally with NaOH solution. The regenerated resin is directed back into the sorption section and the cycle continues.

Despite the combination of high purity and high recovery with ecological safety in this method [35], it is commercially attractive only when efficient production of auxiliary reagents (HCl and NaOH) is possible.

Another possibility for recovering magnesium with carboxylic acid resins appeared with the discovery of isothermal supersaturated solutions of magnesium carbonate in the resin bed [36,37]. This phenomenon was observed upon elution of the mixture of sodium carbonate and sodium bicarbonate solutions through the magnesium-loaded resin bed (KB-4 resin). During elution, the effective desorption of magnesium is observed but the $MgCO_3$ does not precipitate in the column. It remains a 0.5 N solution (with a supersaturation degree of about 5) for an extended period. After removal from the column, the product ($MgCO_3 \cdot 3H_2O$) crystallizes spontaneously in the form of well-shaped nesquegonite crystals that can be easily separated from the supernatant by filtration or sedimentation. Unlike magnesite ($MgCO_3$), nesquegonite crystals are calcium free. Hence, a high-purity magnesium product is obtained. A large pilot plant with output of 300 tons magnesium carbonate per year is erected in the Vladivostak region of Russia, in the Japanese Sea.

B. Potassium

1. *Precipitation Methods*

Potassium is the fourth most abundant macroelement in the sea. It is produced on a small scale from seawater by evaporation and crystallization methods. During seawater evaporation after precipitation with NaCl (halite) and once the solution density reaches a value of 1.356 g/cm^3, the mixed potassium and magnesium salts (mainly carnallite; KCl · MgCl$_2$ · 6H$_2$O) begin to precipitate. The carnallite crystals are treated with small amounts of water for selective dissolution of magnesium chloride. The potassium salt remains as a crystalline residue. This approach is widely used in countries such as Israel and India [40,41]. Different potassium precipitation methods, applying thiosulfates and perchlorates, have been developed in Egypt [42,43] in tandem with solar evaporation techniques. Methods for potassium precipitation and its extraction with different organic reagents have been developed in Israel [44]. Despite numerous methods of potassium recovery using precipitation, e.g., with borofluorides, silicofluorides, titanium fluorides [41], and phosphates [45,46], large-scale production of potassium salts from seawater is still tilted toward the traditional, basic "carnallite" methods. For instance, the annual production of potassium salts recovered from Dead Sea brines by crystallization in ponds exceeds 2 million tons [40].

Alternative low-energy methods for potassium recovery from the sea were initially developed in countries with cold climates. Holland was the first, where the direct precipitation of potassium from seawater with dipicrylamine (DPA) was successfully carried out [47]. The process consists of two stages:

$$HDPA + K^+ \rightarrow KDPA\downarrow + H^+ \tag{2}$$

$$KDPA + HNO_3 \rightarrow HDPA + KNO_3 \tag{3}$$

The potassium DPA salt precipitate, formed in the first stage, is treated with nitric acid. The DPA released is then directed to the next precipitation cycle. Valuable potassium-nitrogen fertilizer, KNO$_3$, is obtained as the product with the second stage. In 1955, \$2 million were invested in building a pilot plant with a capacity of 300 m^3 of seawater per hour, but operation of the unit was stopped because of the high toxicity of DPA.

2. *Ion-Exchange Methods of Potassium Recovery from Seawater*

The results of studies carried out during the past 10–15 years in Japan, the United States of America, and Russia have shown that sorption methods for potassium recovery from the sea will be the most attractive in the near future. The Japanese investigators have proposed that removal of potassium from seawater be effected by sorbents based on polysulfonic acid treated with nitrophenolic compounds [48,49]. The inorganic sorbents based on zirconium phosphate [50,51], aluminum or iron polyphosphates [52], phosphate derivatives mixed with silicates, tungstates, or zirconium phosphomolybdates have also been shown to be promising materials for potassium extraction [53]. Natural materials, e.g., glauconite, have shown great promise as well [54]. Studies carried out at the Vernadsky Institute of the Russian Academy of Science have demonstrated that clinoptilolite is the zeolite most useful for recovering potassium (as well as several other metals) from seawater. Its advantages include low cost, high physicochemical stability, and the ability to extract other valuable microcomponents from seawater, e.g., strontium and rubidium [55].

The ion-exchange capacity of clinoptilolite in extracting potassium from seawater reaches a value of 30 mg/g, which far exceeds the capacity values quoted for the much more costly zirconium phosphate [55,56].

The distribution coefficient values compiled for different ions in the course of their sorption from seawater on different clinoptilolite sources are presented in Table 2 [54]. The clinoptilolite content in all of the samples was approximately the same (about a mass % of 70). All experiments were carried out at t = 20°C on clinoptilolite in the NH^+ form. The distribution coefficient was defined as follows:

$$D_{NH_4}^{Me} = \frac{\overline{M}_{Me}}{C_{Me,O}} \quad [cm^3/g] \tag{4}$$

Table 2 Distribution Coefficients of Metal Ions Sorbed at 20°C from Seawater by Clinoptilolite in NH_4^+ Form

Clinoptilolite (source)	Metal ions					
	Sr^{2+}	Rb^+	Na^+	K^+	Mg^{2+}	Ca^{2+}
Tedzami (Georgia)	104	230	3.5	76.0	0.9	29.0
Chekhovsky (Sakhalin Island)	34.6	225	3.2	69.0	0.3	13.5
Jagodinsky (Kamchatka)	20.0	180	5.0	84.5	0.4	11.5
White Boir (Bulgaria)	5.8	290	2.5	71.1	0.4	4.1

where $\overline{M}_{Me}$ is the equilibrium concentration of the metal ion in the zeolite phase (mg-equiv/g) and $C_{Me,0}$ is the initial concentration of the same ion in seawater (mg-equiv/cm^3).

As can be seen from the $D_{NH_4}^{Me}$ values listed in Table 2, all the clinoptilolite samples studied were rather highly selective for potassium and rubidium; some of the sources (Tedzami in particular) were selective for strontium as well. The compilation of $D_{NH_4}^{Me}$ values for some nonferrous metal ions through their sorption from seawater have shown that clinoptilolite is highly selective, e.g., for copper and nickel $D_{NH_4}^{Me}$; values as large as 10^3 are reached [57].

The main problem in recovering potassium from seawater by natural zeolites occurs in the regeneration step. It is most appropriately resolved through the use of ammonium salts for treating potassium loaded sorbent, as shown in Fig. 3. The concentrate obtained after the regeneration of the clinoptilolite column is further concentrated in the

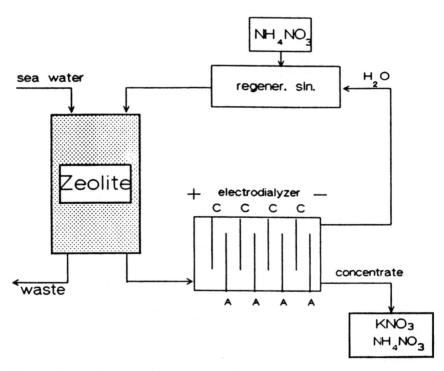

Figure 3 Schematic flowsheet unit producing mixed concentrations of potassium and ammonium nitrates from seawater. C and A denote cationic and anionic membranes, respectively.

electrodialyzer. The final potassium and ammonium nitrate solution that reaches a concentration of 150 g/L can be used as a valuable mineral fertilizer. The main drawback of this process is the risk of possible penetration of seawater by the ammonium nitrate solution. Further development of the "clinoptilolite" technique for potassium recovery from seawater has led to "reagentless" ecologically clean methods of potassium extraction. These methods are based on the strong temperature dependence of clinoptilolite selectivity toward metal ions. For instance, potassium ions are sorbed from seawater at t = 50–75°C by a factor of 2.5–3 times smaller than at t = 20°C, while calcium and magnesium ions respond to a similar decrease in temperature by an increase in sorption [58,59] . This reversal of temperature effect provides a new route for potassium concentration or recovery from seawater without incurring the need for auxiliary reagents. Accordingly, the ambient seawater is passed through the clinoptilolite bed at a low temperature (~20°C) until the equilibrium sorption of K^+ ion is reached; then warm seawater, taken from the cooling cycle of a power plant, is passed through the column. This leads to an increase in the potassium ion concentration of the solution phase and a decrease in its Mg^{2+}, Ca^{2+} (and Sr^{2+}) content. Subsequent repetition of this dual temperature cycle results in the increase of KCl concentration in the final solution to 25–30 g/L and a significant decrease in the concentration of all the other metal ions.

A more attractive variant of this temperature regulated mineral separation process is via potassium loaded clinoptilolite free of agrochemically harmful components. This natural zeolite, containing approximately 7–8% by weight of potassium, represents, in fact, a very valuable chlorine-free potassium fertilizer [58,59] with the following obvious advantages: prolonged action via sustained release, regulation of soil structure via granulation of the zeolite, and neutralizing action toward acidic soils.

A unit producing approximately 4000 tons of zeolite mineral fertilizers per year was designed for the Sakhalin seaside power plant (Sakhalin Island) [59]. Its schematic flow sheet is shown in Fig. 4. Filter F_1 is permanently loaded with 20 tons of zeolite and filter F_2 contains 6 tons of the sorbent, which is periodically removed from the contactor (saturated with potassium ions). The hot potassium concentrate flowing from F_1 and passing through the heat exchanger system (recuperators) R1–R3 is cooled and used for reloading clinoptilolite in F_2 with potassium. The unit is scheduled to process up to 6×10^3 m^3 of cold seawater per year and $\sim 2 \cdot 10^5$ m^3/year of hot water.

The main operational cost of the unit involves heat expenditure, which can be estimated with ~50% heat recovery to be approximately 5×10^9 kcal/year; this corresponds to the heat productivity of about 200 tons of additional fuel.

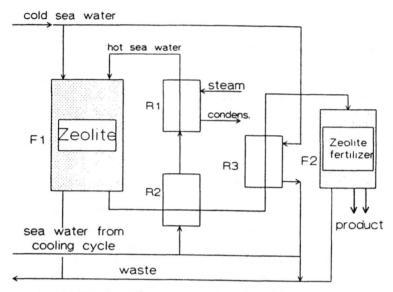

Figure 4 Schematic flowsheet of unit producing clinoptilolite-based potassium fertilizer from seawater. F_1 and F_2 clinoptilolite filters serve for potassium preconcentration and fertilizer production, respectively; R_1–R_3 recuperators provide heating/cooling to solutions under treatment.

C. Bromine from Seawater

1. Air-Stripping Methods

Among the elements recovered from seawater and other sources (using conventional methods) bromine is the most prominent, since 70% of its world production is provided from the sea [12]. Because the ocean contains over 99% of this element, bromine may be referred to as a typical marine element.

Demands for bromine continue to increase throughout the world at a rate predicted to equal to 2% per year. Such an estimate of bromine consumption growth is warranted by increasing production of antidetonators for motor fuel, flameproof plastics, pesticides, and high-density fluids for the deep borehole in the petrol industry.

Bromine is produced directly from seawater by the so-called air-stripping technique [60, 2, p. 31]. The process involves acidifying seawater with H_2SO_4. Treatment with chlorine then oxidizes the bromides in the seawater to volatile elemental bromine. The bromine is removed from the solution in an airstream for reabsorption by different aqueous or organic solutes [60] selected for this purpose. This technology, developed by the Dow Chemical Co., is widely used on a large scale. In one modification of the Dow process [61], bromine, extracted from the air or stream flow is reacted with SO_2 and is then absorbed by water in the countercurrent absorber. Elemental bromine is recovered by secondary chlorination of the emerging liquid phase, Br_2 in HBr solution. One of the most interesting versions of the Dow process was introduced by the Israel Mining Industries Co. [62]. In their approach, bromine is finally extracted with tetrabromoethane.

The air-stripping production of bromine from seawater is effective only under the following conditions: relatively high seawater temperatures and sufficient bromine concentration (usually higher than 50 mg/L). Unfortunately, these two conditions are not always achievable. For instance, in southern parts of Russia where weather conditions are suitable, the concentration of bromine is low, reaching 8–10 mg/L (Kaspyi Sea) or 37 mg/L (Black Sea). In the northern and eastern open seas where the bromine concentration is high (around 65 mg/L [63]) temperatures are far too low.

Currently used technology in Russia (and other states of the former USSR) for the recovery of bromine from seawater involves preliminary concentration of seawater. Seawater brine, usually used for bromine recovery, is brought to the concentration level that corresponds to the point at which NaCl crystallizes (achieved by seawater evaporation [60]). Increasing the density of the bromine from 1.2 to 1.3 g/cm^3 leads to an increase in brine concentration from 0.05 to 0.3 mass%. The development of a complex scheme for processing seawater brine for the production of bromine brines, low-grade sodium chloride, mirabilite (Na_2SO_4), and magnesium oxide [64] was pointed toward treatment of the saline lakes of Krimia and Sivash Gulf (Black Sea). This scheme included bromine stripping from the brine at bromide ion concentrations of about 600 mg/L.

The flowsheet of the air-stripping process for bromine recovery from brines (including seawater brines) is shown in Fig. 5 [60]. The stock brine from a reservoir, mixed with H_2SO_4 and Cl_2, is directed to the top of the desorber. The bromine-free brine is collected at the bottom of the desorber, neutralized with thiosulfate and lime milk prior to disposal. Release of the chlorine/bromine air mixture from the top of the desorber is directed to the dechlorinating tower (1) where the mixture is treated with diluted $FeBr_3$ solution. The halogen exchange is described by the reaction

$$2FeBr_3 + 3Cl_2 \rightarrow 2FeCl_3 + 3Br_2 \tag{5}$$

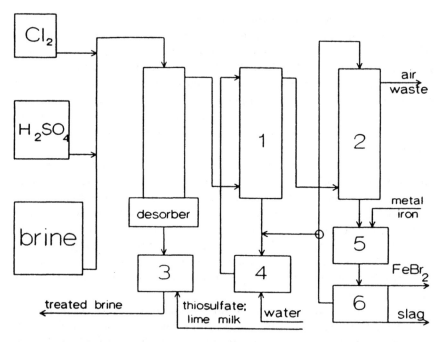

Figure 5 Technological flowsheet for producing bromine from seawater brines by the air-stripping method: 1. dechlorination column; 2. bromine absorber; 3. neutralizer; 4. tank for dechlorinating solution; 5. reducer; 6. settling tank.

Chlorine-free bromine/air mixture is directed to the absorber (2) where Br_2 is reacted with iron (II) bromide:

$$2FeBr_2 + Br_2 \rightarrow 2FeBr_3 \tag{6}$$

Then the iron (III) bromide is reduced with metallic iron to $FeBr_2$:

$$2FeBr_3 + Fe \rightarrow 3FeBr_2 \tag{7}$$

Part of the reduced $FeBr_2$ solution is returned for bromine absorption and dechlorination of the bromine airflow. The rest of this solution is used for producing elemental bromine and different bromine compounds.

Brine processing is carried out at a bromine concentration of 1 g/L and higher by steam stripping after preliminary acidification and by oxidative chlorination in a countercurrent mode. In this case, the raw bromide is obtained in one stage and then further refined.

None of the numerous technological solutions available for bromine recovery from seawater wholly satisfy modern ecological standards. At present, the closest to acceptable methods exclude the use of aggressive and toxic chemicals, thereby contributing to reduced ecological problems and the improvement of economical factors.

2. Ion Exchange and Electrosorption Methods

The first attempts to apply ion exchange in bromine recovery from seawater occurred at the end of the fifties [65, 66] and dealt with the application of ion exchangers for the sorption of bromine from oxidized bromine solutions. The phenomenon of superequivalent sorption of bromine by anion exchangers (Amberlite IRA-400, Dowex-1, AV-17, or the AM-type in the Br⁻ form) through the formation of polybromide complexes in the resin phase ($R-Br_3-R-Br_7$) became the basis for their research. It was determined that the sorption capacity of the anion exchanger toward bromine could achieve more than 6 equivalents per 1 equivalent of the functional groups [65,66]. The technology based on such use of anion exchangers for recovering bromine from seawater was first proposed in 1964 in the United States of America [67,68]. With this technique, seawater is acidified to pH = 3 to 5 and chlorine is passed into the system (approximately 1 g · mol of Cl_2 per 1 g ion of Br), thereby oxidizing bromide into elemental bromine. Following this the bromine is adsorbed on a strong-base anion exchanger. The sorbed bromine can be removed from the resin phase either by hydrolysis with base [69] or by oxidation [67,70]. One advantage of this sorption method is the low dependency of capacity on temperature. This allows processing of the brines during winter at the ambient temperature [71–73].

A number of ecologically acceptable electrosorption methods have been developed at the Vernadsky Institute of The Russian Academy of Science [74, 75]. It has been demonstrated that these methods can compete successfully with existing techniques of bromine recovery from seawater and brines.

The schematic diagram of the double chamber apparatus for the electrosorption recovery of bromine from seawater is shown in Fig. 6. Both anode and cathode chambers are packed with granulated activated carbon and separated from each other by a hydrophilic 0.5-mm diaphragm. Current flow is through 0.5–1.5-mm-diameter titanium wire. A (1 ± 0.25)-mm granular activated carbon with a specific surface area of 500 m²/g fills the anode and cathode regions of the apparatus forming layers of a constant thickness (3 mm) along the full height of the electrolyzer. Seawater upflow at approximately 100 chamber volumes per hour is filtered. A potential of 2.5 V is used.

During the accumulation stage, elemental bromine is formed. Magnesium bromide is obtained after changing the polarity of the electrodes (regeneration stage). Flushing the $MgBr_2$ with two chamber volumes of

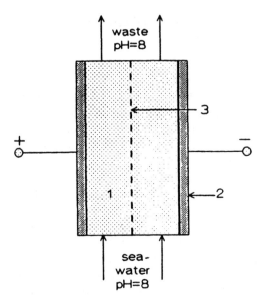

Figure 6 Schematic diagram of double-chamber apparatus for electrosorption recovery of bromine from seawater: 1. activated carbon; 2. titanium current carriers; 3. hydrophilic diaphragm.

water yields an electrolyte containing 95% $MgBr_2$. It is used in the regeneration cycles to increase the concentration of $MgBr_2$ from 20 to $\gg 100$ g/L.

The main electrochemical reactions proceed as outlined via the reduction of H^+ and Br_2 and the oxidation of Br^- and O^- at the electrodes:

1. Accumulation stage:

cathodic process

$$H_2O + 2\bar{e} \rightarrow H_2 + 2OH^- \tag{8}$$

anodic process

$$2H_2O - 4\bar{e} \rightarrow O_2 + 4H^+ \tag{9}$$

$$2Br^- - 2\bar{e} \rightarrow Br_2^{\circ} \tag{10}$$

neutralization reaction

$$Mg^{2+} + 6OH^- + 4H^+ \rightarrow H_2O + Mg(OH)_2 \tag{11}$$

overall reaction

$$Mg^{2+} + 2Br^- + 2H_2O - 2\bar{e} \rightarrow Mg(OH)_2 + Br_2 + H_2\uparrow \tag{12}$$

2. Regeneration stage:

$$2Mg(OH)_2 + 2Br_2 + 4\overline{e} \rightarrow 2MgBr_2 + 2H_2O + O_2\uparrow \qquad (13)$$

The elemental bromine released during the accumulation stage is sorbed by the active carbon in the anodic space. Magnesium hydroxide is released and sorbed in the cathodic chamber. Further regeneration leads to the cathodic reduction of the sorbed elemental bromine and formation of magnesium bromide.

The process is ecologically clean, since seawater leaving the chamber contains no chlorine and the anodic process is applicable only toward bromide ion and the water discharged.

Energy expenditure for this process is 16–20 kWh per 1 kg of bromine (including seawater pumping). The anion impurity of the product does not exceed 5%.

Numerous attempts at sorbing bromine from seawater (as bromide ion) without preliminary oxidation failed to lead to satisfactory results for an extended period [76], mainly because ion exchangers with high selectivity toward Br$^-$ ion are not easily regenerated.

Recent studies including those concerned with the recovery of bromine from seawater [77–79] have shown that the selectivity of strong-base anion exchangers for bromide ion depends heavily on temperature. The value at equilibrium of the separation coefficient for Br$^-$ and Cl$^-$ ions, $\alpha_{Cl^-}^{Br^-}$ (see Eq. (1)) with the strong-base anion exchanger AV-17 (Russian analog of Dowex-1 and Amberlite IRA-400) decreases by almost a factor of 2 for a temperature increase from 10 to 90°C. This allowed the design of reagentless and waste-free processes for concentrating bromine from seawater through predetermined changes in temperature of the ion-exchange system.

3. *Physicochemical Principles of Waste-Free Ion-Exchange Technology for Recovery of Bromine from Seawater*

The main disadvantage of ion-exchange processes for the recovery of valuable mineral components from seawater is attributed to the increase of electrolyte content in waters that originate in the regeneration step. Several ways to overcome this problem have been proposed, e.g., by combining membrane and ion-exchange techniques [81,82], by coupling ion exchange with precipitation [83,84]. Others deal with ion-exchange methods that are based on strong sensitivity of the resin selectivity to an intensive thermodynamic parameter, such as temperature, ionic strength, etc., that facilitate separation of ionic mixtures without recourse to auxiliary reagents. Parametric pumping techniques (see, e.g., [85–89]), dual-temperature ion-exchange processes [90–92], and thermal ion-exchange fractionation [93,94] are typical examples of ion-exchange separation methods of this type. The single feature that makes these separation techniques most attractive is the exclusion of the regeneration step. This leads to ecologically clean and practically waste-free processes.

Most of the element recovery technology has been involved with separating cation mixtures (see, e.g., [93–96]) but separation of anion mixture is also a potential path to this objective, as the following example shows. Consider the case where a solution containing a certain ion to be removed, e.g., in Br⁻ ion recovery from seawater is exposed to a Br⁻ ion-selective exchanger initially in the Cl⁻ form at temperature T_1. The exchange reaction presented below then results in enrichment of Br⁻ in the resin:

$$R - Cl + Br^- \leftrightarrow RBr + Cl^- \tag{14}$$

The equilibrium separation coefficient, α, for this ion-exchange reaction is defined as follows:

$$\alpha_{Cl^-}^{Br^-} = \frac{\overline{M}_{Br}}{\overline{M}_{Cl}} \cdot \frac{C_{Cl}}{C_{Br}} \tag{15}$$

where $\overline{M}$ and C have the same meaning as that in Eq. (1). The value of α usually determined by the experimental conditions (solution composition, ionic strength, etc.).

A thermodynamically meaningful equilibrium constant K can be obtained from α values, determined for different compositions of the contacting phases through the use of the expression [8]

$$\ln K = \int_0^1 \ln \alpha \, d\overline{X}_{Br^-} \tag{16}$$

where $\overline{X}_{Br^-}$ is the fraction of ion exchange groups occupied by Br⁻ ions.

The influence of temperature on the equilibrium constant K can be described with the following fundamental equations:

$$d\frac{\ln K}{dT} = \frac{\Delta H^\circ}{RT^2} \tag{17}$$

$$\ln\frac{K_2}{K_1} = \int_{T_1}^{T_2} \frac{\Delta H^\circ}{RT^2} dT \tag{18}$$

In the simplest case, when ΔH° is independent of temperature, it follows from Eq. (18) that

$$\ln\frac{K_2}{K_1} = \frac{\Delta H°}{R}\left(\frac{1}{T_1} - \frac{1}{T_2}\right) \qquad (19)$$

Enthalpy changes encountered in ion-exchange processes carried out with resins of a conventional type are usually small (see, e.g., [97,102]) when covalent bond formation, association, or complex formation are absent. In systems where association equilibria (or complex formation) prevail in either the solution or the resin phase, the equilibrium is, as a rule, shifted markedly, as a result of the decrease in selectivity with increased temperature (see, e.g., [103–106]).

To examine this particular aspect, consider what happens when the ion-exchange treatment of one of the solutions of interest (seawater) is carried out in a two-sectional countercurrent column that is operated at different temperatures, T_1 and T_2 ($T_2 > T_1$). A diagram of this column is presented in Fig. 7.

The ion-exchange resin moves countercurrent to the upward flow of the seawater. The ion-exchange equilibrium in section 1 at T_1 is described by Eq. (14), while the value defines the equilibrium. Its shift to the left in section 2, where $T_2 > T_1$ and $\alpha_2 < \alpha_1$ as a result, leads to the release of Br⁻ ion from the resin into the solution phase. The Br⁻-enriched solution that results is made available at the outlet near the boundary of

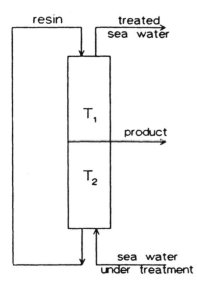

Figure 7 Schematic diagram of two-sectional countercurrent ion-exchange column operating at different temperatures to provide recovery of waste-free bromine from seawater.

the sections. The unloaded resin phase at T_1 is able to sorb Br⁻ once again from the cold seawater. As a consequence, the resin leaving section 2 can be directed into section 1 without any pretreatment. The resin circulates in a practically closed cycle and does not need any regeneration, i.e., the process is absolutely free of waste [79].

The same dual-temperature bromine recovery scheme can be extended to a multistage process using fixed beds of anion exchanger (analogous to the flow sheet presented to depict the potassium concentration scheme shown in Fig. 4 (see Fig. 8). The principle of the dual-temperature multistage bromine concentrating operation is presented in Fig. 9. In a subsequent treatment of cold and hot seawater in a fixed-bed anion-exchange resin column, the concentration of Br⁻ ions in the hot solution exiting the column increases by a factor of 2, while the concentrations of Cl⁻ and SO_4^{2-} ions decrease. The multistage process enriches bromine concentrations from 3 to 5 g/L. This is an acceptable concentration level for further processing. Estimates of bromine production rates indicate that 10 kg can be expected to be formed yearly with approximately 1 kg of ion exchanger [78,79].

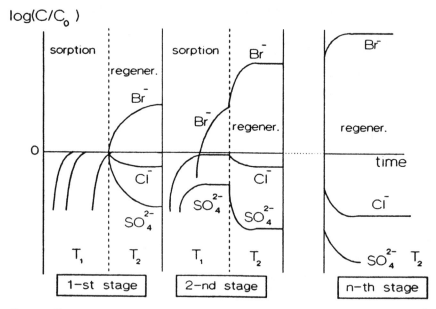

Figure 8   Concentration profiles illustrating the principle of the dual-temperature multistage scheme for bromine recovery from seawater $(T_2 > T_1)$.

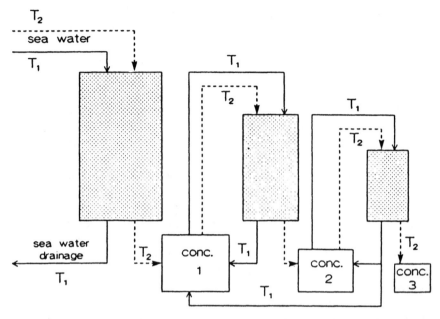

Figure 9 Scheme of dual-temperature multistage unit for bromine recovery from seawater. Storage tanks for bromine concentrates (conc. 1–3) are supplied with facilities for cooling/heating the solution under treatment ($T_2 > T_1$).

Continuously increasing demand for bromine and its compounds provides the stimuli for further development of new methods for its recovery from seawater [80].

IV. RECOVERY OF MICROELEMENTS

A. Lithium

Lithium concentration in seawater does not exceed 0.17 mg/L. Nevertheless, the ocean is considered to be the most promising source of this element in the near future [107]. The overall inventory of lithium in the world's oceans is approximately 2.6×10^{11} tons [2]. With lithium so accessible, continual growth of lithium demand depends solely on new developments and expansion of its recovery from sea.

Lithium is presently widely used for the production of ceramics, glass, in the aluminum industry, in ferrous metallurgy, in the chemical industry and in electrotechnics [108,109]. Significant quantities of lithium are used for producing compact accumulators. But all these quantities are minute compared to the amount that would be needed for regulated ther-monuclear energy synthesis.

Naturally occurring lithium ($\sim 7\%$ of Li^6) needed for one "solid phase blanket" may vary from 5.5 to 46.8 kg/MW; for one "liquid blanket" the need is for 1000 kg/MW [110].

According to predictions for the year 2000, the demand for lithium could reach 50,000–70,000 tons per year, in the Western European countries alone. This amounts to an increase by a factor of 10 from the present demand [111]. According to other estimates, the need for Li^6 isotope in one thermonuclear reactor could be about 5000 tons per year [112].

The most important land-based lithium sources are the following minerals: spodumene ($Li_2O \cdot Al_2O_3 \cdot 4SiO_2$), petalite ($Li_2O \cdot Al_2O_3 \cdot 8SiO_2$), lepidolite ($KLi_{1.5}Al_{1.5}[Si_3AlO_{10}][F,OH]_2$), and amblygonite ($LiAl[PO_4]$-$[OH,F]$) [109]. Only a small part of lithium world production is recovered from hydromineral stocks, which include highly mineralized underground brines, geothermal waters, etc. [113]. The recovery of lithium from hydromineral sources is carried out (on a semi-industrial and industrial scale) in the United States of America [114–116] (salt lakes), in Japan [117–120] (thermal waters), and in Israel [121, 122] (Dead Sea). In Bulgaria and Germany, complex schemes for the recovery of lithium from hydromineral stocks are under development [123, 124]; recovery from diamond deposits in Jakutia and from geothermal waters and brines of the Dagestan and Stavropel regions in Russia [125] are also being investigated.

1. Coprecipitation and Extraction Methods

The problem of lithium recovery from land-based hydromineral sources is very similar to the problem encountered in its recovery from seawater. Coprecipitation, extraction, and ion exchange, the methods used in both instances are practically the same.

Reagents, such as aluminum hydroxide [119–121, 124], potassium, and iron periodates [118] have been successfully applied for lithium coprecipitation. In the last case, ion exchange was used to concentrate lithium after dissolution of the coprecipitate.

Extraction methods in lithium recovery processes have not found as wide use as other techniques. When several extractants, such as C_3–C_5 primary alcohols and C_6–C_8 aliphatic ketones were tried for this purpose, isobutanol, the one that seemed to be the most promising yielded separation coefficient values for Li and Mg^{2+}, and Na^+ and K^+ that were too low. Separation factors were increased by extracting lithium complexes with chlorides of iron, nickel, or cobalt in acidic media [126]. The most interesting extraction method was developed by Japanese scientists at the beginning of the 1980s [127-131] and, in fact it represents the most updated technology. In it, lithium is extracted with cyclohexane mixed with phenolfluoroacetone and trioctyloxyphosphine. The back reaction of lithium with hydrochloric acid is then followed by precipitation of Li with potassium phosphate. The purity of the precipitate obtained is not less than 95%.

2. Ion-Exchange and Sorption Methods

Since the beginning of the 1980s, lithium recovery from seawater has increasingly involved sorption and ion-exchange methods. A number of organic and inorganic sorbents analogous to compounds used for lithium extraction or coprecipitation have been designed for this purpose. Lithium-selective aluminum-containing resins [132–135], prepared by treating the macroporous anion exchanger Dowex-1 type with saturated $AlCl_3$ solution, ammonia, and, finally, with lithium halide solution before heating to obtain a composite of $LiX \cdot 2Al(OH)_3$ microcrystals included in the resin matrix are examples of such products that have been patented in the United States of America.

Sorbents based on tin and antimony compounds have been synthesized in Japan for selective extraction of lithium from seawater. They have raised the enrichment degree in lithium recovery to 4000 [111]. High selectivity toward lithium is demonstrated by charcoal impregnated with tin dioxide [136] as well. One group of inorganic sorbents that was found to show high selectivity toward lithium in alkaline media includes the dioxides of titanium and zirconium, thorium arsenate [137,138] and mixed oxides of titanium and iron [139], titanium and chromium [140], and titanium and magnesium [141]. Although these sorbents cannot be widely used in lithium recovery from seawater, they may be useful for processing lithium concentrates (e.g., for refining and purification).

In the wide spectrum of lithium selective ion-exchange materials, only the cation exchangers based on manganese oxides show promise for direct recovery of lithium from seawater. Originating in Russia, one is based on manganese oxide (ISM-1) and the other on manganese and aluminum mixed oxides (ISMA-1) [142–144]. Similar sorbents, based on hydrated γ-oxides of manganese [145] and mixed oxides of manganese and magnesium [146] have been developed in Japan.

Investigation of the performance of ISMA-1 sorbents, when used to recover lithium from seawater, has yielded the following information: (1) Li^+ ion distribution coefficients of 4×10^4 cm^3 g^{-1} prevail (Eq. (4)); (2) the sorbents are easily regenerated with nitric acid; (3) they exhibit a high capacity for Li^+ ions of about 20 mg/g; and (4) lithium concentrates containing up to 1 g/L of Li^+ can be reached under optimal conditions. These results were applied in the design of a pilot plant with a capacity of 3 m^3 of seawater per hour, using a two-stage scheme for the production of lithium carbonate from seawater [15, 147, 148].

The need to promote stability in the Japanese sorbents of the same type during their use for the recovery of lithium from seawater [146] during five sorption-regenerated cycles [145] lowered their recovery capacity to 8.5 mg/g. Though ISMA-1 sorbents provide higher chemical stability, the degradation of manganese oxide associated with the ion exchangers remains the most serious limitation for their scaled-up applications in lithium recovery processes. Another factor determining the industrial applicability of these sorbents has been the degree of success achieved in the search for flowsheets, which promote increased efficiency and reduce the sorbent inventories. Progress in this direction is observable in the several processes that have been designed to apply ISMA-1 sorbents for lithium recovery from seawater [149, 150]. Their aim has been to increase efficiency or to decrease sorbent inventory [149, 150].

A diagram of the pilot plant unit for two-stage extraction of lithium from seawater is shown in Fig. 10. The first stage of lithium recovery is carried out in the contactor (CL), which contains about 30 kg of the sorbent (ISMA-1). The seawater is passed upward through a fluidized ion-exchange bed to provide equilibrium saturation with Li^+ ions. This step is followed by downflow regeneration of the fixed ISMA-1 bed with HNO_3 solution to yield Li concentrate at the end of the first stage (T3 Li conc. 1) of the operation. The neutralization of excess nitric acid requires alkali or special buffering anion exchangers (columns C2 and C3 in the scheme). The use of buffering anion exchanger permits the neutralization step to occur without an increase in the overall concentration of metal ions in the solution and leads, in turn, to a significant increase in the ion-exchange capacity of the ISMA-1 used in the second stage of the lithium concentration process (columns C4–C10 in the scheme).

Redirecting the flows of the solution (Li conc. I) in the second stage of the process permits both sorption and regeneration steps to be carried out independently and in a continuous mode of operation while assuring complete lithium recovery. The values of α, the equilibrium separation coefficient for Li^+ and the ions accompanying Li^+ in the concentrates obtained after the first and second stages of the process are listed in Table 3 [15]. The unit shown in Fig. 10 operates at a total flow rate of 5 m^3/hr. After preliminary evaporation of the second-stage lithium concentrate (Li conc. II) Li_2CO_3 is precipitated using hot K_2CO_3 solution. The dry product obtained without any additional purification contains more than 95% Li_2CO_3.

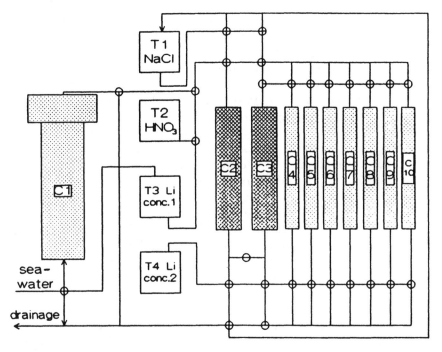

Figure 10 Scheme of pilot unit for two-stage recovery of lithium from seawater (see text).

It has been shown that the economics of the lithium recovery process is determined by the sorbent lifetime in the first stage. During the last few years, a number of ways to prevent sorbent loss and to prolong its useful life were found.

Several alternative methods proposed for lithium recovery from scawatcr usc ion cxchangc aftcr solar cvaporation and fractional crystallization of NaCl, $CaSO_4$ and $KCl \cdot MCl_2$. In these instances, polymeric ion exchangers, such as highly cross-linked Dowex 50 (16% DVB) [110] or Retardion Ag II, A8 (copolymer of styrene and acrylic acid cross-linked

Table 3 Equilibrium Selectivity Coefficients, α_{Me}^{Li} Measured for ISMA-1 in Seawater Sorption Studies

Stage no.	α_K^{Li}	α_{Na}^{Li}	α_{Mg}^{Li}	α_{Ca}^{Li}	α_{Sr}^{Li}
1	15090	21210	6650	1470	270
2	48240	17090	9800	980	165

with DVB and containing both weak-acid COOH groups and strong-base —$CH_2N^+(CH_3)_3$ groups) can be used [152]. These ion-exchange resins pretreated with a 50% water-ethanol mixture have also been shown to be applicable for the recovery of lithium from seawater [153]. A new method for lithium recovery from seawater that uses metallic aluminum has been proposed by Takeuchi [154, 155]. This method is based on the high selectivity of the $Al(OH)_3$ layer (covering the metallic aluminum during its corrosion in seawater) for Li^+ ion. The optimum temperature for the recovery process is 50°C, almost a 70% lithium recovery being reached under batch conditions. The selectivity coefficients of Li^+, Mg^+, Ca^{2+}, and K^+ equal 990, 11, 45, and 90, respectively [154].

In concluding this section dealing with lithium recovery from seawater, we reemphasize that the selectivity for Li^+ ion of all sorbents investigated to date, including metal oxides [137, 138], composite materials based on crystalline aluminum compounds included in polymer resins [133–135, 156, 157], metallic aluminum [154, 155], mixed compounds of tin and antimony, and antimony and titanium [158], and some organic compounds [152], is not competitive with the selectivity for Li^+ ion of manganese-based ion exchangers. Practically all recent publications in this field deal with studies of the structure and sorption properties of MnO_2-based sorbents [147–151, 159–163]. An example of this is provided by the recent publications of thermodynamic parameters for lithium exchange on manganese oxide sorbent from alkaline media [160].

To improve the kinetic properties of manganese oxide sorbents, Japanese investigators have developed a composite material through the inclusion of λ-MnO_2, a fine powder, with a spinal structure, in polyvinylchloride [162]. It is of interest to note that ISM and ISM-1 sorbents synthesized in Russia are also composite materials obtained by applying polymeric binding materials [142, 144]. As a consequence, long-term investigations carried out concurrently by both Russian [142–144, 147–151] and Japanese [145, 146, 159–163] teams in the field of lithium recovery from seawater and other hydromineral sources have led to the development of the same type of sorbents with very similar structural and compositional characteristics.

It is worth emphasizing the fact that the number of publications on lithium recovery from seawater is rising steadily.

B. Uranium

During the past 30 years, investigations of the recovery of uranium from seawater have intensified as a result of expectations that land-based uranium deposits will be exhausted by the turn of the century [164, 165]. The demand for uranium in the year 2000 is estimated to be 2.5×10^5 tons. The world's oceans contain about 5×10^9 tons of this element.

First studies dealing with the extraction of uranium from seawater were initiated in 1953 at the Nuclear Research Centre at Harwell in Great Britain. A number of Japanese research groups have been involved in similar investigations since the beginning of the 1960s. Early in the 1970s a remarkable increase in activity occurred in this field in China, Germany, France, Finland, Israel, India, Italy, Pakistan, and Sweden, as well as in the United States of America and USSR (in spite of their significant reserves of traditional uranium deposits).

The real boom in "uranium from sea" projects arose in the mid-1970s when concerted efforts to develop uranium specific sorbents and to evaluate processes based on their use for uranium recovery from seawater were undertaken in Japan and West Germany. The research in Japan, conducted by the Metal Mining Agency of Japan, was sponsored by the Ministry of International Trade and Industry [166,167]. In West Germany, the studies were carried out at the Nuclear Research Center (KFA) in Jülich [168].

The main focus of all research projects was directed toward solving the following tasks:

1. Knowledge with respect to speciation and distribution of uranium in the ocean
2. Identification and discovery of uranium zones in seawater
3. Synthesis and study of sorption materials for uranium recovery
4. Design and construction of technical facilities required for uranium recovery from the sea

All these problems are very closely interrelated and this aspect is emphasized in the following discussion.

The concentration of uranium in the sea is relatively constant and is equal to 3 µg/l at a salinity of 35%. A slight increase of uranium concentration has been observed to occur upon increase of the general salinity of seawater in numerous deep-water studies [169].

The ratio of the isotopes $^{235}U/^{238}U$ in "marine" uranium samples does not differ from that typical for land-based uranium deposits.

The main ionic forms of uranium compounds and their concentrations in seawater are listed in Table 4 [165,170].

From Table 4, it is apparent that almost all uranium in seawater exists in the form of a highly stable tricarbonate complex $UO_2(CO_3)_3^{4-}$ (its stability constant is approximately 10^{23}). This fact and its very low concentration in seawater dictates the choice of sorbent most suitable for uranium recovery. The sorbent must be available on a large scale and at low

Table 4 Speciation, Concentration, and Relative Content of
Uranium in Seawater at 5 = 25°C and pH = 8.1

Compound, ion	Concentration (mol/L)	Relative content (%)
UO_2^{2+}	1.53×10^{-17}	1.0×10^{-7}
$UO_2(OH)_2$	1.53×10^{-12}	0.01
$[UO_2(CO_3)_2]^{2-}$	5.46×10^{-11}	0.39
$[UO_2(OH)_3]^-$	2.43×10^{-10}	1.75
$[UO_2(CO_3)_3]^{4-}$	1.37×10^{-8}	98.82

cost, for instance, for a unit whose daily production is 1 ton of the uranium concentrate, 1 km^3 of seawater needs to be processed (assuming the
extent of extraction is about 30%). This means that the sorbent investment
must be 10^6–10^7 tons. Only highly stable ion-exchange materials accommodate large-scale technology. Finally, it should be emphasized that apart
from absorption methods no alternative techniques, such as precipitation,
extraction, etc., are acceptable for uranium recovery from seawater, because of ecological damage introduced by the use of chemical agents to
affect such alternative paths to uranium recovery.

1.   Ion-Exchange Materials for Uranium Extraction

Numerous independent cost studies [170] have shown that for a sorbent
to be acceptable, it must be able to concentrate the initial uranium feed
upon contact by a factor (see distribution coefficient, Eq. (4)) of 10^5–10^6.
Such enrichment of the uranium by the sorbent corresponds to a uranium
capacity of 0.3–5.0 mg per gram of dry resin [171].

Those ion exchangers that reach adequate values for the sorption of
uranium from seawater are presented in Table 5.

The first 15–20 years of the "marine uranium" projects were involved
with the development of hydrous titanium oxide–based sorbents. During
the last 10–15 years, directions of this study have changed and the preparation mainly of poly(acrylamidoxime) resins and their modifications has
been sought.

The technological aspects of hydrous titanium oxide production
were developed in Great Britain. This product is obtained in irregular
granular form by precipitation with basic titanium sulfate or chloride
solutions. After the precipitation step further granulation of the precipitate is obtained using different methods. A typical product composition
is 60% TiO_2, 30% H_2O, and 10% a mixture of other components, includ-

Table 5 Uranium Selective Ion-Exchange Materials

Ion exchanger	Functional group	Capacity toward U (mg/g)	$D^a \times 10^{-5}$ (cm³/g)	Reference
Hydrous titanium oxide, $TiO_2 \cdot nH_2O$	Ti with two OH groups	0.55 (a)[b] 0.20 (b)[b]	1.7 0.6	171,172
$TiO_2 \cdot nH_2O$, freshly prepared	Ti with two OH groups	1.55 (a)	4.7	173
Basic zinc carbonate	—	0.54 (a)	1.6	173
Resorcinol–arsenic acid	—As(O) with two OH groups	1.1 (b)	3.3	173
Poly(acrylamid-oxime) (TEGDM-DVB)	C with =N—OH and NH₂	3.2 (b) 4.0 (a)	9.7 12.1	174 174
Duolite ES-346	C with =N—OH and NH₂	3.6 (b)	11.1	175,176
Macrocycloimide resin 508	—NH—CH₂—CH(OH)—CH₂—NH₂ attached to aminophenol ring	0.93 (b)	2.8	177

[1]D values were calculated for uranium concentration in seawater = 3 µg/dm³.
[2]The capacity refers to the metal content (a); or to the dry sorbent (b); TEGDM = tetraethylene glycol dimethacrylate.

ing sodium salts [171]. The uranium sorption on titanium oxide sorbents can be described by the following reaction:

$$[UO_2(CO_3)_3]^{4-} + \begin{bmatrix} HO \\ \ \ \ \ \ \ Ti \\ HO \end{bmatrix}_n \rightleftharpoons \begin{bmatrix} UO_2 \begin{pmatrix} -O \\ \ \ \ \ \ \ Ti \\ -O \end{pmatrix} \end{bmatrix}_n^{(2n-2)-} +3HCO_3^- + (2n-3)H^+$$

$$(20)$$

Numerous publications in the 1960s and 1970s dealt in detail with the description of the mechanism, equilibrium, and kinetics of the uranium sorption reaction on titanium hydroxide [165]. Scaled-up testing of uranium sorption from seawater was carried out in the Soviet Union, United States of America, Great Britain, and Germany. The results were used in the design and construction of units for uranium recovery; approximately 10–100 g of uranium were produced per year [180,181].

The very similar operations common to these units include the following:

1. Uranium sorption by the sorbent (usually in a fluidized bed) with large amounts of seawater
2. Uranium stripping and sorbent regeneration with carbonate solutions (e.g., sodium or ammonium carbonates)
3. Secondary processing of uranium carbonate concentrates on ion exchangers to produce secondary uranium concentrates
4. Precipitation of uranium from the secondary concentrate

The economic aspects of uranium recovery from seawater, using $TiO_2 \cdot nH_2O$ sorbents have been discussed in a number of publications [166, 181–183]. They were based on project production scales of 100 to 1000 tons of U_3O_8 per year, 500 t/year [182], and 180 t/year [183]; the most detailed economic analysis was given by Hirai et al. [166].

The most important problems remaining with titanium oxide sorbents and the attempts being made to overcome them are

1. Low enrichment of uranium concentration ($D \leq 10^5$; see Table 50. An economically viable process for extracting uranium is achievable only with $D \geq 10^5$.
2. Unsatisfactory physicochemical stability of the sorbent. Titanium hydroxide is slightly soluble in seawater (0.1 mg/l [165]) and also shows some degradation upon aging. Mixed titanium and zirconium oxide–type sorbents, known as "thermoxides" with much better stability [178, 179] were developed.
3. Unsatisfactory kinetic characteristics of the sorbent. Several methods were employed to improve the kinetic properties of titanium oxide

sorbents as well. They included the insertion of fine powdery titanium oxide into the highly penetrable supports [184] and also into magnetic supports [185]. Similar improvement was obtained with titanium oxide–active carbon composites [186] or with superfine titanium oxide–coated active carbons [187].

The achievement of an economically viable process for the extraction of uranium from seawater, however, could be achieved only through use of a sorbent with uranium concentrating factors greater than those provided by titanium oxides [181]. Polyacrylamidoxime sorbents, characterized by D value $> 10^5$ ($\sim 10^6$) (see Table 5) made an appearance at the end of the 1970s to remove this impediment to the economic recovery of uranium from seawater. Their appearance reoriented research in the field of uranium recovery toward highly selective organic resins [188, 189].

Detailed information on the research carried out in the field of uranium extraction from seawater up to 1984 are given in reviews [164, 191–198]. The most intensive investigations are being carried out in Japan in institutes dedicated to uranium recovery from seawater [196].

A large number of sorbents containing amidoxime functional groups, in fiber materials, in granulated resins with different cross-linking agents, and in various composite materials (see, e.g., [193, 199–204]), have been synthesized in a number of different countries. The equilibrium and kinetic properties of these sorbents were studied in experiments examining uranium sorption from both natural seawater and model solutions. Uranium sorption behavior has been shown to be described by a Freundlich isotherm with a power index of about 1.6 [205, 206]. The results of enthalpy measurements during the sorption of uranium from model solutions by amidoxime resins suggest a chemisorptional mechanism ($\Delta H = -42.4$ kJ/mol) [206]. Uranium distribution coefficients determined for the different types of amidoxime sorbents from natural seawater were shown to range between 10^5 and 10^6 mL/g, whereas D values for the macroelements corresponded to 0.1 to 10 mL/g [200, 207]. Uranium distribution studies using x-ray microanalysis have shown uranium to be concentrated mainly at the periphery of fibers or beads, while smaller ions, such as Ca^{2+}, are distributed far more homogeneously along the diameter of the sorbent particles. When the matrix incorporates a mixture of hydrophilic (TEGDM) and hydrophobic (DVB) cross-linking agents, the uranium distribution coefficient increase with growth of the TEGDM fraction [203] is accompanied by greater homogeneity of the resin samples. It is still unclear whether these observations can be attributed to quasi-equilibrium in the course of an experiment, which normally does not exceed a period of more than 100 days [200,207], or to sorption specificity. For instance, the commercial amidoxime resin Duolite ES-346 [208] exhibits increased selectivity toward uranium during the first sorption cycle. This is attributed to precipitation of UO_2^{2+} ion on the surface of the sorbent, which occurs due to localization of high pH regions in the production process.

The rate of uranium sorption from seawater is strongly dependent on temperature (activation energy is about 42 Kj/mol [203]). The rate of sorption of polyamidoxime fibers is dependent on their diameter. The magnitude of the inter diffusion coefficient of UO_2^{2+} ion is of the order of 10^{-10} cm^2/s [209], while in the particle itself, the diffusion coefficient is about 10^{-7} cm^2/s [203]. Whether or not the sorption of uranium by macroporous granulated sorbents (of the amidoxime type) cross-linked with DVB/TEGDM is film- or particle diffusion controlled is thus determined by the particle size. Film diffusion across the particle/solution boundary layer is assumed to be rate controlling on this basis in the case of fiber materials [206]. Treatment of fiber sorbents with alkali leads to a significant increase in the rate of uranium sorption (four to five times). This can be attributed to micropores forming within fibers as a result of reaction with the alkali [209, 210]. Increase in cross-sectional area of the solution is affected in this way. Attempts to increase the rate of uranium sorption from seawater by polyamidoxime sorbents have led to the design of unique materials, e.g., irradiation-grafted fibers on polypropylene support [206,207,211] and hollow-fiber sorbents [210, 212]. Practical use of these sorbents with fiber diameters of 40 μm has resulted in a capacity of about 5 μg uranium/g after 30 days contact with seawater [207]. Approximately the same rate of sorption is observed with hollow fiber sorbents [212]. Other amidoxime containing materials, such as fine fibers encapsulated by small, highly penetrable plastic bags (1–5 cm in dimension) [213] exhibit improved kinetics of uranium sorption and are of interest. Other composite fiber sorbents with improved kinetic properties are obtained by incorporating a fine powder of polyamidoxime within SiO_2-based fibers [214], polymer gels [215], or porous membranes [216].

One of the key problems with many ion-exchange materials, including the polyamidoxime sorbents used for uranium extraction is encountered in the desorption (elution) stage. With the polyaminodimer, the HCl elution of uranium is usually carried out in two steps. The removal of calcium and magnesium with dilute HCl solution (~ 0.01 M) is followed by uranium elution with 0.5–1.0 M HCl [212, 213]. The concentrated effluent contains tens to hundreds of milligrams of uranium per liter and the removal of uranium reaches 95%. However, the elution with HCl seriously reduces the sorbent capacity [198, p. 213; 165, 207, 208], e.g., a six-day treatment of the fibrous amidoxime sorbents with 1 M HCl causes up to a 50% capacity loss [207]. Sorbents with improved kinetic properties that reduce elution times to less than 1 hr exhibit much higher stability [207,212] and can be used in 100 or more cycles. But even this gain in stability is considered insufficient for large-scale industrial applications. Alternative eluants are being sought to remedy the situation. Duolite ES-346 has been shown to be very stable when 0.10–1.0 M Na_2S or K_2S are mixed with 0.5–1.0 M NaOH and used as eluants, but the uranium concentration in the eluate is less than 1 ppm [217]. The SiO_2-based fiber composite sorbents were shown to be stable during at least 50 cycles when stripping with 1 M $NaHCO_3$ and 0.7 M NaCl mixtures [214].

These studies of polyamidoxime sorbents are being carried out concurrently with those involving titanium hydroxide, the sorbent initially introduced to facilitate uranium extraction [179, 218]. The new generation of titanium hydroxide based sorbents, such as carbon fiber materials containing a fine powder of TiO_2–nH_2O [219] and a combination of sorbents containing titanium hydroxide and amidoxime groups [219, 220], including membranes based on the same combination [221] have appeared during the past five years. A number of new sorbents containing phosphine and phosphonic functional groups [222–224] and porous polyurethane resins impregnated with oxime-type extractants [225] have also been synthesized for uranium extraction. Other studies have focused on a search for natural and inexpensive sorbents, like peat [226], strong-base anion exchangers modified with humic acid extracted from peat [227] and different tannin derivatives, as well as tannin-protein complexes immobilized in amine-type polymeric matrices [228]. These studies have not yet resulted in discoveries of sorbents as effective as the polyamidoxime sorbents.

2. Technical Devices for Uranium Recovery from Seawater

The facilities known to exist for uranium extraction are of two types. The first type requires an external energy source. The second uses seawater motion (waves, tides, streams, etc.) as the natural source of energy. In the first type, the rate of energy expenditure per 1 ton of product (uranium or any other microelement with a concentration level in seawater of around 10^{-6} g/L) was 3×10^5 kWh [229].

First Type   The first type of unit for uranium recovery has the following general features: The granulated sorbent is packed into net containers [230, 231] and seawater is pumped through the sorbent bed by compressors [230] or pumps [231]. When fiber sorbents are used (e.g., of the polyamidoxime type) they are packed in special modules supplied with a turbine which pumps the seawater through them [232]. The units are located either on the sea surface [230, 231] or are placed under water (up to 60 m [232]). The units are serviced by special ships which can also serve as the energy source [231]. The estimated cost of uranium obtained by units of this type is 220–330 USD per kilo of uranium [232]. Other approaches have been used to design devices for uranium recovery as well [192, 233–235]. For instance, an "endless" belt, either covered with sorbent [233] or fabricated from polyamidoxime/polyhydroxamic acid fiber cloth [192] has been employed for continuous extraction of uranium from seawater. The schematic diagram of a unit of this type is shown in Fig. 11.

Second Type The first proposal to extract microelements from the sea using the energy of tide streams was examined in Great Britain. It proved to be uneconomical for industry [181]. An analogous Japanese proposal was rejected because the tide energy in Japan was considered to be insufficient [165].

Two more seriously considered approaches to the design of sorption units for the recovery of microelements (mainly uranium) of the second type have received attention [165, 181]. The first applied energy from sea streams to units drifting along the sea surface during the sorption stage. In the second, the sorbent containers were attached to fixed bases located in an appropriately turbent zone (such as breakwaters). The intensive seaboard waves and streams promoted the movement of seawater through the sorbent bed. A scheme of a unit of this type is presented in Fig. 12 [236].

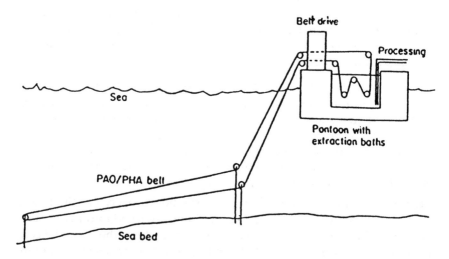

Figure 11 Schematic diagram of continuous extraction of uranium from seawater with polyamidoxime/polyhydroxamic acid fiber cloth used as an "endless belt" (needs permission from Reactive Polymers).

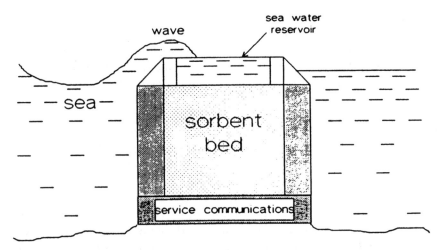

Figure 12 Schematic diagram of module for sorption recovery of uranium using energy of waves for propogation of seawater through the sorbent bed.

A number of additional original technical devices for the second type of uranium recovery from seawater has also been described recently [237–239].

The cost of "marine uranium" is highly dependent on the technical details of the process, e.g., for granulated sorbents it is in the range of 284–750 USD/kg, while with the fiber sorption materials it costs in the range of 132–300 USD/kg. Analysis of the cost of uranium produced from seawater using different units have shown that devices of the first type are approximately 100 USD more costly than units of the second type. But even the lowest average cost of uranium production from the sea are 5–10 times higher than the cost of uranium on the world market at the present time [164]. This situation is expected to change by the turn of the century, since the depletion of land-based uranium deposits, on one hand, and improvements in technology for recovery from seawater, on the other, will make the process for "marine uranium" production economically feasible.

C. Strontium

Strontium is widely used in ferrous and nonferrous metallurgy as a deacidifier and as an antifrictional material for producing glasses and some special optic materials. It exists in seawater at a concentration of 8 mg/L.

Several methods for its extraction as a by-product in uranium processing from seawater when applying titanium oxide have been proposed [175, 240]. Distribution coefficients in the range from $D = 3000$ to 6000 cm^3/g were observed. The strontium enrichment obtained was about a factor of 120 and further concentration was needed to produce technologically suitable solutions from which the relatively pure strontium carbonate could be precipitated. Barium fluorosilicate was applied for sorptive extraction of strontium from seawater for analytical purposes [241]. Composite sorption materials based on a mixture of Al_2O_3, MgO, and active carbon have been proposed for joint recovery of uranium and strontium from seawater [242]. Inorganic cation exchangers based on antimonic and arsenic acids exhibit very high selectivity for strontium that lead to its recovery from highly mineralized solutions containing large excesses of calcium and magnesium [243]. Manganese and aluminum oxide–based sorbents are also highly selective toward strontium. For instance, the ISMA-3 cation exchanger, synthesized and produced on a semi-industrial scale in Russia (at the Perm Polytechnic Institute) has the composition $Na_{0.06} \cdot Sr_{0.1} \times MnO_2 \cdot 0.025, Al_2O_3 \times (0.6–0.7)H_2O$ and is characterized by a structure of the psilomelane type. The equilibrium separation coefficients for Sr-Ca,Sr-Mg, α_{Ca}^{Sr}, and α_{Mg}^{Sr} in the course of the sorption of these metal ions from seawater by this cation exchanger are 23 and 650 [244].

Unfortunately, the relatively high α values are not as meaningful as they could be, the ISMA-3 sorbents being characterized by rather low chemical stability in the HNO_3 regeneration stage. Capacity loss of approximately 0.5–10% is observed after each regeneration cycle. Composite carbomineral sorbents containing MnO_2, selective toward Sr, and organic phase-bearing carboxylic acid groups, which preferentially adsorb calcium have been employed in a two-step desorption procedure to enhance their separation [245].

Synthetic and natural zeolites have also been intensively investigated. They are inexpensive and widely available sorbents that should be employable for strontium recovery. Some of these sorption materials, such as type A zeolites and natural chabazite are known to be effectively employed for the removal of radioactive strontium from technological solutions [246, 247] so contaminated.

Clinoptilolite has been demonstrated to be the most appropriate sorbent for strontium recovery from seawater, as well as from other natural waters and brines [248–250]. This most common zeolite is characterized by the composition $(Na, K)_4 CaAl_6Si_{30}O_{72} \cdot 24 H_2O$. The main advantages of applying clinoptilolite for recovery of strontium from seawater is attributed to the ease with which it is regeneratable by ammonium salts and to the greater likelihood of potassium (see below) and rubidium [251,19] coextraction that it promotes.

It has been shown that clinoptilolites from divergent deposits differ from each other in their selectivity toward the alkali earth metal ions (see Table 2) [57, 250, 251, 19]. We see from Table 2 that α_{Na}^{Sr} values may vary from 30 to 3. This behavior is attributed to both structural and cationic composition features of certain zeolites. The nonexchangeable fraction has been considered to influence not only the overall capacity of natural clinoptilolites but also their sensitivity toward strontium [250]. The pretreatment of natural forms of clinoptilolites with hot concentrated ammonium salt solutions and their "weathering" during 20–50 cycles that include periodic regeneration with 0.5 M NH_4NO_3 solution have been shown to provide uniformity to the properties of clinoptilolites from different sources [250]. The Sr distribution coefficient values most often encountered in sorption from seawater by clinoptilolites with full ion-exchange capacities of about 2.1 mequiv/g are in the range of 70–200 cm^3/g.

The kinetic properties of strontium sorption on clinoptilolite from both model solutions and natural seawater have been investigated in detail [252, 253]. These properties are determined by diffusion in both the zeolite microcrystals and the intercrystalline porous space [252]. Methods for determining the characteristic size of microcrystals by an ion-exchange technique are given in [253]. Such estimates provided a determination of their size-based contribution to the diffusion process.

The additional presence of 50–100 mg/L of strontium in the ammonium stripping solution remains one of the major problems in applying clinoptilolite for the recovery of strontium. Selective separation of strontium and calcium before precipitating $SrCO_3$ as a pure final product also remains a sizable problem. A possible solution to this problem may be provided by applying "cascade schemes" [15]. Such treatment of strontium concentrate could include columns with fixed beds of "selective" and "auxiliary" sorbents, e.g., [KU-2—Cli] → [KU-2—Cli] → etc., where brackets denote one stage of the process; Cli is clinoptilolite and KU-2 is a sulfonate exchanger (the Russian analog of Dowex 50). In such a "cascade scheme," the selectivity toward the cation of the stripping agent should be sizably smaller with the auxiliary sorbent than it is with its companion sorbent. The distribution of Sr, Ca, and Mg in eluates obtained with such a "cascade scheme" is presented in Fig. 13 [15]. It has been shown that application of 0.5 M NH_4NO_3 as a stripping solution for Cli and 2 M NaCl for KU-2 allows the production of strontium concentrates containing up to 600 mg Sr per liter after two subsequent stages (two pairs of sorbents) has been employed [15]. This concentration is appropriate for precipitat-

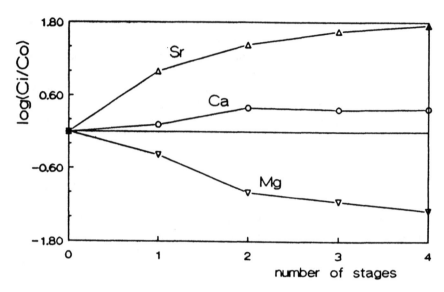

Figure 13 Concentrations of alkali earth metals at different stages of the cascade process in the course of strontium recovery from seawater (see text).

ing $SrCO_3$, but the purity of the product appears to be insufficient because of the high level of calcium carbonate also present in the product.

Another method for recovering strontium from NaCl stripping concentrates as well as from natural underground brines of similar composition has been examined in the laboratory of Professor V. I. Gorshkov at Moscow State University [254–256].

Selective separation of strontium from calcium at significant $SrCl_2$ concentration levels up to 2 g/L was obtained with KB-4 carboxylic acid exchanger employed in a Higgins-type contactor [246]. The pilot unit for strontium recovery from seawater in the closed (and practically waste-free) processing scheme was constructed in the Okhotsk Sea region (Sakhalin power station). The data obtained with the new pilot plant have shown that several components will be recovered simultaneously from seawater. The unit is estimated to produce more than 150 kg of $SrCO_3$, more than 5000 kg of KNO_3, and about 2 kg of $RbNO_3$ [15]. However, recovery of strontium is still uneconomical. Its cost is expected to become comparable to that of strontium produced from traditional, land-based sources.

D.   Rubidium

Rubidium and its compounds are used in electronics, electro-, and radio-, and x-ray techniques. Rubidium compounds are of particular importance in designing IR signal devices. The demand for rubidium may increase significantly in the near future, because of the further development of MHD generators and ion jet engines.

Even though the concentration of rubidium in seawater (about 0.11 mg/L) is comparable to the concentration of lithium in seawater, its price exceeds that of lithium by almost a factor of 200.

The number of publications involved with the recovery of rubidium from seawater is very limited. Most of the work in this field is by Russian scientists, who have proposed several schemes for the combined recovery of rubidium, strontium, and potassium with natural zeolites [15, 19, 250–253, 257]. A number of inorganic sorbents with high selectivity toward rubidium were also synthesized for the recovery of rubidium from natural hydromineral sources, including seawater. Ferrocyanides of the transition-metal ions were shown to exhibit the best properties for this purpose [258, 259]. Mordenite (another natural zeolite) has recently been proposed for selective recovery of rubidium from natural hydromineral sources as well [260]. A review of the properties of inorganic sorbents applicable for the recovery of rubidium from hydromineral sources has been published [261]. Studies of rubidium recovery from seawater [15, 19, 250–253] have shown that the final processing of rubidium concentrates, especially the selective separation of Rb^+ –K^+ mixtures remains the major problem. A report was recently published showing that this problem can be successfully solved by countercurrent ion exchange on phenolic resins [262].

E. Boron

This element is widely employed in the production of glass and glass fibers, fluxes, antiseptics, and other products. Boron compounds are also widely used in nuclear technology [264]. Boron is an element that occurs at a relatively high concentration level in seawater (4.5 mg/L). Yet, economically acceptable processes for boron extraction from the sea do not exist, despite the fact that methods for its recovery from highly mineralized brines have been available since the beginning of the 1960s [253]. With the development of such methods, attempts were made to determine the lowest concentration levels of the element, at which economical processes could be developed [256, 266]. This "critical" concentration of boron was at that time estimated to be around 20 mg/L. Currently, the "critical" concentration of boron is estimated to be 15 mg/L or even somewhat less.

Several Russian projects, by focusing on multielement processing from hydromineral sources, have now assured economically reasonable processes for boron recovery [267–269].

Three alternative approaches to boron recovery are known: precipitation, extraction, and sorption. The coprecipitation of boron with $Mg(OH)_2$ in seawater made basic is one of the most popular and well-known examples of the first approach [270]. A number of extraction methods have been designed. Among them are extraction with β-aliphatic diols [271], with isoamyl alcohol [272] alone, and with a mixture of salicyclic acid derivatives [273, 274]. The sorption methods for boron recovery, currently considered to be the most competitive, are mainly based on inorganic sorbents, such as hydroxides of Zr, La, Ti, Fe, Al, Ce, and V [275,276] or synthetic sorbents, such as weak-base anion exchangers [277]. Several highly selective ion exchangers have been synthesized in Russia, e.g., the MNG-type sorbents based on magnesium and nickel hydroxides [278, 279] and the SB-1 anion exchanger containing aminooxyethylene groups [280, 281], as well as some polymeric complexones [282].

Pilot plant studies (flow rates, 1 cm/s) with the SB-1 anion exchange resin (column diameter, 0.5–0.7 cm) yielded distribution coefficients of the order $D = 400$ cm^3/g. The boron sorption process was shown to be film diffusion controlled. The equilibrium values of boron loading were reached in 6–8 hr [280]. Boron elution and resin regeneration were carried out with 0.1 M NaOH. The complete elution of boron required 10 column volumes at 10 BV and yielded concentrates of 100 mg/L. This facilitated the eventual reduction to solid concentrates of alkali metal borates [281].

F. Molybdenum, Vanadium, and Other Nonferrous Microelements

Molybdenum concentration in seawater varies from 1 to 10 μg/L, according to different literature sources [2, 3, 165]. The ion-exchange resins Chelex-100, Permutit S-1005, and strong-base anion exchangers containing quaternary ammonium groups have been proposed for the sorption of molydbenum from seawater. Their use for this purpose proved successful. Good results were obtained by using iron hydroxide in the colloid state for the sorption of molydbenum [284]. The optimum pH for binding molybdate anion is 4. Further molybdenum concentration is carried out by foam flotation with dodecyl sulfate [285]. The nonaqueous mixture of 4-benzoyl-3-methyl-1-penyl-15-pyrazoline in isoamyl alcohol has also been applied for extractive recovery of molybdenum from seawater [286]. Ion flotation methods for recovering molydbenum and vanadium from seawater have also been proposed [287]. Secondary concentration and separation after flotation is subsequently carried out on the cation exchanger Amberlite IR-120 (for vanadium capture) and the anion exchanger Amberlite IRA-420 (for molydbenum capture).

Vanadium concentration in seawater is around 2 ng/L [2, 3, 165]. It can also be recovered (as noted before) by ion flotation after coprecipitation with colloidal iron hydroxide [288] or by extraction techniques using chloroform or acetone solutions of 4-(2-pyridilazo) resorcinol as a complexing agent [289]. Simultaneous extraction of vanadium, manganese, copper, and zinc by chloroform solution of ammonium pyrrolidildithiocarbomate has been proposed for recovery of these elements from seawater [290]. The recovery of vanadium together with other nonferrous metals by sorption of Chelex-100 and Permutite S-1005, followed by selective elution with ammonium salts has been demonstrated [283]. Vanadium is one of the elements that accumulate in the bodies of some marine organisms and animals. The concentration factor has reached a value of 3×10^5 [2, 288]. Biological methods of vanadium recovery from seawater may, as a result, become valuable.

Generally, the removal of molydbenum, vanadium, and other nonferrous elements (e.g., copper, zinc, etc.) from seawater is accompanied at a relatively high concentration level by other metal components. Simultaneous recovery of nonferrous metals has been observed in all processes where weak-base anion exchangers or chelating resins have been employed for the sorption of a particular element from seawater, e.g., in boron recovery. Similar results were obtained with natural zeolites [57, 291, 292] during their employment for the recovery of, e.g., strontium and rubidium from seawater (see above). It has been shown that stripping with NH_4NO_3 solutions leads to complete elution of Sr and Rb, but fails to remove nonferrous metals from the sorbent phase due to slow kinetics. This results in the gradual accumulation of Cu, Zn, Ni, and other elements on the zeolite. They can be periodically stripped (e.g., every 10–20 sorption -salt regeneration cycles) with HNO_3 solution for separation on an iminodiacetic acid resin [15].

The concentration of microelements from seawater on sorbents is widely used as a preconcentration step for their analytical determination by different instrumental methods. Several recent publications [293–298] and reviews [299–301] are recommended to those who are further interested in this subject.

ACKNOWLEDGMENTS

A visiting professorship to D. M. from the Department of Analytical Chemistry, Autonomous University of Barcelona (Bellaterra) Spain (Programa de Professors Visitants a le Universitat Autonoma de Barcelona

136

(Bellaterra)) is gratefully acknowledged. D. M. is indebted to Professor Manuel Valiente (UAB) for his interest in this work. Financial support from the Ministry of Immigrant Absorption (Israel) in the form of a grant to D. M. is gratefully acknowledged.

REFERENCES

1. Sverdrup, H. U., Johnson, M. W., and Fleming, R. H. *The Oceans, Their Physics, Chemistry, and General Biology,* Prentice-Hall, Englewood Cliffs, NJ, 1942.
2. Mero, J. L. *The Mineral Resources of the Sea,* Elsevier, New York, 1965, Chap. III.
3. Horne, R. A. *Marine Chemistry. The Structure of Water and the Chemistry of the Hydro-sphere.* Wiley-Interscience, New York, 1969, p. 444.
4. Riley, J. P. and Chester, R. *Introduction to Marine Chemistry,* Academic Press, New York, 1971, p. 5.
5. Ford, G., Niblett, C. and Walker, L. The future for ocean technology, Future Sci. Technol. Ser., Frances Pinter, Wolfeboro, NH, 1987.
6. Schultze, L. E. and Bauer, D. J. Operation of a mineral recovery unit on brine from the Salton Sea known geothermal resource area. Tech. Rep.BUMIENES-RI-8680, Bureau of Mines, Reno Research Center, Reno, NV, 1982.
7. Ehrlich, P. R. and Ehrlich, A, H. *Population, Resources, Environment: Issues in Human Ecology.* W. H. Freeman, San Francisco, 1970.
8. Hogfeldt, E. in *Ion Exchangers* (K. Dorfner, ed.), Walter de Gruyter, Berlin, 1991.
9. Tallmadge, J. A., Butt, J. B., and Solomon, H. J. *Ind. Eng. Chem,* 56:44 (1964).
10. Hanson, C. and Murthy, S. L. N. *Chem. Eng., Aug.,* 295: (1972).
11. McIlhenny, W. F. in *Chemical Oceanography,* (J. P. Riley and G. Skirrow, eds.), 2nd ed., Academic Press, New York, 1975; Vol. 4. pp. 155–218.
12. Seetharam, B. and Srinivasan, D. *Chem. Eng. World, 13:* 63 (1978).
13. Senyavin, M. M. *Ion Exchange in Technology and Analysis of Inorganic Compounds,* Khimia, Moscow, 1988, p. 236, (Russian).
14. Massie, K. S. in *The North-West European Shelf Seas: The Sea Bed and the Sea in Motion.* II: *Physical and Chemical Oceanography and Physical Resources,* (F. T. Bauner, M. B. Collins and K. S. Massie, eds.), Elsevier, New York, 1980, Chap. 19, p. 569.
15. Senyavin, M. M. Khamizov, R. Kh., Bronov, L. V. and Venetsianov, Ye. V. in *Theory and Practice of Sorption Processes,* VGU, Voronezh, 1988, Vol. 20, pp. 58–72. (Russian)
16. Waldichuk, M. *Mar. Pollut. Bull., 18:* 378 (1987).
17. Gaskell, T. F. *Chem. Ind.,* Oct.: 1149 (1971).
18. Skinner, B. J. *Earth Resources,* Englewood Cliffs, NJ, 1989.
19. Khamizov, R. Kh., Senyavin, M. M., Butenko, T.Yu, Bronov, L. V. and

Novikova, V. A. in *Zeolites for Nineties, Recent Res. Rep. 8th Int. Zeolites Conf.*, Amsterdam, p. 167, 1989.

20. Khamizov, R. Kh., Butenko, T. Yu., Bronov, L. V., Skovyra, V. V. and Novikova, V. A. *Izv. AN SSSR, Ser. Khim.*, p. 2461, 1988. (Russ.)
21. Gorshkov, V. I., Nikolaev, N. P., Ivanov, V. A., Kovalenko, Ju. A., Muraviev, D. N., Timofeevskaja, Staina, I. V. and Tavasov, V. N. *A study of dynamics of sorption and desorption in countercurrent systems for recovery of metal ions from sea water*, Project report No. 0290.0040815, 1989, MGU, Moscow, 55 pp. (Russian)
22. Kanno, M. *Sep. Sci. Technol.*, 16: 999 (1981).
23. Warshawsky, A. in *Ion Exchange and Sorption Processes in Hydrometallurgy* (M. Streat and D. Naden, eds.), Wiley, New York, 1987, p. 208.
24. Bondarenko. S. S., Popov, V. M. and Strepetov, V. P. *Main Types of Resources and Seals of Processing Hydromineral Raw Materials in Developed and Developing Countries*, Review VIEMS, Moscow, 1986. (Russian)
25. Armstrong, E. F. and Miall, L. M. *Raw Materials from the Sea*, Chemical Publ., Brooklyn, NY, 1946.
26. Kryzhanovsky, R. A. *Effectiveness of Processing Resources of the Continental Shelves of World Ocean*, Nedra, Leningrad, 1989. (Russian)
27. Shnjukov, E. F., Beloed, R. M. and Tsemko, V. P. *Valuable Resources of the World Ocean*, Naukova Dumka, Kiev, 1974. (Russian)
28. Gilpin, W. C. and Heasman, N. *Chem. Ind.*, 14: 567 (1977).
29. Bauman, W. C., Pat. USA. No 3615181, 1971.
30. Hellferich, F. *Ion Exchange*, McGraw Hill, New York, 1962.
31. Lee, J. M. and Bauman, W. C. Pat. USA No 4116857, 26.09.78. (B01D 15/04)*.
32. Lee, J. M. and Bauman, W. C. Pat. USA No 4183900, 15.01.80 (B01D 15/04).
33. Lee, J. M. and Bauman, W. C. Pat. USA No 4243555, 6.01.81 (C01B 9/00).
34. A. A. Zhdanova, B. P. Nikolsky, F. A. Belinskaya, E. D. Makarova, E. I. Bobrikova and O. M. Berezhkovskaya, Pat. USSR No 738651, 8.06.80 (B01 1/22).
35. R. Kh. Khamizov, T. E. Mitchenko, L. V. Bronov, M. M. Senyavin, A. A. Uzbekov and L. E. Postolov. Russ. Pat. No 1678771, 23.09.91 (C 02F 1/42).
36. Myasoedov, B. F., Senyavin, M. M., Khamizov, R. Kh. and Rudenko, B. A. in *The Perspective Approaches in Design and Technology for Processing Mineral and Technogenious Raw Materials*, (I. Sh. Sattaev, ed.), Mekhanobr, St. Petersburg, 1991, p. 218–231. (Russian)
37. Mironova, L. I. and Khamizov, R. Kh. Russ. Pat. Pos. Dec. of 22.06.92 to the claim No 5049130/26, 1992.
38. Barba, D., Brandony, V. and Foscolo, P. H., *Desalination, 48*: 133 (1983).
39. Khamizov, R. Kh., Mironova, L. I. and Serebrenikova, O. V. Russ. Pat. Pos. Dec. of 28.03.92 to the claim No 4869581, 1992.
40. Chari, K. S., *Chem. Age India, 28*(9): 788 (1977).

*International Pat. Index.

41. Epstein, J. A. *Chem. Ind., 14*: 572 (1977).
42. Bark, M. Y. and Zatont, A. A. *Chem. Age India, 26*(3):197 (1975).
43. Bark, M. Y. and Zatont, A. A. *Chem. Age India, 26*(3):200 (1975).
44. Epstein, J. A., Altaras, D., Feist, M. and Rosenzweig, *J. Hydrometall., 1*(1): 39 (1975–1976).
45. Sudzuki, T. *Nippon Kaisui Gakkaishi (Bull. Soc. Sea Water Sci. Japan), 19* (5): 274 (1966). (Japanese)
46. Nagatani, Z. and Goto, T. *J. Nat. Chem. Lab. Ind., 79*(11): 437 (1975).
47. Kielland, J. *Chem. Ind.,* 1309 (1971).
48. Sanno Saburo and Tokyoto Itachashi. Jap. Pat. No 46-43082, 1971 (B01 1/ 08).
49. Sanno Saburo and Tokyoto Itachashi. Jap. Pat. No 46-43204, 1971 (B01 1/ 08).
50. Matsushita Hiroshi and Takayanagi Toshiko, *Nippon Kaisui Gakkaishi (Bull. Soc. Sea Water Sci. Japan), 24*(3): 96 (1970). (Japanese)
51. Matsushita Hiroshi, Jap. Pat. No 48-23269, 1973 (C01 3/00).
52. Kobalshi Etzuro and Tokyoto Ota, Jap. Pat. No 47-44151, 1972 (B01 1/04).
53. Matsushita Hiroshi and Takayanagi Toshiko, *Nippon Kaisui Gakkaishi (Bull. Soc. Sea Water Sci. Japan), 25*(4): 269 (1972). (Japanese)
54. Thomas, Ch. L. Pat. USA No 3497314, 1970 (C01 3/06).
55. Khamizov, R. Kh., Skovyra, V. V., Senyavin, M. M., Slinko, G. M., Bronov, L. V. and Butenko, T. Yu. in *Extraction. Processing and Application of Natural Zeolites* (G. V. Tsitsishvili, ed.), Sakartvello, Tbilisi, 1989, p. 219–223. (Russian)
56. Khamizov, R. Kh., Novikova, V. A., Butenko, T. Yu., Mironova, L. I., Bronov, L. V., Senyavin, M. M. and Skovyra, V. V. in *Proc. USSR Allunion Conf. Utilization of Natural Zeolites*, Kemerovo, 1990, Siberian Branch of Russian Academy Publ., Novosibirsk, 1991, Vol. 2, p. 147–153. (Russian)
57. Mironova, L. I. and Nikashina. V. A. *Izv. ANSSSR, Ser. Khim., 7*: 1452 (1984). (Russian)
58. Khamizov, R. Kh., Novikova, V. A. and Melikhov. Russ. Pat. Pos. Dec. of 08.04.92 to the claim No 4953782/26, 1992.
59. Khamizov, R. Kh., Novikova, V. A. and Melikhov. in *Proc. Russian Republic Conf.* (Natural Zeolites of Russia), Novosibirsk, 1991, Inst. Mineral. Siberian Branch of RAS Publ., Novosibirsk, Vol. 1, 1992, p. 160–164. (Russian)
60. Ksenzenko, V. I. and Stasinevitch, D. S. *Chemistry and Technology of Bromine, Iodine and Their Compounds*, Khimia, Moscow, 1979, p. 303. (Russian)
61. Yaron, D. in *Bromine and Its Compounds* (Z. E. Jelles, ed.), Ernest Benn, London, 1966, p. 250.
62. Ingham, J. J. *Chem. Ind.,* 1863 (1966).
63. Popov, N. I., Fyodorov, K. N. and Orlov, V. M. *Sea Water*, Nauka, Moscow, 1979, p. 327. (Russian)
64. Taran, Yu. A., Dubovik, N. A., Koifman, M. L., Volovik, T. M. and Mukhina, L. I. *Chemical Technology and Engineering of Bromine, Iodine and Manganese Compounds Production.* Khimia, Moscow, p. 5, 1989. (Russian)

65. Aveston, J. and Everest, D. A., *Chem. Ind., 37*: 1238 (1959).
66. Ziegler, M., *Angew. Chem., 71*: 283 (1957).
67. Hein, R. F. Pat. USA No 3037845, 1961.
68. Gradishar, F. J. and Hein, R. F. Pat. USA No 3116976, 1964.
69. Nakamura, H., Katoo, H., Minejima, N., Shimizu, H., Satoo, A., Nozaki, M. and Doochi Y. Pat. USA no 3352641, 1967.
70. Mills, E. J. Pat. USA No 3050369, 1962.
71. Zalkind, G. R. and Koifman, M. L. *Iodo-Bromine Industry*, No. 11–12: p. 59 (1967). (Russian)
72. Ksenzenko, V. I. and Stasinevitch, D. S. in *Ion Exchange* (M. M. Senyavin, ed.), Nauka, Moscow, 1981, p. 201. (Russian)
73. Ksenzenko, V. I. and Stasinevitch, D. S. in *Technology of Bromine and Iodine*, Khimia, Moscow, 1960, p. 237.
74. Abramov, E. G. Russ. Pat. No 1673643, 1992.
75. Abramov, E. G. *Dokl. AN SSSR, 313(3)*: 653 (1990). (Russian)
76. Koifman, M. L., Zalkind, G. R. and Gurevitch, R. N. in *Chemistry and Technology of Iodine, Bromine and Their Derivatives* (S. I. Yavorsky, ed.), Khimia, Leningrad, 1965, p. 104.
77. Khamizov, R. Kh., Fokina, O. V., Ivanov, V. A. and Gorshkov, V. I. Russ. Pat. No. 1728133, 1992.
78. Khamizov, R. Kh., Fokina, O. V. and Senyavin, M. M. Russ. Pat. No. 17226387, 1992.
79. Khamizov, R. Kh., Fokina, O. V. and Senyavin, M. M. in Abstracts of Lectures and Posters of *6th Symp. Ion Exchange*, Balatonfured, Hungary, 1990, p. 140.
80. Lyday, P. A. *Mining Eng. (USA), 41*(6): 433 (1989).
81. Bursik, A., Reitz, H. and Spindler, K. *VGB Kraftwerkstechnik, 63*(6): 325 (1983).
82. Dobias, J., Schutze, R. and Stahl, U. *VGB Kraftwerkstechnik, 65*(10): 959 (1985).
83. Kadlec, V. and Hubner, P. in *Ion Exchange for Industry* (M. Streat, ed.), Ellis Horwood, Chichester, 1988.
84. Muravicv, D. N., Svcrchkova, O. Yu., Gorshkov, V. I. and Voskresensky, N. M. *Reactive Polym. 17*: 75 (1992).
85. Chen, H. T. in *Handbook of Separation Techniques for Chemical Engineers* (P. A. Schweitzer, ed.), McGraw-Hill, New York, 1979, p. 467.
86. Grevillot, G. in *Handbook for Heat and Mass Transfer* (N.P. Cheremisinoff, ed.), Gulf Publ., West Orange, NJ, 1985, Chap. 36.
87. Tondeur, D. and Grevillot, G. in *Ion Exchange: Science and Technology* (A. E. Rodrigues, ed.), Martinus Nijhoff, Dordrecht, 1986, p. 369.
88. Rice, R. G. *Separ. Purif. Methods, 5*: 139 (1976).
89. Wankat, P. C. in *Percolation Processes, Theory and Applications* (A. E. Rodrigues and D. Tondeur, eds.), Sijthoff and Noordhoff, Alphen aan den Rijm, 1978, p. 443.
90. Andreev, B. M. Boreskov, G. K. and Katalnikov, S. G. *Khim. Prom. (Chem. Ind.), 6*: 389 (1961). (Russian)

91. Gorshkov, V. I., Kurbanov, A. M.: and Apolonnik, N. V. *Zhur. Fiz. Khim. (Russ. J. Phys. Chem.), 45*: 2969 (1971). (Russian)
92. Gorshkov, V. I., Ivanova, M. V., Kurbanov, A. M. and Ivanov, V. A. *Moscow Univ. Bull., 32*: 23 (1977).
93. Bailly, M. and Tondeur, D. *J. Chem. E. Symp. Ser., 54*: 111 (1978).
94. Bailly, M. and Tonduer, D. *J. Chromat., 201*: 343 (1980).
95. Timofeevskaya, V. D., Ivanov, V. A. and Gorshkov, V. I. *Russ. J. Phys. Chem., 62*: 1314 (1988).
96. Ivanov, V. A., Timofeevskaya, V. D. and Gorshkov, V. I. *React. Polym., 17*: 101 (1992).
97. Boyd, G. E., Schubert, J. and Adamson, A. W. *J. Am. Chem. Soc., 69*: 2818 (1947).
98. Bonner, O. D. and Swith, L. L. *J. Phys. Chem., 61*; 1614 (1957).
99. Cosgrove, J. D. and Strickland, J. D. H. *J. Chem. Soc.,* 1845 (1950).
100. Gregor, H. P. and Bregman, J. J. *J. Colloid. Sci., 6*: 323 (1951).
101. Kressman, T. R. E. and Kitchener, J. A. *J. Chem. Soc.,* 1190 (1949).
102. Groszek, A. J. in *Ion Exchange for Industry* (M. Streat, ed.), Ellis Horwood, Chichester, 1988.
103. Bonner, O. D., Dickel, G. and Brummer, H. *Z. Phys. Chem. N. F., 25*: 81 (1960).
104. Glass, R. A. *J. Am. Chem. Soc., 77*: 807 (1955).
105. Krauss, K. A. and Raridon, R. J. *J. Am. Chem. Soc., 82*: 3271 (1960).
106. Warshawsky, A. and Kahana, N. *J. Am. Chem. Soc., 104*: 2663 (1982).
107. Kapitsa, P. L. *Experiment, Theory and Practice*, Nauka, Moscow, 1980, pp. 110, 460. (Russian)
108. Kogan, B. I. *Rare Metals*, Nauka, Moscow, 1979, p. 9. (Russian).
109. *Chemistry of Rare and Dispersed Elements*, (K. A. Bolshakova, ed.), Vysshaya Shkola, Moscow, 1965, Vol. 1, p. 11. (Russian)
110. Steinberg, M. and Dang, V-D. Geological Survey Professional Paper, No 1005, p. 79, 1976.
111. Narita, E. *Chem. Chem. Ind., 37*(1): 149 (1984).
112. Symons, S. A. *Separ. Sci. Technol., 20*(9–10): p. 633 (1985).
113. Averill, W. A. and Olson, D. in *Proc. Symp. Corniug, NY,* 1977, Oxford E. A. 1978, pp. 305–313, 1978.
114. Macey, J. G. Pat. USA No 3268289, 1966.
115. Macey, J. G. Pat. USA No 3342548, 1967.
116. Deuti Tosimitsu and Honda Tzugitoku, Pat. USA No 4291001, 1981. (C 01 15/00).
117. Hayashi Hiroshi and Ueno Hiroshi. Jap. Pat. No 56-50113 (C 01 15/02).
118. Hayashi Hiroshi and Ueno Hiroshi. Jap. Pat. No 56-5319, 1981 (C 01 15/04).
119. Yanagase, K., Yoshinaga, T., Kawano, K. and Matsuoka, T. *Bull. Chem. Soc. Jpn., 56*(8): 2490 (1983).
120. Yoshinaga, T., Kawano, K. and H. Imoto, *Bull. Chem. Soc. Jpn., 59*(4): 1207 (1986).
121. Pelley, I. J. *Appl. Chem. Biotechnol., 28*(7): 469 (1978).

122. Epstein, J. A., Feist, E. M., Zmora, J. and Marcus, Y. *Hydrometallurgy, 6*(3): 269 (1981).
123. Neikova, E. and Dimitrova, C., *Rudobiv, 32*(11): 22 (1977). (Bulgarian)
124. Holldorf, H., Kropp, E. and Mosler, H. Pat. GDR No 233549, 1986, (C01 15/00).
125. Bondarenko, S. S., Lubensky, A. A. and Kulikov, G. V. *Geologiko-Economical Estimation of Underground Brine Resources*, Nedra, Moscow, 1988, p. 203. (Russian)
126. Joseph, B. and Theodor, E. A. Pat. USA No 3537813, 1970 (C01 15/02).
127. Kitamura, T. and Wada, H. Jap. Pat. No 956532, 1978 (C01 15/02).
128. Kitamura, T. and Wada, H. *Nippon Kaisui Gakkaishi (Bull. Soc. Sea Water Sci. Jap.), 32*: 78 (1978). (Japanese)
129. Kitamura, T. and Wada, H., Ooi, K., Takagi, N., Katoh, S. and Fujii A., *Rep. Cov. Ind. Res. Inst., Shikoku, 12*(1): 1 (1980).
130. Ooi, K., Wada, H., Kitamura, T., Katoh, S. and Sugasaka, K., *Rep. Cov. Ind. Res. Inst., Shikoku, 12*(1): 6 (1980).
131. Wada, H., Kitamura, T., Fujii, A. and Katoh, S. *Chem. Soc. Jpn*, 1156 (1982).
132. Dang, V. - D. and Steinberg, M., *Meyer Energy, 3*(3): 352 (1978).
133. Lee, J. M. and Bauman, W. C. Pat. USA No 4116856, 1978. (B01 15/04).
134. Lee, J. M. and Bauman, W. C. Pat. USA No 4116858, 1978. (B01 15/04).
135. Repsher, W. J., Jackson, L. and Rapstein, K. J. Pat. USA. No 4291001, (C01 15/00).
136. Ho, P. C., Nelson, F. and Kraus, K. A. *J. Chromatogr., 147*: 78 (1978).
137. Alberti, G. and Massuci, M. A. *J. Inorg. Nucl. Chem., 32*: 1719 (1970).
138. Kratchak, A. N., Nikashina, V. A., Khainakov, S. A. and Stepanchenko, T. V. *Izv. AN SSR Ser. Khim.*, No. 2, 263 (1986). (Russian)
139. Khodyashev, N. B. Volkhin, V. V., Onorin, S. A. and Sesynina, E. A. Pat. USSR No 790424, 1980.
140. Zilberman, M. V., Ioffe, A. D. and Volkhin, V. V. Pat. USSR No 716578, 1980.
141. Khodyashev, N. B., Volkhin, V. V., Onorin, S. A. and Sanlin, D. B. Pat. USSR No. 1012486, 1982.
142. Volkhin, V. V., Leontieva, G. V., Onorin, S. A., Khodyashev,N. B., Kudryavtsev, P. G. and Shvetsova, T. I. *Chemistry and Technology of Inorganic Sorbents*, P. P. I/ Publish., Perm, 1980, p. 67. (Russian)
143. Leontieva, G. V., Volkhin, V. V., Tchirikova, L. G. and Mironova, E. A. Pat. USSR. No. 735296, 1980.
144. Leontieva, G. V., Volkhin, V. V., Tchirikova, L. G. and Mironova, E. A. *Zh. Prikl. Khim., 55*: 1306 (1982). (Russian)
145. Ooi, K. and Miyai, Y. *Separ. Sci. Technol., 21*(8): 755 (1986).
146. Miyai, Y., Ooi, K. and Katoh, S. *Separ. Sci. Technol., 23*(1–3): 179 (1988).
147. Senyavin, M. M. and Khamizov, R. Kh., *Priroda (Nature), 7*:25 (1990). (Russian)
148. Senyavin, M. M. and Khamizov, R. Kh. *Fundamentalnye Nauki Narodnomu Khozysistvu*, Nauka, Moscow, 1990, p. 246.
149. Khamizov, R. Kh., Kratchak, A. N., Mironova, L. I., Bronov, L.V., Senyavin,

M. M., Melikhov, S. A. and Zilberman, M. V. Pat. USSR No 1462566, 1988 (C01 15/00).

150. Kratchak, A. N., Khamizov, R. Kh., Nikashina, V. A., Mironova, L. I., Zilberman, M. V. and Melikhov, S. A. Pat. USSR No. 1524253, 1989.

151. Melikhov, S. A., Khamizov, R. Kh., Senyavin, M. M., Mironova, L. I., Kratchak, A. N., Zilberman, M. V. and Bronov, L. V. Russ. Pat. No. 1762379, 1992.

152. Koyanaka, V. and Yasuda, Y. *Suiyo Kaishi, 18*: 523 (1977).

153. Koyanako Iosio, Jap. Pat. No. 52-133898, 1977 (C01 15/00).

154. Takeuchi, T. J., *Nucl. Sci. Technol., 17*(2): 922 (1980).

155. Takeuchi, T. J., *Rust. Prev. Countr., 26*(10): 369 (1982).

156. Burba, J. L. Pat. USA No 4.461714, 1984.

157. Burba, J. L. Pat. USA No 4.472362, 1984.

158. Mitsuo, A., Chitrakov, R. and Kenji, H. in *Chem. Separ. Select. Pap. 1st Int. Conf. Separation Science and Technology*, New York, 1986, Deuver, Viena, 1986, Vol. 1, pp. 15–17.

159. Katoh, S., Miyai, Y., and Ooi, *K. Kagaku Kogyo (Chem. Chem. Ind), 42*(10: 1987, (1989). (Japanese)

160. Ooi, K., Miyai, Y., Katoh, S., Maeda, H., and M. Abe, *Langmuir, 6*(1): 289 (1990).

161. Ooi, K., Miyai, Y. and Katoh, S. *Kemikaru Eudjiniyaringu (Chem. Eng.), 35*(2): 151 (1990). (Japanese)

162. Miyai, Y., Ooi, K., Sakibara, J. and Katoh, S. *Nippon Kaisui Gakkaishi (Bull. Soc. Sea Water Sci., Jap.), 45*(4): 193 (1991). (Japanese)

163. Ooi, K., Miyai, Y. and Katoh, S., *Shikoku Kogyo Gijutsu Slikenso Kenkyu Hokoku, 20*: 1 (1991).

164. Sibata Hayadji, *Sigen Enerugi Kyunkyukai Kenshyu Kanshai Daigaku Koge Gidzyutshu Kenkyudje, 8*, Osaka, 1990, pp. 19–28. (Japanese)

165. Schwochau, K., *Topp. Curr. Chem., 124*: 91 (1984).

166. Hirai Tatsuo, Sakata Norihoko, Yamata Mihimasa, Tsujitani Tunichi, Okazak, Morio and Tamon Hajime, *Bull. Soc. Sea Water Sci., Jpn., 42*(1): 7 (1988).

167. Katoh, S., Sugasaka, K., Sakane, K., Takai, N., Takahashi, H., Umezawa, Y. and Itagiri̇, K., *Nippon Kagaku Kaishi, 9*: 1449 (1982).

168. Astheimer, L., Schenk, H. J., Witte, E. G. and Schwochau, K. *Sep. Sci. Technol., 16*: 999 (1981).

169. Ku, T. L., Knauss, K. G. and Mathieu, G. G., *Deep Sea Res., 24*: 1005 (1977).

170. Ogata, N., Inoue, N. and Kakihana, H. *Nippon Genshiryoku Gakkaishi (J. Atom. Energy Soc. Jap.), 13*: 560 (1971). (Japanese)

171. Keen, N. J., *Brit. Nucl. Energy Soc., 7*: 178 (1968).

172. Yamashita, H., Ozawa, Y., Nakajima, F. and Nurata, T. *Bull. Chem. Soc. Jpn., 53*: 3050 (1980).

173. Davies, R. V., Kennedy, J., Peckett, J. W. A., Robinson, B. K. and Streeton, R. J. W., U. K. At. Energy Auth. Res. Group Rep., AERE-R-5024, 1965.

174. Sugasaka, K., Katoh, S., Takai, N., Takahashi, H., and Umezawa, Y., *Sep. Sci. Technol., 16*: 971 (1981).

175. Schwochau, K., Astheimer, L., Schenk, H. J. and Witte, E. G., *Chem. Ztg.*, *107*: 177 (1983).
176. Schwochau, K., Astheimer, L., Schenk, H. J. and Witte, E. G., *Z. Naturforsch.*, *37b*: 307 (1982).
177. Chen, Y. F. *Nippon Kaisui Gakkaishi (Bull. Soc. Sea Water Sci. Jpn.)*, *36*: 24 (1981).
178. Novikov, Ju. P., Komarevsky, V. M. and Myasoedov, B. F., *Radiochem. Radioanal. Lett.*, *48*(1): 45 (1981).
179. Nuriev, A. N., Akperov, G. A., Komarevsky, V. M., Mamedov, R. M. and Novikov, Ju. P., *Radiokhimia, 33*(2): 93 (1991). (Russian)
180. Sugano Masayoshi, *Isot. News, 391*: 12 (1987).
181. Kolmers, A. D., *Separ. Sci. Technol., 16*(9): 1019 (1981).
182. Cambell, H. M. and Binney, S. E., *Trans. Amer. Nucl. Soc., 33*: 464 (1979).
183. Best, F. R. and Driscoll, M. J. *Trans. Amer. Nucl. Soc.,34*: 380 (1980).
184. Kokubu Toshinore, *Miyakonojo Koge Koto Sammon Gakko Keukyu Hokoky (Res. Rept. Miyakonojo Techn. Coll.), 20*: 33 (1986). (Japanese)
185. Ozawa Yoshihiro, Murata Toshifumi, Yamashita Nisao and Nakajima Fumito. *J. Nucl. Sci. Technol., 17(3)*: 204 (1980).
186. Ogata, N. *Genday Kagaku (Chem. Today), 112*: 16 (1980). (Japanese)
187. Pritshepo. R. S., Pershko, A. A., Vasilevsky, V. A. and Betenenkov, N. D., *Radiokhimiya, 27*(5): 626 (1985). (Russian)
188. Kyffin, T. W. *Chelating Ion-Exchange on Amidoxime and Hydroxamic Acid Resins*, Ph.D. thesis, Salford, 1976.
189. Vernon, F. and Shah, F., *React. Polym., 1*: 301 (1983).
190. Kanno, M. *J. Nucl. Sci. Technol., 21*(1): 1 (1984).
191. Szefer, P., *Stud. i. Mater. Oceanol. PAN. Chem. Morza, 6*: 121 (1985). (Polish)
192. Vernon, F., *Curr. Eng. Pract., 29*(1): 1 (1986).
193. Katoh Shunsaku and Sugasaka Kazuhiko, *Nippon Kaisui Gakkaishi (Bull. Soc. Sea Water Sci., Jpn.), 40*(5): 17 (1987).
194. Hotta Hitoshi, *Oceanus, 30*(1): 44 (1987).
195. Kobuke Yoshiaki, *Chemen., 26*(7): 461 (1988). (Japanese)
196. Ogata, N., *Nippon Kaisui Gakkaishi (Bull. Soc. Sea Water Sci, Jpn.), 40* (4): 247 (1987). (Japanese)
197. Warshawsky, A. in *Ion Exchange and Sorption Processes in Hydrometallurgy* (M. Streat and D. Naden, eds.), Wiley, Chichester, 1987, p. 210.
198. Kantipuly, C., Katragadda, S., Chow, A. and Gesser, H. D. *Talanta, 37*(5): 491 (1990).
199. Dia-Prosium, Vitry, France, Technical Data Sheet 0100A, 1977.
200. Takagi Norio, Hirotsu Takahiro, Sakakibara Jitsuo, Kotoh Shunsaku and Sugasaka Kazukiko, *Nippon Kaisui Gakkaishi (Bull. Soc. Sea Water Sci. Jpn.), 42*(6): 279 (1989). (Japanese)
201. Moroka Shideharu, Kato Takafumi, Inada Mitsutoshi, Kago Tokichiro and Kusakabe Katsuki, *Ind. Eng. Chem. Res. 30*(1): 190 (1991).
202. Katyn Syunsaku and Hirotzu Takahio, Jap. Pat. No. 61-219718, 1986.

203. Hirotzu Takahiro, Katoh Shunsaku, Sugasaka, Kazuhiko, Takai Nobuharu, Seno Manabu, and Stagaki Takaharu, *Sep. Sci. Technol., 23*(1–3): 49 (1988).

204. Nakayama Morio, Nonaka Takamasa and Egawa Hiroaki, *J. Appl. Polym. Sci., 36*(7): 1617 (1988).

205. Zheng Baugding, Cai Shuiyuan, Zhuang Miugjiang and Jiang Zongheng, *Acta Oceanol. Sci., 3*: 417 (1985).

206. Omichi, H., Katakai, A. and Okamoto, *J. Separ. Sci. Technol., 23*(10–11): 1133 (1988).

207. Omichi, H., Katakai, A., Supo, T., Okamoto, J., Katoh, S., Sakene, K., Sugasaka, K. and Itagaki, T., *Separ. Sci. Technol., 22*(4): 1313 (1987).

208. Lieser, K. H., Eroboy, M. and Thybusch, B., *Angew. Makromol. Chem., 152*; 169 (1987). (German).

209. Shirotsuka Tadashi, Onoe Kaoru and Mochizuki Seiichi, *Nippon Kaisui Gakkaishi (Bull. Soc. Sea Water Sci., Jpn.), 41*:(1): 15 (1988). (Japanese)

210. Egawa Hiroaki, Kalay Nalau, Nonaka Takamasa and Shuto Taketomi, *Nippon Kaisui Gakkaishi (Bull. Soc. Sea Water Sci., Jpn.), 45*:(2): 87 (1991). (Japanese)

211. Uezu Kazuya, Saito Kyoichi, Furusaki Shintaro, Sugo Takanobu and Okamoto Jiro, *Nippon Genshiroku Gakkaishi (J. Atom. Energy Soc. Jpn.), 32*(9): 75 (1990). (Japanese)

212. Takeda Toshiya, Saito Kyoichi, Uezu Kazuya, Furusaki Shintaro, Sugo Takanobu and Okamoto Jiro, *Ind. Eng. Chem. Res., 30*(1): 185 (1991).

213. Morooka Shiceharu, Kusakabe Katsuki, Kago Tokihiro, Inada Mitsutoshi and Egawa Hiroaki, *J. Chem. Eng. Jpn., 23*(1): 18 (1990).

214. Kobuke Yoshiaki, Aokitakao, Tanaka Hiromitsu and Tabushi Iwao, *Ind. Eng. Chem. Res., 29*(8): 1662 (1990).

215. Vatanable Sumiya, Hirotzu Takahiro, Kato Sansaku, Takagi Keniyo, Takai Nobuharu, Banjen Takahary and Outi Hideyosi, Jap. Pat. No 60-210532, 1985.

216. Saito Kyochi, Hori Takahiro, Furusaki Shintaro, Sugo Takanobu and Okamoto Jiro, *Ind. Eng. Chem. Res., 26*(10): 1977 (1987).

217. Kataoka Yushiu, Matsuda Masaaki and Aoi Masahiro, Pat. USA No 4786481, 1988.

218. Nuriev, A. N., Akrepov, G. A., Komarevsky, V. M., Mamedov, R. M., and Novikov, Yu. P., *Radiokhimia, 33*(2): 89 (1991). (Russian)

219. Kato Shunsaku and Hirotsu Takahiro, *Sugen Syori Gidjuzu (Resour. Process.), 34*(1): 62 (1987).

220. Nakamura, S., Mori, S., Yoshimuta, H., Ito, Y. and Kanno, M., *Separ. Sci. Technol., 23* (6–7): 731 (1988).

221. Takase Hisao and Yoshimura Goshihide, *J. Chem. Eng. Jpn., 24*(4): 500 (1991).

222. Kobayasi Suro, Tanabe Takaku, Sayegusa Takeo and Masuo Fudjio, Jap. Pat. No. 61-247623, 1986.

223. Egawa Hiroaki, Nonaka Takamaza and Makayama Morio, *Nippon Kaisui Gakkaishi (Bull. Soc. Sea Water Sci. Jpn.), 44*:(5): 316 (1990). (Japanese).

224. Egawa Hiroaki, Nonaka Takamaza and Makayama Morio, *Ind. Eng.Chem. Res., 29*(11): 2273 (1990).
225. Akiba, K. and Hashimoto, H., *J. Radioanal. Nucl. Chem. Art., 130*(1): 13 (1989).
226. Tchistova, L. R., Khamizov, R. and Sokolova, T. V., *Vesti AN BSSR, Ser. Khim. Nauk, 3*: 91 (1991). (Russian)
227. Heitkamp, D. and Wagener, K., *Separ. Sci. Technol., 25*(5): 535 (1990).
228. Nakajima Akira and Sakaduchi Takashi, *J. Chem. Technol. Biotechnol., 47*(1): 31 (1990).
229. Svanidze, A. G. *Technological Schemes and Devices for Sorption Recovery of Microelements from Sea Water*, Deponent VINITI USSR No. 8030-B85, Moscow, 1986.
230. Sanno Saburo and Siki Sikasimati, Jap. Pat. No. 54-89977, 1979.
231. Koske, P. and Ohlogge, K., Offenlegungsschrift DE 2936399 A1, 1981. (German)
232. Driscoll, M. J., *Meeting on Recovery of Uranium from Sea Water, Japan, 1983.*
233. Hess, P. D. Pat. USA No 3436213, 1969.
234. Lesser, R. Offenlegungsschrift 2900966, 1980 (German)
235. Nakamura, S., Ito, Y., and Kanno, M., *Nippon Kaisui Gakkaishi (Bull. Soc. Sea Water Sci., Jpn.), 41*:(1): 38 (1987). (Japanese)
236. Lagstrom, G. E., Taby Och and Forberg, S., Swed. Pat. No. 417913, 1981.
237. Katoh Shunsaku, Hirotzu Takahiro, Sakana Kodji and Nobukawa Hisashi, Jap. Pat. No. 60-46929, 1985.
238. Okazaki Morio, Tamon Hajime and Yamamoto Takigi, *Nippon Kaisui Gakkaishi (Bull. Soc. Sea Water Sci., Jpn.), 41*:(5): 257 (1988). (Japanese)
239. Nobukawa Hisashi, Muchimoto Junichi, Kobayashi Masanory, Nakagawa Hiroyuki, Sakakibara Jitsuo, Takagi Norio and Tamehiro Masayuki, *Nippon Dzosen Gakkai Rombunshyu (J. Soc. Nav. Archit., Jpn.), 168*: 319 (1990). (Japanese)
240. Saburo, S., Haswyoshi, O. and Hido, K. *Nippon Kaisui Gakkaishi (Bull. Soc. Sea Water Sci., Jpn.), 38*:(4): 227 (1984). (Japanese)
241. Krylov, V. N. and Trofimov, A. M., Pat. USSR No 405562, 1973.
242. Hommura Isamu, Jap. Pat. No 55-132635, 1980.
243. Lazarev, K. F., Kovalev, G. N. and Ushatsky, V. N. Pata. USSR No 490 493, 1976.
244. Leontieva, G. V., Tchirokova, L. G. and Volkhin, V. V., *Zhur. Prikl. Khimii, 6*: 1229 (1980). (Russian)
245. Tolmacheva, E. I., Senyavin, M. M. and Rilo, R. P. Pat. USSR No 1079277, 1982.
246. Mimura Hitoshi and Kanno Takuji, *J. Nucl. Sci. Technol., 22*(4): 284 (1985).
247. Tchernyavsky, N. B., Zhdanov, S. P., Andreeva, N. R., Shubaeva, M. A., and Ostretzova, L. N. Pat. USSR No 1173.600, 1986.
248. Tchelitshev, N. F., Volodin, V. F. and Kryukov, V. L. *Ion Exchange Properties of Natural Highly Silicic Zeolites*, Nauka, Moscow, 1988, p. 128. (Russian)

249. Nikashina, V. A., Senyavin, M. M., Mironova, L. I. and Tyurina, V. A. in *Proc. 7th Int. Zeolite Conf.*, Japan, 1986, Kodansha, Tokyo, 1986, p. 283.
250. Khamizov, R. Kh., Butenko, T. Yu., Bronov, L. V., Skovyra, V. V. and Novikova, V. A. *Bull. Acad. Sci. USSR, D. Chem. Sci., 37*(11): 2215 (1988).
251. Khamizov, R. Kh., Novikova, V. A., Mironova, L. I., Butenko, T. Yu., Bronov, L. V. and Senyavin, M. M. Abstracts of lectures and Posters of *6th Symp. Ion Exchange*. Balatonfured, Hungary, 1990, pp. 163–165.
252. Khamizov, R. Kh., Butenko, T. Yu., Veber, M. L. and Zaitseva, R. V. *Bull. Acad. Sci., USSR, D. Chem. Sci., 39*(1): 2 (1990).
253. Butenko, T. Yu., Khamizov, R. Kh., Kaplun, R. V., Golikov, A. P., Kurilenko, L. N. and Zaitseva, E. V. *Bull. Russ. Acad. Sci., 41*(2): 273 (1992).
254. Nikolaev, N. P., Ivanov, V. A., Gorshkov, V. I., Muraviev, D. N., Saurin, A. D. and Ferapontov, N. B. Pat. USSR No. 1590441, 1988.
255. Nikolaev, N. P., Ivanov, V. A., Gorshkov, V. I., Nikashina, V. A. and Ferapontov, N. B., *Reactive Polymn., 18*: 25 (1992).
256. Ivanov, V. A., Gorshkov, V. I., Ferapontov, N. B., Nikashina, V. A. and Nikolaev, N. P. Pat USSR No. 1473835, 1989.
257. Senyavin, M. M. Venitsianov, E. V., Kolosova, O. M., Nikashina, V. A. and Khamizov, R. Kh., *Zhur. Prikl. Khimi: 62*: 989 (1989). (Russian)
258. Shulga, E. A. and Volkhin, V. V. *Rare Alkali Elements,* PPI Perm.,1969, p. 331. (Russian)
259. Kozlova, G. A., Volkhin, V. V. and Zilberman, M. V., *Zhur. Prikl. Khimii, 52*: 1411 (1979).
260. Onadera Yoshio and Iwasaki Takashi, *Nippon Koge Kaisui (J. Mining and Met. Inst. Jpn.), 104* (1203): 277 (1988). (Japanese)
261. *Chemistry and Technology of Inorganic Sorbents*, (E. S. Boitchinova, ed.), PPI Perm. 1989. p. 5. (Russian).
262. Gorshkov, V. I., Ivanov, V. A. and Staina, I. V. Russ. Pat. No. 1781313, 1992.
263. Kloppfenstein, P. K. and Arnold, D. S., *J. Metals, 18*: 1195 (1966).
264. Pilipenko, A. T., Grebenyuk, V. D. and Melnik, L. A., *Khimia i Tekhnologiya Vody (Chemis. Technol. Water) 12*(3): 195 (1990). (Russian)
265. Murthy, S. L. N. and Hanson, C., *Chem. Ind.,* 669 (1969).
266. McIlhenny, W. F. and Ballard, D. A. in *Selected Papers on Desalination and Ocean Technology* (S. N. Levine, ed.), Dover, New York, 1986, p. 250.
267. Mukazhavo, M. B., Filippova, Z. O. and Dzhurambaev, A. M. *Complex Utilization of Mineral Resources, 7*: 81 (1988). (Russian)
268. Nuriev, A. I. and Dzhabbarova, Z. A., *Azerb. Khim. Zhurnal, 2*: 112 (1982). (Russian)
269. Khodzhamamedov, A. *Physico-Chemical Fundamentals and Design of Methods for Processing Brines*, D. Sc. thesis, Mendeleev Institute of Chemical Technology, Moscow, 1987. (Russian)
270. Chisholm, H. E. Pat. USA No. 3232708, 1966.
271. Shiappa, G. A., Place, J., Hudson, R. K., and Grinstead, R. R. Pat. USA No. 3493349, 1969.

272. Hanson, C. and Murthy, S. L. N., Ph.D. thesis, University of Bradford, 1971.
273. Grannen, E. A. Pat USA No. 3839222, 1974.
274. Peterson, W. D. Pat. USA No. 3741731, 1971.
275. Nikolaev, A. B. and Ryabinin, A. I., *Dokl. ANSSSR, 207*(1): 149 (1972). (Russian)
276. Nomura, J. and Ishibashi, J. Pat USA No. 4596590, 1986.
277. Samborsky, I. V., Vakulenko, V. A., Potarenko, L. P., Petrov, B. A. and Derkatch, L. P. *Theory and Practice of Sorption Processes*, VGU, Voronezh, 1980, No. 14, p. 90. (Russian)
278. Leontieva, G. V., Tomtchuk, T. K. and Volkhin, V. V. Pat. USSR No. 946647, 1982.
279. Volkhin, V. V., Tomtchuk, T. K. and Leontieva, G. V., *Zhur. Prikl. Khimi., 56*(4): 767 (1983). (Russian)
280. Serova, I. B., Sorotchan, A. M. and Vakulenko, V. A., *Zhur. Prikl. Khimi., 62*(2): 293 (1989). (Russian)
281. Serova, I. B. and Vakulenko, V. A., *Zhur. Prikl. Khimi., 65*(4): 721 (1992). (Russian)
282. Krasnikov, V. A., Leikin, Yu. A., Ostryakov, I. Yu. and Svitsov, A. A. in *Proc. All-Union Conf. USSR on Chemistry of Complexones*, Tchelyabinsk, 1988, Tchelyabinsk, 1989, pp. 208–210. (Russian)
283. Riley, J. P. and Taylor, D., *Anal. Chim. Acta, 40*: 479 (1968).
284. Kim, Y. S. and Zeitlin, H., *Separ. Sci. Technol., 6*: 505 (1971).
285. Kim. Y. S. and Zeitlin, H., *Chem. Commun.,* 672 (1971).
286. Akami, Y., Nakai, T. and Kawamura, F., *Nippon Kaisui Gakkaishi (Bull. Soc. Sea Water Sci., Jpn.) 33*: 180 (1979). (Japanese)
287. Tori, N. and Hiroshi, J., *Nippon Kaisui Gakkaishi (Bull. Soc. Sea Water Sci., Jpn.) 42(4)*: 186 (1988).
288. Hagadone, M. and Zeitlin, H. *Anal. Chim. Acta, 86*: 289 (1976).
289. Monien, H., Stangel, R. and Fresemus, Z., *Anal. Chim., 311*: 209 (1982).
290. Kusaka, Y., Tsuji, H., Tamari, Y., Sagawa, T., Ohmorti, S., Imai, S. and Ozaki, T., *Radioanal. Chem., 37*: 917 (1977).
291. Nikashina, V. A., Tyurina, V. A. and Mironova, L. I., *J. Chromatogr. 201*: 107 (1980).
292. Sorotchan, A. M., Saldadze, K. M., Katz, E. M., Komarova, I. V., Stroganova, N. N. and Pron, S. V., *Radiokhimiya, 23*: 505 (1981). (Russian)
293. Caroli, S., Alimonti, A., Petrucci, F. and Horvath, Z., *Anal. Chim. Acta., 248*(1): 241 (1991).
294. Daih, B. J. and Huang, H. J. *Anal. Chim. Acta, 258*(2): 245 (1992).
295. Fadeeva, V. I., Tichomirova, T. I., Kudryavtsev, G. V. and Yuferova, I. B., *Zhur. Anal. Khimii, 47*(3): 466 (1992). (Russian)
296. Liu, Z. S. and Huang, S. D., *Anal. Chim. Acta, 267*(1): 31 (1992).
297. Andreev, G. and Simeonov, V., *Toxicol. Environ. Chem., 36*(1–2): 99 (1992).

298. Baffi, F., Cardinale, A. M. and Bruzzone, R., *Anal. Chim. Acta, 270*(1): 79 (1992).
299. Myasoedova, G. V. and Savvin, S. B., *Crit. Rev. Anal. Chem., 17*: 1 (1986).
300. MacCarthy, P., Klusman, R. W. and Rice, J. A., *Anal. Chem., 59*: 308R, (1987).
301. Kantipuly, C. J. and Westland, A. D., *Talanta, 35*: 1 (1988).

4

Investigation of Intraparticle Ion-Exchange Kinetics in Selective Systems

A. I. Kalinitchev

*Institute of Physical Chemistry, Russian Academy of Science,
Moscow, Russia*

I. INTRODUCTION

Ion-exchange processes and the chemical reactions associated with them are receiving much attention. During the last decade they have been the subject of extensive research. These investigations are reviewed in a number of books [1–12]. The ion exchangers employed in these studies have been widely used for water treatment, in the field of heat power engineering, for extraction of metals from waste and natural waters [6–12], as well as for nuclear technology [8,13–15]. There has been an increase in the use of ion-exchange technology for environmental protection as well [9–12].

Numerous studies on the kinetics of ion exchange (IE) have been summarized elsewhere as well [1–12,16,17]. They demonstrate that in some instances the diffusion mechanism is complicated by additional interactions. One such complication arises from the interdependence of the fluxes of diffusing ions. The interference effects of diffusing components are governed by the modifying action of the variable electric field in the ion exchanger [1–12,16–23]. The electrodiffusion potential of the field can be expressed in terms of the chemical potential gradients (or the concentration gradients) of the exchanging ions [16, 24–28]. Use of irreversible thermodynamics [29] in its formulation for the multicomponent system yields terms with cross phenomenological coefficients of the type L_{ij} describing the interference of the i and j components.*

The phenomenologically-based theory of diffusion processes for binary ion exchange has been examined in detail by F. Helfferich [2,5,16,18,19] and N. Tunitsky with co-workers [22,23]. Their theoretical considerations are combined with the assumption that the exchanging ions retain their individual nature and that the process consists of the redistribution of the invading (B) and displaced (A) ions by diffusion between the ion exchanger and the contacting solution. Such an approach is valid for a variety of ion-exchange systems including unlike ions and free-acid or free-base forms of the exchangers where the exchange rate is controlled by the interdiffusion of B and A counterions. Formation of a very sharp boundary for the forward B–RA exchange and one that is spread for the reverse A–RB exchange obeys the "minority role" [5,16]. This rule is applicable not only to binary exchange but to the general case of coupled diffusion processes as well [5].

However, the above approach (based only on the diffusion mass-transfer equation where the ion fluxes are described by the Nernst-Planck law) becomes invalid for ion exchange involving a chemical reaction between the fixed R groups of the exchanger and A and B counterions (or the counterions and Y co-ions) [2,21,30]. In putting forward the fundamental considerations for the theory of IE kinetics in such ion exchangers Helfferich [2,30] gives 11 typical examples of accompanying IE reactions, namely dissociation, neutralization, hydrolysis, and complex formation. In addition the reactions are classified into three major groups. According to Helfferich [5] the effect of the chemical reaction on the mass-transfer process is likely to lead to an appreciable decrease in the exchange rate (see also Ref. 31), a change in the type of dependence of the process on the solution bulk concentration (c_0), and appearance of distinct mechanistic features.

A moving boundary mechanism, with the sharp boundary of B/A exchange in the resin bead, was predicted for a number of reactions carried out in solution at high concentration levels. Analytical expressions corresponding to this expectation were obtained [5,30]. The existence of such a boundary was documented by photographs taken through a microscope of transparent resin used in exchanges involving metals (Me):Me–RH and Me_2–RMe_1 [32,34]. The mathematical description of the

---

*In the following discussion B is the ion entering the resin and A is the ion displaced from it so that the ion exchanger, initially in form RA, is transformed to form RB during the ion-exchange process.

process was carried out using a progressive shell mechanism combined with the pseudo-steady-state concept of diffusion [35] which is applied to noncatalytic fluid-solid reactions [36,37], mass-transfer processes in sorption systems [38,39], heat or mass transfer [40], and IE systems [5,15,26,27,30,32,41]. In these "shell-core" models [14,30,42–44] a chemical reaction occurs initially on the bead surface and then on the moving boundary between the reacted shell RB and the core RA.

With few exceptions discussed elsewhere [14,44] and also mentioned by Helfferich [5], the reaction rate is much more rapid than the diffusion process so that the effect of the reaction on the kinetic process becomes apparent through the reaction parameters measured in the approach to equilibrium [7–12,25–28,44–47].

The purpose of this review is to discuss IE kinetics in selected systems using the kinetic model developed in this author's research [26–27,45–50]. It is an attempt to demonstrate, in particular, that phenomenological regularities and criteria describing intraparticle diffusion kinetics for conventional ion exchange are not applicable for the selective ion-exchange systems.

A. Notation

The dissociation constants K_{Ri} of the complexes Ri (i = B or A) are assumed to be independent of the degree of conversion F. In the case of weakly dissociated components (when $K_{Ri}/C_0 \ll 1$) the selectivity factor K_{RA}/K_{RB} is responsible for the shape of the exchange isotherm $a_i = f_i(c_B, c_A)$ (where i = B or A) in the resin phase. The exchange isotherm is favorable (or "convex") for invading ion B if the resin is selective for this ion and unfavorable (or "concave") for it if the resin is selective for the counterion A being displaced. The relationship $K_{RA}/K_{RB} > 1$, holds in the first case and the relationship $K_{RA}/K_{RB} < 1$, is applicable in the second case.

In the notation of this paper, forward exchange corresponds (as a rule) to the exchange of the ion with the favorable isotherm, i.e., the invading ion B where the selectivity of the ion exchanger is higher for ion B than for displaced ion A. Reverse exchange in the notation of this paper refers to the exchange with the unfavorable isotherm of invading ion B where the selectivity of ion exchanger is lower for counterion B than for displaced ion A.

In our experiments the commercial type of ion exchangers CB-4 (carboxylic), VPC (vinylpyridine carboxylic), ANKB (ampholyte), and AN-13 (ampholyte) were used.

The cation exchanger, CB-4, consists of a copolymer of methyl methacrylic acid and divinylbenzene (DVB).

The vinylpyridine ampholyte VPC used in the experiments is an oxymethylated copolymer of 2-methyl-5 vinylpyridine and divinylbenzene with picolinic acid as the functional groups.

ANKB-35 resin is an iminodiacetic acid exchanger; this multidentate chelating functionality is capable of forming metal coordination complexes with different stabilities [10,51].

The results of static and kinetic investigations with these resins are presented in Refs. 6,10,and 51. The relationship of these resins to their analogs outside of Russia is noted [6,10].

II. THEORETICAL KINETIC MODEL

Earlier [26,27,45,46] a phenomenological approach, based on the premise that the thermodynamics of irreversible processes [29] joined with Nernst-Planck equations for ion fluxes, would be useful was applied to the solution of intraparticle diffusion controlled ion exchange (IE) of fast chemical reactions between B and A counterions and the fixed R groups of the ion exchanger. In the model, diffusion within the resin particle, was considered the slow and sole controlling step.

Most of the known IE kinetic problems have been solved by the use of a single mass-balanced diffusion equation [1–5,7–11,14–24,34–43]. They are, on this basis, identified as one component systems and the diffusion rate for the invading B ion is controlled by the concentration gradient of this ion alone. In these cases the effective interdiffusion coefficient depends on the ion concentrations and the equilibrium constants of the chemical reaction between both ions in the ion exchanger [2–5,7–12,16–22,23,25,30,32,34,42,52–54].

In some studies [24,55,56] the effect of mutually diffusing ions is allowed for by taking into account the interdependence of ion fluxes in the ion exchanger. Any effect of chemical reaction is omitted, however.

Helfferich [2,5,30] states that in addition to the mutual interference of substances i and j, characterized by the phenomenological cross coefficients of the type L_{ij}, one should take into account the presence of a co-ion in the ion exchanger as well. As a result, the simplified solution is inappropriate, even to the problem of ordinary IE. By use of only one diffusion mass-transfer equation, as in this case, account for the presence of co-ion has been neglected. It is, as a consequence, necessary to consider the Nernst-Planck relation for the co-ion also.

In the model to be presented, the Nernst-Plank relations are applied to multicomponent systems and two nonlinear differential equations have to be solved simultaneously.

A. Premises of Diffusion-Controlled Model

The diffusion model for examination of the kinetics of weakly dissociating and complex-forming ion exchangers that is proposed in our studies [26,45,46] describes these exchangers as multicomponent systems to distinguish between this model and the other published models [32,34,42,43,52–54,57]. The effect of the interference of B and A components in the process of interdiffusion in the ion exchanger is accounted for by the interdependence of the fluxes J_i and J_j of the diffusing i and j components. The mathematical formulation of the IE kinetic process in this case is described by a system of two nonlinear diffusion equations expressed as partial derivatives where the effective self-diffusion D_{ij} coefficients and the cross-interdiffusion D_{ii} coefficients depend on the concentration of the diffusing counter-(B and A) and co-(Y) ions as well as on the equilibrium constants K_{RB} and K_{RA} describing the stability of the RB and RA complexes in the ion exchanger phase. With the model presented, the presence of co-ions is taken into account as recommended by Helfferich [2,5,30]. The solution involves three types of ions, namely B, A and Y so that the problem cannot be reduced to a single mass-transfer diffusion equation.

It should be noted here that in a state-of-the-art report [44], it was pointed out by Petruzelli, Helfferich, and others that another approach is possible. It includes mass balance equations with Nernst-Planck relations for fluxes of the "free" species and additional terms that are determined by the quantities formed in various reactions with species that include co-ions, neutral molecules, and neutral fixed groups (Hwang-Helfferich model [28]).

1. *Premises*

For the system the following assumptions and restrictions, which in the majority of papers on ion-exchange kinetics are considered to be satisfied [1–5,7–12,14–28,32,34,41–50,52–57] are proposed:

The exchanger (resin) is considered as a quasi-homogeneous phase.
The diffusional exchange process is assumed to be isothermal.
Convective transport is neglected.
Sorption and desorption of the solvent as well as swelling and shrinking of the exchanger are not considered.
Ion fluxes in the exchanger phase are described by Nernst-Planck relations and coupling of ionic fluxes other than by electric forces and reactions is disregarded.

The individual diffusivities D_i of mobile species in these relations are
   assumed to be constant.
Activity coefficient gradients are disregarded.
The dissociation coefficients K_{Ri} of complexes Ri are assumed to be constant and to be independent of the degree of conversion F.
Local equilibrium of the reactions is assumed to be maintained at all times
   at any location in the exchanger.

The final assumption is valid in almost all cases since the vast majority of ionic reactions is very rapid compared to the diffusion process
in an ion exchanger. Moreover in this model, any ions bound to the fixed
groups or the matrix are considered to be completely immobile.

Thermal effects are neglected in all theoretical analyses and such
neglect is justifiable for all fluid media.

All interpretations based on the assumptions with respect to use of
the Nernst-Planck relationships are, however, subject to sizable uncertainty because the constancy attributed to the diffusion coefficients used
in these relationships is susceptible to sizable variability. Bead volume
variations are ignored. This leads to variations in ion mobilities which
dictate changes in the diffusion coefficients of the Nernst-Planck relationships. Sizable divergence of measurements from prediction can be expected on this basis. Even in rigid ion exchangers such as zeolites the
difference in size of the counterions exchanged usually affects their mobility, and so leads to variations of the diffusivities [5].

Moreover the electrodiffusion potential gradient is likely to cause
electroosmotic transfer of the solution, whose local content is not in
equilibrium with that of the counterions [5]. In this case, as it is pointed
out in Ref. 5, the ion mobility and concentration depend on the prior
history of the process which can bring about "non-Fickian diffusion." The
application of Nernst-Planck equations to the real system may require
inclusion of additional terms that account for the effect of activity coefficient gradients which may be important in IE with zeolites [4,5].

Except for the few early stage theoretical investigations of IE kinetics that were based on the Stefan-Maxwell equations [58,59], the majority of such studies [1–5,7–12,14–28,30,42–50,52–57] were based mainly
on the Nernst-Planck equations. The fact is that the Nernst-Planck equations opposed to Fick's law, account for electric field coupling with the
ion fluxes with the result that a significant improvement over Fick's law
is achieved with a relatively minor increase in complexity. It is equally
important that the Nernst-Planck relations permit inclusion of the accompanying reaction which, being superimposed on the effect associated with
the electric field may significantly alter the exchange rate [5].

B. Equations of the Model

In the model it is proposed that during the exchange RA + B $\Leftrightarrow$ RB + A involving a chemical reaction of the counterions, B and A, with the resin site, R, they exist in the exchanger in two states, namely "free" and as the species formed by their combination with R. Counterions B and A are considered to be immobile while "bound" and to move unimpeded while "free." Ideally, ion concentrations in the "free" and "bound" states (c_i and a_i, respectively) are related to the dissociation constants, K_{Ri}, of the complexes Ri by the equilibrium relations for the complexation reactions. These dissociation-association equilibria have the form

$$\left(\frac{a_R}{a_0}\right) K_{Ri} = \left(\frac{c_i}{a_i}\right)^{1/z_i}, \qquad i = B, A \tag{1}$$

where $a_R = a_0 - Z_A a_A - Z_B a_B$ and a_0 is the exchanger capacity.

An approach with consideration of the two states (i.e, "free" with concentration c_i and "bound" with concentration a_i) of the B and A counterions is consistent with the "loose quasi-crystal" molecular model [41] that states that mass transfer in the exchanger may not occur by transfer of the complete counterions mass, as suggested by the classical quasi-homogeneous exchanger model [1–5,7–12,16,17], but solely by the free counterions that are not bound to the fixed exchanger groups.

The IE system, with the above premises, is described as follows [26,27,45,46] with concentrations in the reduced form shown.

material balance equations:

$$\frac{\partial(U_i + C_i)}{\partial t} = -\text{div } J_i, \quad U_i = \frac{a_i}{c_0}, \quad C_i = \frac{c_i}{c_0}, \qquad i = B, A \tag{2}$$

electroneutrality relation:

$$\sum_i z_i C_i - U_R = 0, \qquad i = B, A, Y, \quad U_R = \frac{a_R}{c_0} \tag{3}$$

Nernst-Planck relation for ion fluxes:

$$J_i = -D_i(\text{grad}(C_i) + z_i C_i (\mathscr{F}/RT) \text{ grad } \phi), \qquad i = B, A, Y \tag{4}$$

absence of electric current:

$$\sum_i z_i J_i = 0, \qquad i = B, A, Y \tag{5}$$

The system is confined to exchange between an ion exchanger initially containing ion A as the only counterion and continually renewed solution ("infinite solution volume" condition). The initial conditions are

$$c_A = c_A^0 \quad \text{or} \quad C_A^0 = 1 \quad \text{and} \quad C_B = 0 \tag{6}$$

Boundary conditions for a spherical bead

$$\text{at } r = r_0 \quad c_B = c_B^0 = c_0 \quad \text{or} \quad C_B = 1 \quad \text{and} \quad C_A = 0; \tag{7}$$

$$\text{at } r = r_0 \quad r^2\left(\frac{\partial C_i}{\partial r}\right) = 0$$

As opposed to the conventional neglect of Donnan invasion of the ion exchanger by salt, Eq. (3) includes the concentration, C_Y, of co-ion Y that is present in the exchanger phase. By the use of Eq. (5) to eliminate the electrical potential ϕ [24–28], the Nernst-Planck relation (4) can be reformulated before being substituted into the mass balance Eq. (2) so that that the ion fluxes can be expressed in terms of concentrations and their gradients. Only two out of three gradients (grad(C_i, i = B,A and Y) in (4) are independent. The relation between the three functions (grad(C_i)) can be readily obtained by differentiation of Eq. (3). One of the C_i values is eliminated by use of the electroneutrality relation (3). The most symmetrical form of the final equation is reached by elimination of the C_Y value.

Finally, substitution of the reformulated expressions for ion fluxes in material balance Eq. (2) results, for the spherically symmetric particle, in the two nonlinear parabolic type partial differential equations that follow in reduced form:

$$\frac{\partial(U_i + C_i)}{\partial Fo} = \frac{1}{R^2} \cdot \frac{\partial}{\partial R}\left\{ R^2\left[D_{iA}\frac{\partial C_A}{\partial R} + D_{iB}\frac{\partial C_B}{\partial R} \right] \right\} \tag{8}$$

$$U_i = V_i\left(\frac{a_0}{c_0}\right), \quad V_i = \frac{a_i}{a_0}, \quad i = B,A$$

where the Fourier number Fo = $D_0 t/r_0^2$ = reduced time

D_0 = scale diffusion coefficient

R = r/r_0 = reduced radial coordinate in the spherical bead (0 < R < 1)

U_i = a_i/c_0 = reduced concentration (0 < U_i < a_0/c_0)

Concentration U_i, expressed through C_B and C_A values from Eq. (1) and as an example for equally charged ions, reduced Eq. (9)

$$U_i = \frac{U_0 c_i g}{K_{Ri}}, \qquad i = B, A \tag{9}$$

where factor $g = K_{RA} K_{RB} / (K_{RA} K_{RB} + K_{RA} C_B + K_{RB} C_A)$.

The effective diffusion coefficients D_{ij} form a 2×2 matrix where each component depends on concentrations C_B and C_A:

$$D_{ij} = \frac{D_i}{D_0} \delta_{ij} + \frac{C_i D_Y (1 + a_0 g^2 / K_{Rj}) - D_Y}{C_B (D_B + D_Y) + C_A (D_A + D_Y) - D_Y a_0 / c_0 g} \tag{10}$$

where δ_{ij} is the Kroneker symbol ($\delta_{ij} = 1$, $i = j$; $\delta_{ij} = 0$, $i \neq j$).

It would be sufficient to prescribe only three individual diffusivities (D_A, D_B, D_Y) to determine the matrix $\| D_{ij} \|$.

Effective diffusivities D_{ij} control the diffusion flux of component i, coupled with its concentration gradient. The effective diffusion cross-efficients D_{ij} ($i \neq j$) characterize the cross effect (or the interdiffusion interference of the two components) thereby defining the diffusion of given component i associated with the concentration gradient of the other component j. The cross diffusivities are small compared to the diagonal ones and may, in principle, be negative and tend to zero while C_A, C_B tend to zero.

C. Results of Computer Simulation

The set of diffusion equations (8) along with the appropriate initial and boundary conditions (6), (7), solved using a computer, gives transient concentration profiles, C_i, as functions of r and t values. The simplifying assumption that the concentration, c_B^0, of free B ions in solution and at the boundary, r_0, in the surface layer of the associated exchanger bead is the same is justified by the absence of the Donnan exclusion of electrolyte in the associated layer RB. The same assumption has been adopted for calculations elsewhere [32,34,42,43,52–54,56].

1. Approximate Analytical Solution

In a particular limiting case where the formation of the stable RB complex, on an "absolute" basis, is formally consistent with the "rectangular" isotherm of B ion invasion of the B–RA exchange, the sharp boundary of this exchange advancing toward the bead center. The existence of such a boundary has been demonstrated elsewhere [32–34]. In this chapter it is presented in the microphotographs (Fig. 1) obtained for the bead of the

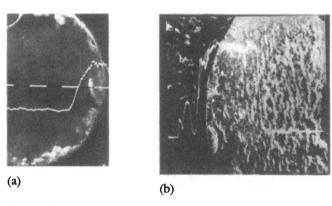

(a)

(b)

Figure 1   Scanning electron microscope micropictures of resin VPC beads with Ni ion distributions (lines) in radial direction for (a) Ni–RH exchange, (b) Ni–RNa exchange.

complex forming vinilpyridinecarboxylate exchanger (VPC) by the method of x-ray spectral electronic sond microanalyses with the ISM scanning electron microscope [60].

An approximate analytical solution of the system with Eq. (8) was obtained [26,27] for the "rectangular" isotherm of the invading B ion by using the integral relations method [61,62]. In this case the conversion (F) versus time (t) dependence is too close to what it is for conventional ion exchange [2,16,30,41]; however, the effective diffusion coefficient D_{ef}, is determined by the somewhat different relation [26,27] presented below:

$$F = 3\left[\frac{2D_{ef}}{a_0}\frac{c_0}{r_0^2}t\right]^{1/2}, \qquad D_{ef} = \frac{(z_B - z_Y)D_B D_A}{z_B D_B - z_Y D_Y} \tag{11}$$

Equation (11) is valid for small fractional conversions (F < 0.5). Its main advantage is that the effective diffusivity, D_{ef}, is obtained from the individual diffusivities D_B and D_A. The linear dependence of the fractional conversion F versus $(c_0 t)^{1/2}$ values in the initial stage is typical for intraparticle kinetic processes. Within the approximate solution, the moving boundary velocity is independent of the diffusivity D_Y. There is dependence on charge, z_Y, and this aspect is considered.

2.   Ion-Exchange Rate and Transient Concentration Profiles

The numerically implicit finite difference method was used to solve the set of nonlinear differential equations (8) for a wide range of model parameters such as diffusivities, D_B, D_A, and D_Y, dissociation constants, K_{Ri}, exchanger capacity, a_0, and bulk concentration, c_0, of the solution.

For weakly dissociated ion exchangers (when $K_{Ri}/c_0 \ll 1$) the selectivity factor K_{RA}/K_{RB} is responsible for the shape of the exchange isotherms for counterions ($U_i = f(C_A, C_B)$) in the exchanger phase. The exchange isotherm is favorable if the resin is selective for the invading B ion ($K_{RA}/K_{RB} > 1$) and unfavorable if it is selective for the A ion being displaced ($K_{RA}/K_{RB} < 1$).

Reduced (dimensionless) parameters for some of the IE systems subjected to numerical calculations are shown in Table 1. Variants I through III (high selectivity, $K_{RA}/K_{RB} > 1$, of invading B ion or favorable isotherm) and variants I.e through III.e (low selectivity, $K_{RA}/K_{RB} < 1$, of B ion or unfavorable isotherm) are listed in this Table.

Figures 2a and b presents fractional conversion F versus reduced time Fo kinetic curves for favorable (variants I through III(a) and I.2 through I.6(b)) exchange isotherms.

Table 1 Values of Reduced Parameters of IE Systems Used in Computer Simulation, $a_0/c_0 = 25$

var. N	D_B/D_0	D_A/D_0	D_Y/D_0	K_{RB}/c_0	K_{RA}/c_0	z_B	z_A	z_Y
			Influence of diffusivity factor D_A/D_B					
I.	1.0	1.0	0.1	0.005	0.001	1	1	−1
I.e				0.001	0.005			
II.	1.0	0.1	0.1	0.005	0.001	1	1	−1
II.e.				0.001	0.005			
III.	0.1	1.0	0.1	0.005	0.001	1	1	−1
III.e.				0.001	0.005			
			Influence of counterion charges ($z_B \neq z_A$)					
I.2	1.0	1.0	1.0	0.005	0.001	2	1	−1
I.2e				0.001	0.005	1	2	−1
I.3	1.0	1.0	1.0	0.005	0.001	2	1	−1
I.3e				0.001	0.005	1	2	−1
			Influence of coiosn charge and diffusivity D_Y					
I.4	1.0	1.0	0.01	0.005	0.001	1	1	−1
I.4e				0.001	0.005			
I.5	1.0	1.0	1.0	0.005	0.001	1	1	−1
I.5e				0.001	0.005			
I.5b				0.005	0.001	1	1	−2
				0.001	0.005			
I.6	1.0	1.0	1.0	0.005	0.001	1	1	−1
I.6e				0.001	0.001			

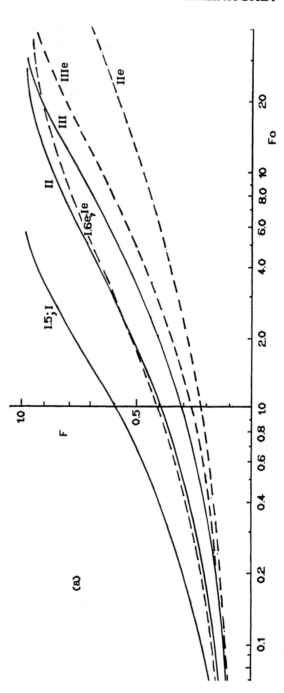

(a)

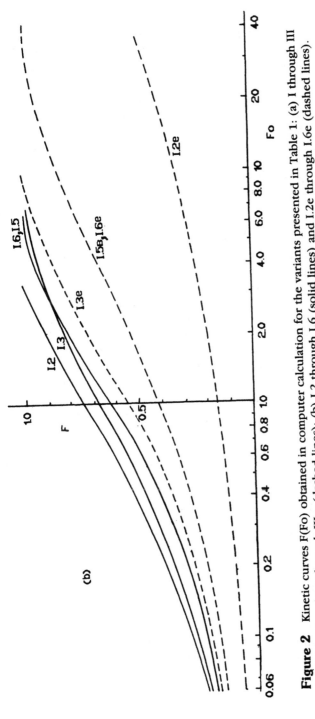

Figure 2 Kinetic curves F(Fo) obtained in computer calculation for the variants presented in Table 1: (a) I through III (solid lines) and I.e through III.e (dashed lines); (b) I.2 through I.6 (solid lines) and I.2e through I.6e (dashed lines).

The radial distribution of ion concentrations corresponding to various conversions, F, are given in Fig. 3 and 4. It follows from these figures that the concentration profiles U_B, C_A, C_Y depend on the combined relationship between two factors, selectivity factor K_{RA}/K_{RB} and diffusivity factor D_A/D_B.

The ion-exchange kinetics accompanied by association-dissociation in this model differ, on the one hand, from nonlinear sorption processes where the isotherm shape is responsible for the concentration profiles in the bead and for the kinetic rate [63] and, on the other, from conventional ion exchange where the kinetic rate and concentration profiles C_B, C_A are governed by the diffusivity factor D_A/D_B [16]. In the model both the selectivity (K_{RA}/K_{RB}) and the diffusivity (D_A/D_B) factors play a role in the IE process.

3. Ion-Exchange Kinetics for Unequally Charged Ions

The equations of the model are now extended to IE systems that include multivalent exchanging ions [64]. Figures 4a–f show the V_B and C_A concentration distributions obtained by computer calculations for I.2 through I.6 and I.2e through I.6 variants (Table 1) including unlike ions. These results confirm the earlier conclusions with respect to the influence of the two factors mentioned. For all the variants in Fig. 4, the diffusivity factor is the same and the relation $D_B/D_A = 1$ applies (Table 1).

The sharpest counterion profiles (shell-core behavior) are obtained for variant I.2 ($z_B = 2$, $z_A = 1$), where the selectivity of the divalent B ion is very high. The most spread counterion profiles are obtained for the variant I.2e of exchange ($z_B = 1$, $z_A = 2$), where the selectivity of monovalent B ion is very low. For the intermediate variants whether the concentration profiles are comparatively sharp or not depends on the value of the selectivity factor, K_{RA}/K_{RB}. Accordingly the half-time of conversion ($T_{0.5}$) values bear the following relationship:

$$T_{0.5}^{1.2} < T_{0.5}^{1.3} < T_{0.5}^{1.6} < T_{0.5}^{1.3e} < T_{0.5}^{1.6e} < T_{0.5}^{1.2e} \tag{12}$$

(compare kinetic curves in Fig. 2b). It can be seen from such a comparison in Fig. 2b that the kinetic rate is not affected by the charge and diffusivity of co-ion Y (variants I.5b, I.6 and I.4, I.5).

4. Selectivity and Diffusion Factors Determining the Ion-Exchange Kinetic Rate

The physical reality of the process becomes evident when considering the features of the C_i and U_i concentration profiles for the free and bound ions, respectively (Fig. 3). The sharpest profiles appear when the two

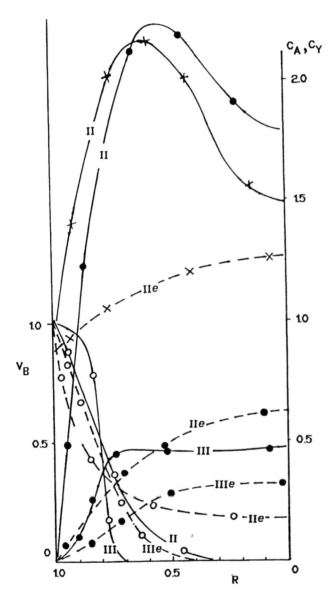

Figure 3 Computerized concentration distributions $V_B(\bigcirc)$, $C_A(\bullet)$ and $C_Y(x)$ in radial direction within exchanger bead for variants: II and II.e ($D_A = 1.0$, $D_B = 0.1$); III and III.e ($D_A = 0.1$, $D_B = 1.0$); II and III (forward exchange, solid lines); II.e and III.e (reverse exchange, dashed lines).

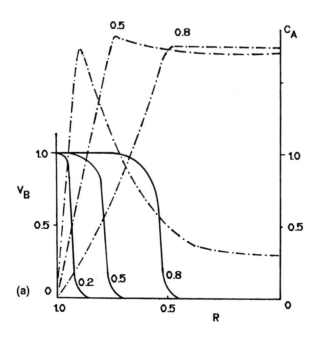

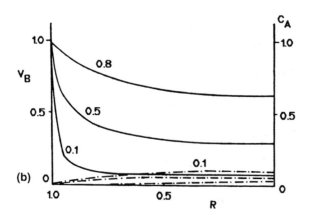

Figure 4 Computerized V_B (solid lines) and C_A (dashed lines) concentration distributions during IE kinetic process for heterovalent exchange (unlike ions): —— V_B, —·—··— C_A: (a) variant I.2 (forward IE), $z_B = 2$, $z_A = 1$, $z_Y = -1$; (b) variant I.2e (reverse IE), $z_B = 1$, $z_A = 2$, $z_Y = -1$; (c) variant I.3 (forward IE); $z_B = 2$,

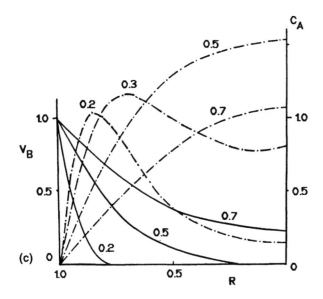

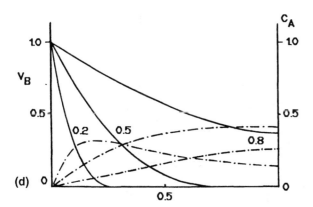

$z_A = 1$, $z_Y = -1$; (d) variant I.3e (reverse IE); $z_B = 1$, $z_A = 2$, $z_Y = -1$; (e) variant I.6 (forward IE); $z_B = 1$, $z_A = 1$, $z_Y = -1$; (f) variant I.6e (reverse IE); $z_B = 1$, $z_A = 1$, $z_Y = -1$. Numbers near curves correspond to the fractional conversion F.

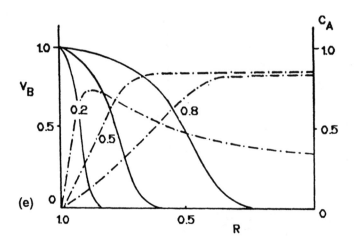

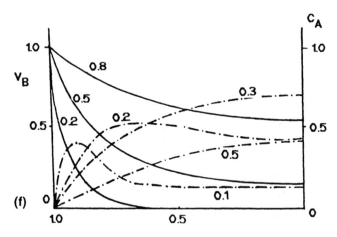

Figure 4   Continued

factors $D_A/D_B > 1$ and $K_{RA}/K_{RB} > 1$ sharpen the concentration fronts (Fig. 3, V_B, curve III). The sharp exchange boundary advances toward the bead center, the counterions are transferred to the exchange boundary and to the boundary of the bead with participation of co-ions Y penetrating from the solution into the exchanger bead. Meanwhile, ions A have time to diffuse into and escape through the surface of the bead due to the fact that its diffusivity is high ($D_A > D_B$). As a consequence, the C_A value is small (Fig. 3, C_A, curve III).

With the favorable isotherm and at the relation $D_A < D_B$, zone of most intensive exchange accumulation of displaced A ions occurs due to their low mobility (Fig. 3, C_A, curve II). The C_A value increases so that the equilibrium of reaction RA + B $\Leftrightarrow$ RB + A shifts toward the left-hand side and profile V_B spreads (Fig. 3, V_B, curve II).

According to the electroneutrality condition accumulation of ion A becomes practicable owing to diffusion of a large number of coins Y from the solution into the bead (Fig. 3, C_Y, curve II) under the effect of the electric field.

With the unfavorable isotherm and at the the the relation $D_A > D_B$, rapid migration of A ions out of the bead into the solution favors the sharpening of profiles (Fig. 3, V_B, curve III.e). However, for variant III.e the profiles are spread due to the effect of the selectivity factor ($K_{RA}/K_{RB} > 1$).

At the initial stage the dependence of fractional conversion F on $(Fo)^{1/2}$ is linear. The exchange rate is dependent on the value of a_0/c_0 and independent of D_Y. This is in accordance with the analytical solution (11) for the "rectangular" isotherm of the B counterion. The greatest spread of concentration profiles appears in situations where both factors act together to spread the concentration distribution of the B ion: i.e., at $D_A/D_B < 1$ and $K_{RA}/K_{RB} < 1$ (Fig. 3, V_B, curve II.e). In this instance the diffusion of the A ion is slower than the diffusion of the B ion ($D_A < D_B$) and this results in accumulation of the A ion (Fig. 3, C_A, curve II.e). The accumulation is partially attenuated due to the effect of the selectivity factor when $K_{RA}/K_{RB} < 1$ continues to prevail (compare C_A concentration profiles in variants II and II.e). Co-ion Y also enters the bead (Fig. 3, C_Y, curve II.e) although not as vigorously as in the case with variant II.

In the comparison of concentration profiles it should be noted that the combined influence of diffusivity and selectivity factors result in the situation where the selective adsorption of the invading substance is not sufficient for sharp V_B, C_B concentration profiles to appear in the bead. Formation of a sharp concentration sorption front is typical when $D_A/D_B > 1$ and the isotherm is favorable (Fig. 3, V_B, curve III). In the case of a favorable isotherm and with $D_B > D_A$ (Fig.3, V_B, curve II) or of an unfavorable one and with $D_A > D_B$ (Fig. 3, V_B, curve III.e) factors responsible for a sharpened sorption profile are partially compensated for by those factors spreading the profile. As a result the concentration profiles for these variants (II and III.e) are spread more than those for variant III with a favorable isotherm (Fig. 3).

When $a_0/c_0 \gg 1$ applies the rate of the process is virtually independent of the diffusivity D_Y (Table 1, variants I.5 and I.4). This can be attributed to the fact that in this case the boundary of the B/A exchange advances slowly and the co-ions have time to diffuse and redistribute under the influence of the electric field so that electroneutrality is maintained. The numerical solution that is obtained for a number of variants with different K_{RA} and K_{RB} values also demonstrates that if $K_{Ri}/c_0 \ll 1$, the rate of the kinetic process is strongly affected by the selectivity factor K_{RA}/K_{RB} with practically no effect attributable to the separate K_{RA} and K_{RB} values. Theoretical-based calculations of ion distributions in the bead permit determination of the applicability limits of the approximate solution of Eq. (11). The solution is applicable to the case of a favorable isotherm for the B ion: at $D_A > D_B$ for quantitative deductions, and at $D_A < D_B$ for semiquantitative estimates and qualitative deductions.

Comparison of the computerized results obtained for the kinetic curves (Fig. 2a) reveals a very interesting feature of the IE process under discussion. Kinetic curves F (Fo) for variants I and III, resolved when a favorable isotherm of the B ion and the relation $D_A > D_B$ apply, are described by the kinetic equation of diffusion into a spherical bead with constant diffusivity D_{ef}. In other words the kinetic curves F(Fo) coincide with the isotope exchange kinetic curves if a sharp profile appears in the bead and do not coincide if the exchange zone is greatly spread. The remaining kinetic curves in Fig. 2 formally correspond to the exchange process where D_{ef} varies in value. This is especially evident when comparing kinetic curves II and I.e (Fig. 2a).

The exchange rate can be conveniently estimated using the reduced parameter $T_{0.5}$ which corresponds to the half-time of exchange (F = 0.5) and is defined by the relation $T_{0.5} = Fo_{0.5}(c_0/a_0)$. Figure 5 shows the dependence of this parameter on the selectivity factor. This dependence is demonstratable by comparing the kinetic curves calculated for D_B, D_A, and D_Y diffusivities and a number of variants (including I–III and I.e–III.e variants of Table 1) as the selectivity factor is varied between 100 and 0.01. It is evident from Fig. 5 that the increase of half-time, $T_{0.5}$, with decreasing selectivity, K_{RA}/K_{RB}, occurs more strongly the smaller the diffusivity factor is (Fig. 5, curve II–II.e, $D_A = 0.1D_B$).

In the favorable isotherm region for the B ion (where $K_{RA}/K_{RB} > 1$) the slowest process is that for which the relation $D_B = 0.1D_A$ is applicable (Fig. 5, curve III, solid line). For an unfavorable isotherm the slowest

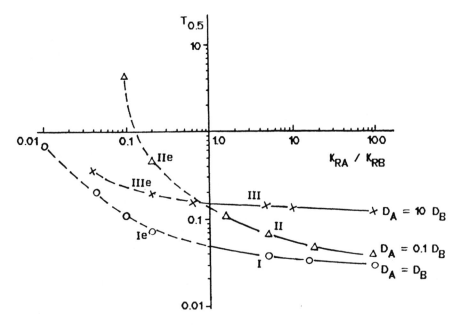

Figure 5 Calculated dependencies of exchange rate ($T_{0.5}$) versus selectivity factor (K_{RA}/K_{RB}) at various diffusion factors (D_A/D_B) including variants I through III and I.e through III.e: $K_{RA}/K_{RB} > 1$—favorable isotherm (solid lines); $K_{RA}/K_{RB} < 1$—unfavorable isotherm (dashed lines).

process occurs in those systems where the relation $D_A = 0.1D_B$ applies (Fig. 5, curve II.e, dashed line, $D_A = 0.1D_B$).

The different effect of diffusivity ratio on kinetic rate for favorable and unfavorable isotherms can be visualized as follows: the low mobility of displaced A ions leads to its accumulation in the resin bead and a lower ratio of free-ion concentrations (C_B/C_A). The relative variation of concentration as the C_B/C_A ratio varies is considerably smaller in the case of the favorable isotherm than the unfavorable isotherm and in the long run may result in variation of the kinetic rate as shown in Fig. 5.

Accordingly numerical comparison of the variants for different combinations of dissociation constants and diffusivities reveal that the kinetic behavior of the system is strongly affected by the mobility of the preferred ion (i.e., by the diffusivity D_B when K_{RA}/K_{RB} and by the diffusivity D_A when $K_{RA} < K_{RB}$ [48]. Of course the preferred ion is more strongly complexed, and in its "free" state is thus more frequently present in the minority so that the above conclusion is in accordance with the well-known "minority rule" [44].

The results demonstrate that selectivity for one ion over another (stronger complex formed by that ion) is not a sufficient condition for shell-core behavior. This is the case for high selectivity of B ion (K_{RA}/K_{RB} > 1) and high diffusivity of the A ion ($D_A > D_B$).

Whether such systems may or may not exhibit shell-core behavior depends on the relative diffusivities and on whether the entering B counterion forms the stronger or weaker complex. In both cases the presence of co-ions in the exchange can be neglected because the Donnan effect (normally excluding co-ions) is greatly reduced by counterion association with the fixed charges.

The author and co-workers have made an attempt to use this intraparticle kinetic model [66] for the explanation of results obtained by L. Liberti et al. for the IE kinetics of forward and reverse sulfate-chloride exchange with complex-forming ion exchangers [53]. It was shown later [67] that in this case the kinetic process was determined by combined film and particle diffusion control. In this connection it was noted [44] that despite their refinement the models often do not allow a mechanism to be identified unambiguously on the basis of experimental observations.

With suitable adjustment of the parameters, essentially the same rates and even the same concentration profiles were obtainable in the two physically different situations that follow: in (a) where some ions of the species are immobolized by bonding while the others move freely as in this author's model [45–50], and in (b) where all ions of the species are mobile but slowed by interactions as they are in the Helfferich-Hwang model [44].

This result demonstrates that experimental observations of rates and even transient concentration profiles can be explained with either set of assumptions so that no conclusive information concerning the actual physical mechanism can be obtained in this way [44].

This conclusion applies especially to kinetic curves because as composite representations of their characteristics, they provide insensitive and rather poor insights for distinguishing between various kinetic mechanisms [66,44].

For all variants of the intraparticle IE process in selective ion exchangers the shape of the concentration distribution pattern in the exchanger bead depends on values of both selectivity ($K_{RA}/K_{RB} \gtrsim 1$) and diffusivity ($D_A/D_B \geq 1$) factors (Figs. 3 and 4). The same conclusions with respect to the simultaneous qualitative influence of diffusional transport and the selectivity process in the cation exchange resin bead were reached later through the use of Fick's law and the same approach with respect to the assignment of two ("free" and "bound") states of the exchanging of ions in the resin bead [65].

The methods of identifying the kinetic mechanism [2,16], effective for nonselective IE, are applicable only to a limited extent in selective systems. This conclusion is based on the results of this author's (and co-workers') numerical and experimental analyses of the kinetic model in weakly dissociating and complex forming ion exchangers [26,27, 45–50,68].

III. EXPERIMENTAL CONCENTRATION PROFILES AND KINETIC CURVES

Usually the IE kinetics in real systems is estimated from the experimental dependence of conversion (F) on time (t) and from bead section microphotographs taken at various intervals [32,34,53]. These methods do not provide any idea of how the concentration distributions of "free" ions (c_i and C_i) and those of the "bound" ones (a_i or V_i) are related in the bead.

Profiles of C_i and V_i concentrations are simulated in these studies by the use of a special kinetic cell [46,47,50,69] that is designed to represent a model of the conic sector of a single bead. The cell made up of separate ceramic rings is filled with fine wet beads of resin (1–20 μm). After a predetermined time interval during which the solution is in contact with the cone, IE between the bulk solution and the resin is interrupted, and the resin in the cone is divided into separate layers (along the line of the contacting rings). The resin in each layer is treated in such a way that the ions not bound to fixed groups and thus simulating ions with C_i concentration are removed. Then the immobile counterions in the beads and thus simulating ions with a_i concentration are removed by displacement [69].

The a_i and c_i concentration distributions along the layers in the cone are similar to their distributions along the radius of the resin bead. This is confirmed below by comparison of the effect of various factors (separation coefficients, bulk solution concentration c_0, diffusivity D_i) on the concentration profile a_i in the bead (from the microphotographs) and in the conical cell. This comparison, exemplified by metal (Me) exchanges such as Me_1–RMe_2 and Me–RH in the ampholyte, VPC, shows that mass transfer in the bead and cell respond to stimuli in the same way, namely the boundary of the B/A exchange is either sharp or spread depending on the type of exchange isotherm and the displacement rate dependence of the B/A moving boundary on bulk concentration c_0.*

---

*Compare Me concentrations distributions in Fig. 1 with Figs. 7 and 14 for the Me/H and Me/Na ion-exchange reactions.

Note that the D_{ef} value of the Me^{n+} ion in the cone is about three times larger than in the exchanger bead. This is attributed to the fact that the resin filling the cone has additional porosity, i.e., the space between the resin beads. More information with respect to the various approaches to the study of concentration profiles a_i and c_i, as well as to measurements of the exchange rate in a thin layer are disclosed elsewhere [46–48, 50,68,69].

Such data permit utilization of information on concentration profiles in the cone to be made for comparison with the theoretical ion distributions in the bead. Information of this kind is presented in Figs. 7 through 14.

Experimental values of c_0, a_0, and $D_H = 10^{-5}$ cm²/s, $D_{Ni} = 10^{-6}$ cm²/s [70], $D_{Pr} = D_{Nd} = 0.5 \times 10^{-7}$ cm²/s were used in the computer simulations. The D_{Me} value was resolved from the experimental kinetic curves of the Me–RNa exchange using analytical expression (11) (Sec. II) for examination of the formation of weakly dissociating RNi, RPr, RNd complexes in the exchanger bead as these reactions take place. Ion distribution data between exchanger and solution at equilibrium, that are essential for esti-

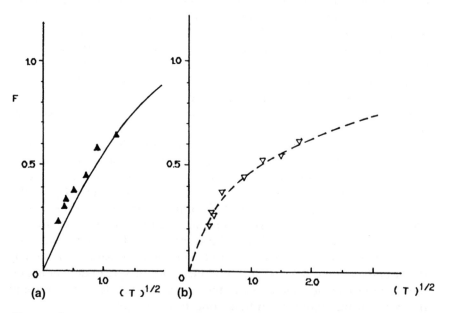

Figure 6 Kinetic curves $F((T)^{1/2})$ for H/Ni exchanges: (a) forward Ni–RH (solid lines) and (b) reverse H–R₂Ni (dashed lines) exchanges; lines—computerized calculation, $K_{RNi}/K_{RH} = 0.2$, $D_{Ni} = 0.2D_H$, $D_H = 10^{-5}$ cm²/s, $r_0 = 0.04$ cm; ▲—experiment for forward exchange, △—experiment for reverse exchange.

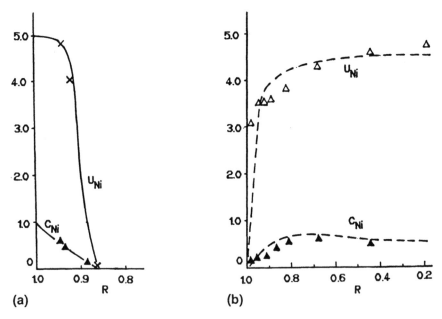

Figure 7 Simulated Ni^{2+} concentration radial distributions (U_B ($\times$), C_B($\blacktriangle$) and U_A($\triangle$), C_A ($\blacktriangle$) for Ni/H exchanges: (a) forward exchange Ni-RH (solid lines), F = 0.2; (b) reverse exchange H–R_2Ni (dashed lines), F = 0.3; lines—computerized calculation; $\triangle$, $\times$, $\blacktriangle$ - experiment with cone cell.

mating the selectivity factor K_{RA}/K_{RB} were compiled in independent static experiments with the vinylpyridinecarboxylic complex-forming ampholyte, VPC [47,48], the carboxyl cation exchanger CB-4 and the weak-base anion exchangers AN-18 and AN-18-P (macroporous) [49]. The exchange rate in exchangers VPC and AN-18 was measured using a thin-layer technique [16]. The filtrate was analyzed with a Spectromom 204 spectrophotometer (Hungary).

Table 2 lists parameters for some of the IE systems examined in this study. Figures 6 through 11 are presented to illustrate the applicability of the model to the exchange of unlike ions ($z_B \neq z_A$). Ion distribution data compiled at equilibrium and the relation between the individual ion diffusivities were used in calculations to facilitate preparation of Table 2 and Figs. 6 to 14.

The effect of the type of exchange isotherm on the U_i, V_i, C_i concentration distributions is illustrated in Figs. 7 through 11, 13, and 14 on the exchange rate in Figs. 6 and 12.

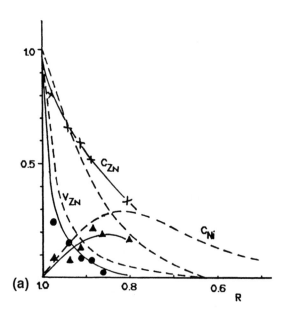

(a)

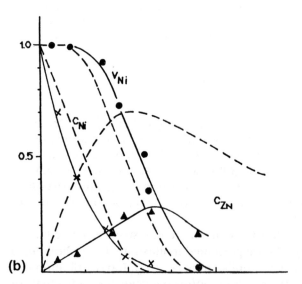

(b)

Figure 8 Simulated Ni^{2+} and Zn^{2+} concentration radial distributions $V_B(\bullet)$, $C_B(\times)$ and C_A (▲) for forward Ni-R_2Zn (a) and reverse Zn-R_2Ni (b) ion exchanges: dashed lines—computerized calculation; solid lines—experiment obtained with the cone cell filled by VPC resin; $c_0 = 0.25$ mmol/cm³: (a) time of contact 66 hr; $F = 0.25$; (b) time of contact 264 hr; $F = 0.18$.

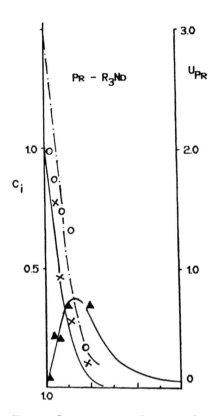

Figure 9 Simulated Pr^{3+} and Nd^{3+} concentration radial distributions --·--·-- U_B —— C_B and C_A: U_{Pr}(o), C_{Pr}(×) and C_{Nd}(▲) (c_{OPr} = 0.55 mg-eq/cm^3, contact time 120 hr): lines—computerized calculations with K_{RNd}/K_{RPr} = 2; o, ×, ▲—experiment obtained with the cone cell filled by VPC resin beads in Nd form.

It can be shown, as well, that the dependence of the mass exchange rate in the exchanger phase on the bulk solution concentration (c_0) and selectivity parameters that is revealed by theoretical studies is also observed in the real IE systems during exchange accompanied by formation of weakly dissociating compounds RMe or RH.

Comparison of the experimental and theoretical results shows that agreement of the kinetic curves (Figs. 6a, b and 12) and the radial concentration distributions (Figs. 7 through 11, 13, and 14) for the H–RMe, Me–RH, and Me_1–RMe_2 exchange reactions is reasonably quantitative. The shape of the kinetic curves in Figs. 6 and 12 is attributable to the features of ion distribution along the bead in the IE process. The ion distribution

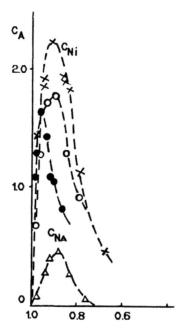

Figure 10 Simulated C_A concentration distributions of displaced Ni (C_{Ni}) or Na (C_{Na}) ions (dashed lines) in the cone cell: × H–R$_2$Ni, CB-4 resin, F = 0.16; ● and ○ H–RNa, CB-4 resin, F = 0.1 and 0.22; △ Ni–RNa, VPC resin, F = 0.2.

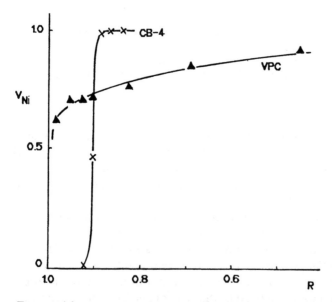

Figure 11 Simulated Ni concentration radial distributions V_{Ni} in the cone cell with resins CB-4 (×) and VPC (▲) for H-R$_2$Ni exchange.

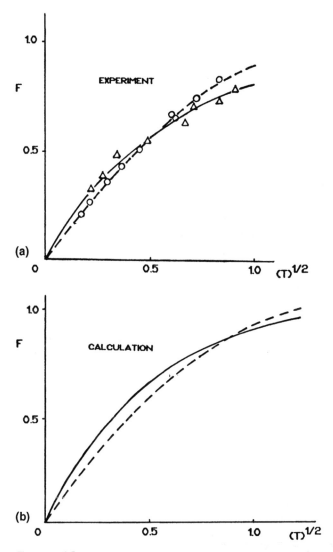

Figure 12   Kinetic curves $F((T)^{1/2})$ for the forward H–RAg (solid lines) and reverse Ag–RH (dashed lines) exchanges: (a) experiment $\triangle$ for the forward H–RAg exchange; $\bigcirc$ for the reverse Ag–RH exchange; (b) computerized calculation with parameters $K_{RAg}/K_{RH} = 1.5$ and $D_{Ag} = 0.1D_H$.

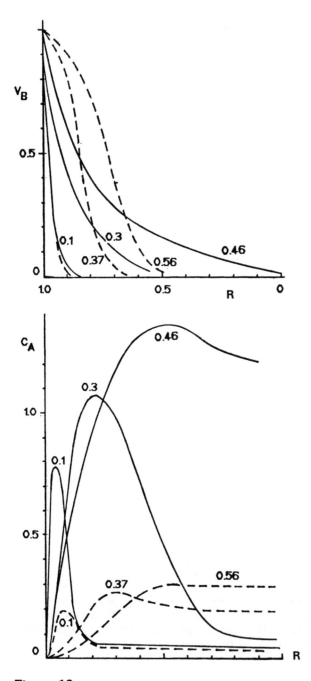

Figure 13 Computerized V_B and C_A concentration distributions in radial direction: for forward H-RAg (solid lines) and reverse Ag-RH (dashed lines) exchanges at various fractional conversions F (numbers near the lines).

178

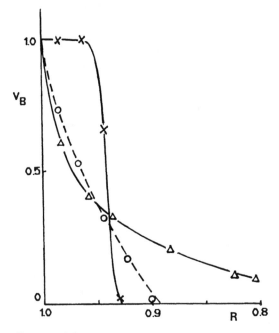

Figure 14 Simulated V_B concentration radial distributions obtained experimentally with the cone cell filled by VPC resin beads for various IE systems: forward exchange Ag–RNa (×, "rectangular" isotherm); forward exchange H–RAg, (Δ, solid line) and reverse exchange Ag–RT (○, dashed line) at fractional conversion F = 0.1.

features, in turn, depend on the selectivity (the type of ion B isotherm) and diffusivity factors.

The experimental kinetic curves in Fig. 6 for the Ni–RH and H–R$_2$Ni exchange reactions in VPC indicate that the IE rate decreases on passing from favorable (Fig. 6a) to unfavorable (Fig. 6b) isotherms and correspond to the calculated curves when the exchange isotherm and diffusivity factor are used as starting parameters for these calculations. The favorable Ni ion isotherm of the forward Ni–RH exchange favors a sharpening of the Ni/H exchange boundary in the bead, whereas the unfavorable H$^+$ ion isotherm for the reverse H–R$_2$Ni exchange leads to spreading of the H/Ni exchange boundary (Fig. 7b). During the reverse H–R$_2$Ni exchange the concentration of free Ni^{2+} ions in the bead increases as they are displaced by H$^+$ ion because of their relatively low mobility. The reaction equilibrium, R$_2$Ni + H ⇔ RH + Ni, shifts toward the left and the H/Ni exchange boundary spreads (Fig. 7b). During the forward Ni–RH exchange rapid migration of desorbed H$^+$ ion from the resin bead to the solution favors

Table 2 Experimental Parameters of the Investigated IE Systems

IE system (B – R$_\eta$A)	c$_0$ (mol/L)	r$_0$ (cm)	t$_{0.5}$ (s)	Type of resin	Type of ion B exchange isotherm
	Effect of bulk concentration, c$_0$, on exchange rate				
Nd–RNa (at 90°C)	0.01	0.04	1,700		Rectangular
	0.03		580	VPC	Rectangular
	0.1		200		Rectangular
Ni–RNa (at 20°C)	0.025	0.04	1,020		Rectangular
	0.05		500	VPC	Rectangular
	0.15		160		Rectangular
H–R$_2$Sr	0.05	0.005			Favorable
Sr–RNa	0.025	0.005		VPC	Favorable
Ni–R$_2$Sr		0.005			Favorable
Zn–RNa	0.0025	0.005			Favorable
Pr–RNd	0.55 mg-eq/ml (cone cell)				Slightly favorable
Cl–R$_3$Ag(S$_2$O$_3$)$_2$	from 0.05 until 0.1		from 60 until 4000	AN-18	From favorable to unfavorable
Cl–R$_3$Cu(S$_2$O$_3$)$_2$	from 0.5 N until 4 N				From favorable unfavorable
	Effect of IE direction				
Forward IE Ni–RH	0.15	0.04	460		Favorable
Reverse IE H–R$_2$Ni (at 20°C)	0.3		1,080	ANKB, VPC, CB-4	Unfavorable
Forward IE Ni–R$_2$Zn	0.025	0.005	63	VPC	Favorable
Reverse IE Zn–R$_2$Ni (at 25°C)	0.025		3,080		Unfavorable
Forward IE Ni–RNd	0.15	0.04	1,100	VPC	Favorable
Reverse IE Nd–R$_2$Ni (at 25°C)	0.1		400,000		Unfavorable
Forward IE H–RAg	0.06	0.04	630	VPC	Slightly favorable
Reverse IE Ag–RH (at 20°C)	0.06		680		Slightly unfavorable
Forward IE Cu–RH		from 0.05 until 0.1	from 60 until 5,000	ANKB	From favorable to unfavorable
Reverse IE H–R$_2$Cu (at 20°C)	from 0.0007N until 0.3N				From favorable to unfavorable

the displacement of the equilibrium toward the side where the RNi complexes are formed and the Ni/H boundary sharpens (Fig. 7a). As a result the Ni sorption-desorption rates that are very much the same at the beginning of the process differ markedly at F > 0.4 (Fig. 6).

The reasonably quantitative agreement between the theoretically predicted and the experimentally measured concentration distributions is illustrated in Figs. 7 through 11, 13, and 14 for various types of IE isotherms. It is evident that the concentration distribution profile depends on the type of the invading ion B isotherm, which is favorable for the forward Ni–RH and Ni–R_2Zn exchanges, unfavorable for the reverse H–R_2Ni and Zn–R_2Ni exchanges (Figs. 7, 8, 11, and 12) and close to linear for the Pr–RNd (Fig. 9) and Ag–RH exchanges (Figs. 13 and 14) (see also Table 2).

It is interesting to consider the IE kinetics of invading Ni^{2+} or Zn^{2+} ions for the forward Ni–R_2Zn and the reverse Zn–R_2Ni exchanges within ampholyte VPC (Fig. 8). Computer simulation of the IE kinetics for this system are carried out for this purpose, with the ratio K_{RA}/K_{RB} = 12 for the favorable isotherm (forward exchange Ni–R_2Zn) and K_{RA}/K_{RB} = 1/12 for the unfavorable isotherm (reverse exchange Zn–R_2Ni).

The value of the selectivity factor K_{RA}/K_{RB} is identical with the relationship of the distribution coefficients of Ni and Zn ions between the solution and the ion-exchanger phase with the conditions $K_{RA}/K_{RB}/c_0 \ll 1$ and $c_0 \ll a_0$. This relationship is obtained from the experimental isotherm of the Ni/Zn exchange system.

As observed from this author's data [45] experimental kinetic curves for the systems Ni–RNa and Zn–RNa can be used to show through use of Eq. (11), that $D_{Ni} = D_{Zn} = 2 \times 10^{-7}$ cm^2/s.

The concentration distribution curves calculated for the IE kinetics of Ni/Zn ions are similar to the experimental ones (Figs. 8a and b).

Some disagreement that is observed between the theoretical and experimental results can be attributed to the variability of ion activity coefficients and their effect on the assessment of dissociation constants. However, it is difficult and sometimes inadvisable to blame these factors because disagreement in the concentration distribution curves lead to only small differences in the kinetic rates [46]. For example satisfactory agreement between the experimental and calculated values was observed on comparing the rates of the forward Ni–R_2Zn and reverse Zn–R_2Ni exchanges. The $T_{0.5}$ (forw)/$T_{0.5}$ (rev) ratios are 7.8 and 6.5 for the experimental and calculated values, respectively.

Figure 10 compares C_A, the radial distribution concentrations of displaced ions A for ion-exchange reactions, with different diffusivity factors (D_B/D_A). The exchange isotherms in all three cases are strongly favorable since the hydrogen ions in carboxylic cation exchanger CB-4 and nickel ions in VPC ion exchanger yield weakly dissociating compounds ($K_{RH} = 10^{-5}$, $K_{RNi} = 10^{-7}$). It is seen from Fig. 10 that maximum concentrations of displaced A ions increase as their relative mobility decreases ($D_{Ni} < D_{Na} < D_H$) and this is consistent with the theoretically based predictions (Sec. II).

The above dependencies confirm the conclusions based on calculations for two types of isotherm with selectivity factors $K_{RA}/K_{RB} = 5$ and $K_{RA}/K_{RB} = 0.2$ as projected in Table 1 (Sec. II). These results also show that the theoretical calculations are in qualitative agreement, with experimental data for the IE kinetics of Ni,$^{2+}$ Zn^{2+}, Nd^3, Na$^+$, and H$^+$ in the weakly dissociating carboxyl (CB-4) and vinylpiridinecarboxyl (VPC) ion exchangers.

To enable one to use the results of computer simulation for quantitative evaluation of the interdiffusion rate of ions in real systems, it is essential to determine as effectively as possible the accuracy with which one has to specify the magnitude of dissociation constants K_{RA}, K_{RB} and diffusion coefficients D_B, D_A, D_Y. To accomplish this, computer simulations were performed to compare results when K_{RA} and K_{RB} differed by 10-fold and when K_{RA} and K_{RB} were taken to be equal [48,68]. The calculated kinetic curves coincide if the relations $K_{RA}/c_0 < 10^{-3}$ and $K_{RB}/c_0 < 10^{-3}$ are applicable. In this case the concentration of the free ions C_i is so low that its variation probably does not affect the electric field in the ion exchanger so that the distribution of the ions in the resin depends only on the ratios K_{RA}/K_{RB} and D_A/D_B. On this basis, an approximate estimate of the order of K_{RA} and K_{RB} can be tolerated. As shown above, their ratio is accessible from the experimental exchange isotherm. Also it is evident from Fig. 5 that in the case where the ratio D_A/D_B taken for calculation purposes does not correspond to the real system, an error in determining the process rate will be larger for the unfavorable (relation $K_{RA}/K_{RB} < 1$) than it is for the favorable (relation $K_{RA}/K_{RB} > 1$) isotherms.

For the H/Ag exchange agreement between the predicted and experimental kinetic results has been demonstrated to be quite acceptable in Figs. 12 through 14 [48,68]. The relation between the theoretically and experimentally based data is particularly demonstrable by reference to the exchange of Ag and H in the VPC ion exchanger. When selecting the basic parameters data reported elsewhere in the literature [70] and static experimental data [68] from this author and co-workers were included in the calculations. According to Helfferich [16,21,70], the dissociation constant of —COOH groups in the carboxylic cation exchanger $K_{RH} = 1.5 \times 10^{-5}$ m/L, the diffusion coefficient $D_H = 10^{-5}$ cm^2/s, and the ratio $D_H/D_{Na} = 7$. Owing to the fact that Ag has a slightly lower mobility than

Na the approximate relations $D_H/D_A = 1.0$ is considered to be acceptable. The static equilibrium measurements [68] lead to resolution of a selectivity factor of 1.5 for K_{RAg}/K_{RH} as well.

The experimental kinetic curves for Ag/H exchange in the ampholyte VPC were obtained by thin-layer chromatography [48,68] under the following conditions:

$$a_0 = 1.5 \text{ mg-equiv/mL}, \quad c_0 = 0.06 \text{ mg-equiv/mL}, \quad r_0 = 0.4 \pm 0.02 \text{ mm}$$

The experimental and calculated curves can be conveniently represented with the same coordinates. The dimensionless parameter, $T_{ex} = D_0 t_{exp} c_0/r_0^2 a_0$, where $D_0 = D_H = D_{Ag} = 10^{-9}$ m²/s, was introduced.

Comparison of the kinetic curves in Figs. 12a and b shows that there is reasonable agreement between the experimental and calculated data for the case under discussion. Also the data in Fig. 12 provide support for the theoretically based conclusion with respect to the equality of the rates of the forward and reverse exchange reactions (Fig. 12) when the isotherm is close to linear. An interesting feature of the kinetic curves (solid and dashed lines in Figs. 12a and b) is that they cross at $F = 0.85$ for computerized dependencies (Fig. 12a) and at $F = 0.55$ for the experimental ones (Fig. 12b). The reason for the different response to time of the forward (solid lines) and reverse (dashed lines) IE can be established through analysis of the V_B and C_A concentration distribution curves.

Computer- and experimentally simulated V_B concentration distributions (Figs. 13 and 14) employed for modeling the distribution of the complexes RB (B = H) and the displaced ions A (A = Ag) (Fig. 13) demonstrate that even though the isotherm of the H-RAg IE is a favorable one ($K_{RA}/K_{RB} = 1.5$; >1) the V_H concentration profile along the bead radius is very broad (Figs. 13 and 14) owing to an increase in the C_A concentration of the less mobile displaced ion A (A = Ag) (Fig. 13) and a shift of the exchange equilibrium toward formation of the RAg complexes. This result agrees with the data reported in Sec. II showing that selective sorption of the invading B ion is not a sufficient condition for the formation of a steep concentration distribution profile in the ion-exchanger particle. In the case of an unfavorable isotherm (V_{Ag} and C_{Ag} curves, Figs. 13 and 14) for the reverse Ag–RH exchange the V_B, C_B (B = Ag) concentration profiles are sharper (Figs. 13 and 14). The sharpening of the profiles in this case is favored by a fast migration of the displaced ions (A(A = H) from the particle to the solution (C_H curves, Fig. 13). The difference in the kinetic curves (solid and dashed lines) of Fig. 12 and their crossing are likely to be governed by the concentration distribution behavior of the ions within the bead during the forward H–RAg and reverse Ag–RH exchanges (Figs. 13 and 14).

It is evident that the experimental distribution curves for the invading B component in a cone-shaped kinetic cell [48] used to model a spherical element of the ion-exchanger bead were obtained for the H ions entering ampholyte VPC in the RAg form and for the Ag ions entering the same ampholyte in the RH or RNa forms. The sorption V_H front of hydrogen ions is very broad compared to that of the Ag ions (Fig. 14, curves V_H and V_{Ag}), agreeing with the data compiled for the numerical study (Sec. II). For the comparison the very sharp distribution curve V_{Ag} for Ag-RNa exchange in Fig. 14 represents the distribution of the invading Ag ions in the case of a "rectangular" (strongly favorable) IE isotherm.

The satisfactory agreement between the experimental and the calculated (theoretical) kinetic curves and concentration profiles suggests that the values of $D_H = 10^{-5}$ cm^2/s, $D_{Ag} = 10^{-6}$ cm^2/s, and $D_{Na} = 1.5 \times 10^{-6}$ cm^2/s adopted for the calculations do indeed characterize the mobility of H^+ and Ag^+ ions in ampholyte VPC. By using these values and the above ratio, $D_H/D_{Ag} = 7$, it was possible to evaluate the rate of the Ag/Na exchange. A rectangular isotherm characterizes the sorption of Ag^+ ions by the VPC ampholyte and this warrants the use of the approximate analytical solution (11, Sec. II) [26,27] for the above computations.

In this system $D_{ef} = 2D_{Na}D_{Ag}/(D_{Ag} + D_{Na})$ and the values of r_0c_0 and a_0 are equal to those adopted for experiments with Ag^+ and H^+ ions. Equation (11) gives $F/t^{1/2} = 0.024$. Experimental measurements of the kinetic rate for Ag/Na exchange give $F/t^{1/2} = 0.022$.

Use of the relation $D_A > D_B$ is suitable for sharpening the exchange boundary in the resin bead. For favorable isotherms, these variations of $T_{0.5}$ for curves I and III in Fig. 5 are relatively small. As a consequence, it can be concluded that the so-called shell-core model [5,14,16] is suitable not only for selective IE but also for exchange isotherms close to linear when $D_A > D_B$.

It follows that the analytical scheme (Eq. (11) in Sec. II can be used to study ion exchange kinetics in selective IE systems.

The theory developed (Sec. II) also anticipates, with reasonable accuracy, results of a study of the exchange of monovalent ions H/Ag accompanied by association with fixed ion-exchange groups of picoline acid in the ion exchanger.

The demonstrated capability of the theory to provide reasonably accurate anticipation of experimental results with the various examples presented attests to its value. Anticipation of ion-exchange processes with the thin layer and cone models shows their value also. The formation of the sharp boundary B/A in the case of strongly favorable isotherms of invading B counterion in the bead and in the cone model (Figs. 7, 8, 10,

11, and 14) shows that structural heterogeneity does not effect the formation of the a_i, C_i concentration distributions.

IV. SELECTIVITY AND BULK CONCENTRATION FACTORS IN INTRAPARTICLE ION-EXCHANGE KINETICS

The set of experiments presented in Sec. III for the examination of IE kinetics in selective systems has confirmed the projections of the theoretical analyses in Sec. II. Selectivity and diffusivity factors have great influence on kinetic rate and concentration distributions in complex forming and weakly dissociating ion exchangers. With polyvalent ion exchange the dependence of separation factor upon solution concentration complicates explanation of the IE kinetics results.

It is equally interesting to compare the Ni ion distribution in ion exchanger VPC which is highly selective to Ni ions and in weak-acid carboxylic exchanger CB-4 with no selectivity for Ni ion (Fig. 11). It is evident that Ni concentration distribution depends on the type of exchange isotherm, rectangular for CB-4 resin and unfavorable for VPC resin.

In the discussion of the results, it is especially of interest to note that many of the dependencies typical of the complex forming exchangers are identical to those exhibiting film exchange kinetics [49]. This observation is pertinent to the effect of the solution concentration, selectivity parameters, and the dependence of the individual counterion diffusivities during the forward and reverse IE on the rate of the kinetic process. In this context, using conventional methods for the identification of a kinetic mechanism for the selective ion exchangers becomes unreasonable. Erroneous results can also result from use of the Helfferich-Tunitsky criterion [2,16,22,23] whose calculation does not take into account the dependence of the effective diffusivity in the exchanger phase on the exchange direction and bulk concentration of the solution.

The problems encountered in determination of the limiting kinetic mechanism are exemplified by desorption of thiosulfate complexed silver anions from the weak-base anion exchanger AN-18 with a NaCl solution. In the concentration range from 0.1 to 2 N where an intraparticle diffusion mechanism is most probable, the rate of the desorption of the selectively retained $Ag(S_2O_3)_2^{3-}$ ions from the porous anion exchanger AN-18-P increases by about 600-fold (Fig. 15, curve 1). In the same range, the rate of the nonselective CNS^-/Cl^- exchange increases by no more than sixfold (Fig. 15, curve 3). The growth of the desorption rate, in the first case, is likely due not only to increase in the relation between exchanger

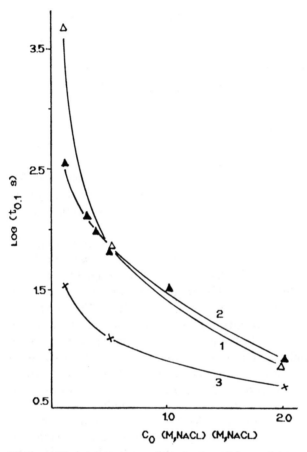

Figure 15   Effect of NaCl solution concentration on the desorption rate ($t_{0.1}$) of $Ag(S_0O_3)_2^{3-}$ ions (curves 1 and 2) and CNS^- ions (curve 3) for anion exchanger AN-18 of a porous (Δ, × curves 1 and 3) and gel ($\blacktriangle$, curve 2) structure.

volume and solution concentration but to a decrease in $Ag(S_2O_3)_2^{3-}$ ions as well [50].

    It is shown [49] that the dependence of exchange rate on the solution concentration and selectivity are typical, both for film and particle diffusion, in the selective exchangers. Figure 15 (curves 1 and 2) depicts the influence of the inner structure of the anion exchanger on the kinetic rate. The ultimate conclusion that particle diffusion is controlling in this case can only be confirmed by more rigorous studies of the $Ag(S_2O_3)_2^{3-}$ / Cl^- exchange with the AN-18 resin provided varying inner structures (Fig. 15, curves 1 and 2).

The effect of concentration and selectivity factors is illustrated here for various polyvalent IE systems in selective ion exchangers. The IE kinetics in selective ion exchangers were studied on the complex-forming resin ANKB-35 for H/Ni, H/Cu, and Ni/Na IE systems (Figs. 16 through 18) and on a weak-base ion exchanger AN-18-10P for the $Cl/[Ag(S_2O_3)_2]^{3-}$ and $Cl/[Cu(S_2O_3)_2]^{3-}$ IE systems (Figs. 19 and 20).

Stoichiometric mixtures of dilute silver nitrate, cupric sulfate, and sodium thiosulfate solutions were used to saturate the AN-18-10P ion exchanger with $[Ag(S_2O_3)_2]^{3-}$ and $[Cu(S_2O_3)_2]^{3-}$ ions. The range of eluent normalities was 0.0007–0.3 N for the ANKB-35 exchanger and 0.5–4.0 N for the AN-18-10P exchanger. The analytical tests were performed using standard titrations and atomic absorption spectrophotometry (AAS-4 Germany) [71].

The half-time ($t_{0.5}$) of exchange was selected to provide estimate of rates in the ANKB-17 exchanger (Fig. 35). The time, $t_{0.2}$, for 20% fractional conversion (Fig. 20) of AN-18-10P exchanger was selected for this purpose because of the much slower $[Ag(S_2O_3)_2]^{3-}$ and $[Cu(S_2O_3)_2]^{3-}$ ion desorption rates in AN-18-10P exchanger when the NH_4Cl solution concentration was very low.

Besides the diffusivity factor solution concentration (c_0) and selectivity factors affect the rate of the kinetic process. In addition selectivity of the IE is influenced by the variation of the solution concentration.

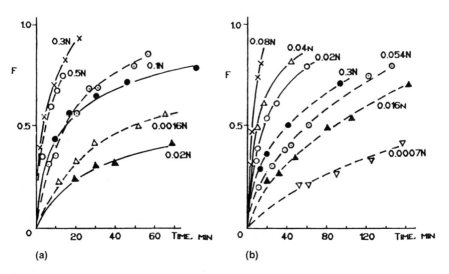

Figure 16 Experimental kinetic curves for H/Cu (a) and H/Ni (b) exchanges in ANKB resin at various solution (c_0) concentrations: forward exchanges (solid lines) and reverse exchanges (dashed lines). Numbers near lines correspond to concentrations c_0.

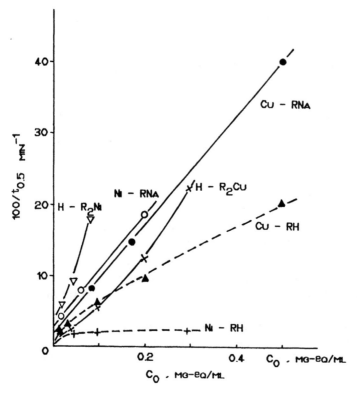

Figure 17 Effect of bulk solution concentration (c_0) on IE kinetic rate ($1/t_{0.5}$) for metal/H and metal/Na exchanges: forward exchanges (solid lines) and reverse exchanges (dashed lines).

In all IE systems studied the exchange rate depended on the c_0 concentration of the external solution (Figs. 16, 17 and 19, 20). This dependence differed for the Ni/H and Cu/H exchanges (Figs. 16 and 17) as well as for the Cl/Cu(S_2O_3)$_2$ and Cl/Ag(S_2O_3)$_2$ exchanges (Figs. 19 and 20).

The rates of the forward (Ni–Rh) and reverse (H–R_2Ni) ion-exchange reactions are very different (Fig. 17) since in these instances the selectivity is changed strongly with the solution concentration (see Ni/H exchange isotherms in Fig. 18). Both concentration and selectivity factors influence the rate of the Ni/H exchange.

The rates of forward (Cu–RH) and reverse (H–R_2Cu) Cu ion-exchange reactions are not very different (Fig. 17) since in these instances the selectivity does not change markedly with the solution concentration c_0 (Cu/H exchange isotherms, Fig. 18). As a consequence only the con-

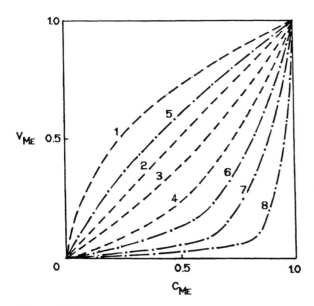

Figure 18 Effect of solution concentration (c_0) on IE isotherms for metal–RH exchanges (N = normality): – – – Cu–RH, 1 – c_0 = 0.02 N, 2 – c_0 = 0.1 N, 3 – c_0 = 0.2 N, 4 – c_0 = 0.5 N, –·–·–·– Ni–RH; 5 – c_0 = 0.0007 N, 6 – c_0 = 0.016 N, 7 – c_0 = 0.05 N, 8 – c_0 = 0.3 N.

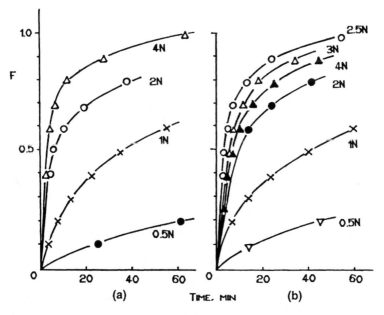

Figure 19 Experimental kinetic curves for $Cl–R_3Cu(S_2O_3)_2$ (a) and $Cl–R_3Ag(S_2O_3)_2$ (b) exchanges with AN-18-10P resin at various bulk solution (c_0) concentrations (numbers near lines).

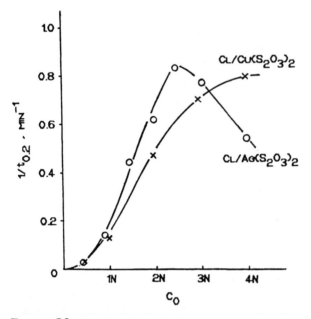

Figure 20 Effect of bulk solution concentration (c_0) on IE kinetic rate ($1/t_{0.2}$) for Cl–R$_3$Cu(S$_2$O$_3$)$_2$ (×) and Cl–R$_3$Ag(S$_2$O$_3$)$_2$ (o) exchanges with AN-18-10P resin at various bulk solution (c_0) concentrations.

centration factor has a strong influence on the rate of the Cu/H exchange reactions. For example in the concentration range of 0.1 N, the Cu/H exchange isotherm is close to linear (curve 2, Fig. 18) and the kinetic curves for the forward and reverse Cu/H exchange differ slightly (Fig. 17) only because of the difference in diffusivity factor. This effect is typical for systems where the exchange isotherm is close to linear (Ag/H exchanges; Secs. II and III).

During metal–RNa exchange (Fig. 17) the selectivity of the resin for the metals Cu, Ni is always high so that the rate of mass transfer depends only on the concentration factor. The rate versus concentration dependencies is close to linear for these cases (Fig. 17). The small difference in rates for Ni/Na and Cu/Na exchange is due to the difference in diffusivity of Cu and Ni ions.

Concentration dependence for Ni–RNa exchange is due to the diffusion of comparable quantities of two kinds of Ni ions: free Ni counterions from the external solution and Ni ions whose source is the dissociation process of the not very stable RNi complex. Increase of external solution concentrations leads to depression of the dissociation process. The rate of Ni–RH exchange is much slower than for H–R_2Ni and Ni–RNa exchanges (Fig. 17) though in all three instances, the least mobile Ni ions are responsible for the rate of the diffusion process.

Figures 19 and 20 show the dependence of rates on the concentration factor for Cl/[Ag(S_2O_3)$_2$]$^{3-}$ and Cl/[Cu(S_2O_3)$_2$]$^{3-}$ exchange in the weak base resin, AN-18. The fact that rate versus concentration dependence exhibited in Fig. 20 is not linear is attributable to a complicated influence of c_0 concentration on the selectivity factor. It follows from the IE isotherms obtained in static experiments [72] that the selectivity factor for the displacement of Ag(S_2O_3)$_2^{3-}$ and Cu(S_2O_3)$_2^{3-}$ ions decreases with increase of c_0 concentrations. The significant decrease of Me(S_2O_3) ion selectivity explains the rate versus concentration c_0 dependence observed in Fig. 20 when c_0 concentration varies from 0.5 N to 2.5 N. But this decrease does not follow a simple pattern for the Ag(S_2O_3)$_2^{3-}$ ion and, as a consequence, the rate versus concentration dependence in Fig. 20 for the Cl[Ag(S_2O_3)$_2$] IE system is not simple as well.

The difference in rates is due to the influence of different equilibrium relations: the H–R_2Ni and Ni–RNa exchange isotherms are strongly favorable but the Ni–RH, Cl–R_3[Ag(S_2O_3)$_2$] and Cl–R_3[Cu(S_2O_3)$_2$] exchange isotherms are unfavorable. In addition the separation factor decreases for the Ni–RH exchange system and increases for the Cl–R_3[Ag(S_2O_3)$_2$] exchange system with increasing external solution concentration.

It is shown for the selective IE systems that the rate of IE depends strongly on bulk concentration (c_0) of the external solution (Figs. 17 and 20). Variation of solution concentration affects the quantity of free ions in the resin phase as well as the shape of exchange isotherm (Fig. 18). Thus the complex and multivaried types of IE kinetic dependence is due to the combination of two factors: solution concentration and selectivity (that is affected in addition by the concentration c_0).

V. CONCLUSION

The theoretically based interpretation of the kinetics of ion exchange that has been developed has been confirmed by experiments that have shown the interdiffusion rates in the exchanger phase to depend, as predicted, on diffusivity and selectivity factors as well as on bulk solution concentrations. For unfavorable ion-exchange isotherms these dependencies are formally similar to those observed in film diffusion [49].

The results from investigations with the intraparticle model developed demonstrate that chemical reaction has a profound effect on IE kinetics in selective systems. Specific binding of counterions to fixed functional groups in selective systems impair their mobility and slow down the exchange rate. Such behavior is defined by Petruzelli and Helfferich et al. as "reaction retarded diffusion control" [44]. The difference in kinetic rate between forward and reverse exchange of two counterions depends significantly on the combined effect of diffusivity and selectivity factors.

It has been pointed out [4,5] that the electric field, while affecting exchange rate and producing a difference between forward and reverse exchange has much less influence.

An interesting observation that has been pointed out is that the difference in rate between forward and reverse exchange of two counterions may arise not only from unequal diffusion coefficients but also from unequal dissociation constants of complexes formed with the fixed groups.

Another significant effect is connected to the bulk concentration of the external solution (c_0). It is shown (Sec. IV) that the bulk concentration effects not only the concentration (c_B) of the invading counterion in the bead but, more importantly, the selectivity of exchanger for the exchanging counterions. According to the theory of mass transfer in sorption systems [63] the motion of the sorption concentration front of substances depend on the curvature of their sorption isotherm. For IE selective systems this effect is connected to the selectivity factor since this factor controls the shape of the ion-exchange isotherm.

In selective systems application of the Helfferich-Tunitsky criterion [2,5,16,22,23] may result in erroneous estimates of the contribution of film and intraparticle diffusion. This is because the ion diffusivities in the resin phase that are included in this criterion can undergo very abrupt changes in magnitude with the reversal of exchange direction. This was not taken into account by Helfferich and Tunitsky in the theoretical expressions as they used them for conventional IE. Moreover the criterion neither includes the influence of the equilibrium parameters characterizing the selectivity in the resin phase nor is there any account for co-ion invasion. The effective diffusivity of the kinetic process may be quite different if these factors are taken into consideration.

Deviation from the ideal exchange kinetic dependencies introduced by selectivity effects can arise in any ion-exchange system in which the resin phase ions can exist in two different states: i.e., relatively free (condensed) and bound (complexed) as assumed in the model projected [45–50]. This is true for complex forming, weakly dissociating, chemically and structurally inhomogeneous ion exchangers.

If local equilibrium is established instantaneously the ionic composition of parallel diffusion fluxes in the bead pores (i.e., "free" ions) and the gel phase of the bead (i.e., "condensed" and "bound" ions) is determined in particular by the selectivity factor, i.e., by the shape of the exchange isotherm. If the mobility of the ions in the pores is much higher than that in the gel portion of the resin bead, then, in terms of the model considered, the kinetic system is analogous to IE accompanied by association of counterions with fixed exchange groups. It is believed, as a consequence, that the kinetic regularities for macroporous ion exchangers are similar to those for weakly dissociating or complex-forming exchangers.

It follows from the experimental results [49,71–73] that the kinetic dependencies on concentration and selectivity effects are analogous in complex-forming and macroporous ion exchangers.

The electrolyte invading the resin bead can also be regarded as a "stagnant mobile phase" (a term from the field of liquid chromatography). It is important, in practice, that the IE process can be strongly retarded in macroporous ion exchangers when the isotherm is unfavorable because of the absence of automatic removal of the displaced ions by convective flow from the "stagnant phase" of the resin bead.

SYMBOLS

B	Counterion entering the resin bead
A	Counterion being displaced from the resin bead
Y	Co-ion
c_i	Solid-phase concentration of "free" ion i (mmol cm^{-3})
C_i	Reduced (dimensionless) solid-phase concentration of "free" ion i ($C_i = c_i/c_0$)
c_0	Bulk concentration of external solution
a_i	Solid-phase concentration of counterion i "bound" to fixed groups R and presumed immobile (mmol cm^{-3})
a_0	Capacity of ion exchanger (total concentration of fixed groups R) (mmol cm^{-3})
U_i, V_i	Reduced concentrations of "bound" counterion (i = B or A, $U_i = a_i/c_0$, $V_i = a_i/a_0$)
D_i	Solid-phase diffusion coefficients of ion i (cm^2/s^{-1})
D_{ij}	Solid-phase interdiffusion coefficient of counterions i and j
D_0	Scale diffusion coefficient (cm^2/s^{-1})
F	Fractional conversion
$\mathcal{F}$	Faraday constant (96500 C mol^{-1})
J_i	Flux of ion i (i = B, A, or Y)
K_{Ri}	Dissociation constant of complex Ri (i = B or A) (mmol cm^{-3})
r	Radial distance from particle center (cm)
r_0	Particle radius (cm)
R	Reduced radial distance (**R** = r/r_0)

194   KALINITCHEV

R   Fixed ionic group
R Gas constant (8.34 J mol^{-1}K $^{-1}$)
t   Time (s)
Fo or T   Reduced time (Fo = $D_0 t/r_0^2$, T = Fo $\cdot$ c_0/a_0)
ϕ Electric potential

REFERENCES

1. J. A. Marinsky, Ed., *Ion Exchange*, Dekker, New York, 1966.
2. F. Helfferich, in *Ion Exchange* (J. A. Marinsky, ed.), Dekker, New York, 1966, Chap. 2.
3. J. A. Marinsky and Y. Marcus, Eds., *Ion Exchange and Solvent Extraction*, Vol. 7, Dekker, New York, 1974.
4. L. Liberti and F. Helfferich, Eds., *Mass Transfer and Kinetics of Ion Exchange*, Martinus Nijhoff, Hague, 1983.
5. F. Helfferich, in *Mass Transfer and Kinetics of Ion Exchange* (L. Liberti and F. Helfferich, eds.), Martinus Nijhoff, Hague, 1983, pp. 157–183.
6. K. M. Saldadze and V. D. Kopylova-Valova, *Complex Forming Ionites*, Khimiya, Moscow, 1980. (Russian)
7. B. P. Nikolsky and P. G. Romankov, Eds., *Ion Exchangers in Chemical Technology*, Khimiya, Leningrad, 1983. (Russian)
8. M. Streat and D. Naden, Eds., *Ion Exchange Technology*, Ellis Horwood, Chichester, 1984.
9. B. A. Bolto and L. Pawlowski, *Wastewater Treatment by Ion Exchange* E&FNspon, 1986.
10. A. I. Voljinsky and V. A. Konstantinov, *Regeneration of Ionites*, Khimiya, Leningrad, 1990. (Russian)
11. K. Dorfner, Ed., *Ion Exchangers*, de Gruyter, Berlin, 1991.
12. M. Abe, T. Kataoka and T. Suzuki, Eds., *New Developments in Ion Exchange, Proc. Int. Symp. Ion Exchange, ICIE 91*), Kodansha-Elsevier, Amsterdam and Tokyo, 1991.
13. M. Streat and G. N. Takel, *J. Inorg. Nucl. Chem.*, 43:807 (1981).
14. M. Streat, *React. Polym.*, 2:79 (1984).
15. W. I. Harris, L. B. Lindy and R. S. Dixit, *React. Polym.*, 4:99 (1986).
16. F. Helfferich, *Ion Exchange*, McGraw-Hill, New York, 1962.
17. N. I. Nikolaev, Ed., *Diffusional Process in Ionites*, NIITEHIM, Moscow, 1973. (Russian)
18. F. Helfferich, *Angew. Chem.*, 68:693 (1956).
19. F. Helfferich and M. S. Plesset, *J. Chem. Phys.*, 28:418 (1958).
20. R. Schlogl and F. Helfferich, *J. Chem. Phys.*, 26:5 1957.
21. F. Helfferich, *J. Phys. Chem.*, 66:39 (1962).
22. O. P. Fedoseeva, E. P. Cherneva and N. N. Tunitsky, *Zh. Fiz. Khim.*, 33:936, 1140 (1959).
23. E. P. Cherneva, V. V. Nekrasov and N. N. Tunitsky, *Zh. Fiz. Khim.*, 30:2185 (1956).

24. R. K. Bajpai, A. K. Gupta and M. G. Rao, *Am. Inst. Ch. E. J.*, *20*:989 (1974).

25. Ju. S. Ilnitsky, *Zh. Fiz. Khim.*, *50*:2132 (1976).

26. A. I. Kalinitchev, T. D. Semenovskaya and K. V. Chmutov, in *Sorptsia and Khromatographia* (P. E. Tulupov, ed.), Nauka, Moscow, 1979, p. 144.

27. K. V. Chmutov, A. I. Kalinitchev and T. D. Semenovskaya, *Dokl. Akad. Nauk SSSR, 239*:650 (1978).

28. Y. L. Hwang and F. Helfferich, *React. Polym.*, *5*:237 (1987).

29. R. Haase, *Thermodynamics of Irreversible Processes*, Mir, Moscow, 1967. (Russian)

30. F. Helfferich, *J. Phys. Chem.*, *69*:1178 (1965).

31. A. Schwarz, J. A. Marinsky and K. S. Spiegler, *J. Phys. Chem.*, *68*:918 (1964).

32. W. Holl and H. Sontheimer, *Chem. Eng. Schi.*, *32*:755 (1977).

33. M. Ju. Hasel and V. P. Meleschko, *Theor. i Prakt. Sorbts. Proc.*, *11*:711 (1976).

34. W. Holl, *React. Polym.*, *2*:93 (1984).

35. K. B. Bischoff, *Chem. Eng. Schi.*, *18*:711 (1963).

36. P. W. Weisz and R. Goodwin, *J. Catal.*, *2*:397 (1963).

37. C. Y. Wen, *Ind. Eng. Chem.*, *60*:34 (1968).

38. P. B. Weisz, *Trans Faraday Soc.*, *63*:1801 (1967).

39. V. F. Frolov and P. G. Romankov, *Tehor. Osn Khim. Technol.*, *2*:396 (1968).

40. J. Crank, *The Mathematics of Diffusion*, Clarendon Press, Oxford, 1975.

41. N. I. Nikolaev, *Kinetics and Katalyses,9*:870 (1968). (Russian)

42. M. Nativ, S. Goldstein and G. Schmuckler, *J. Inorg. Nucl. Chem.*, *37*:1951 (1975).

43. G. Schmuckler, *React. Polym.*, *2*:103 (1984).

44. D. Petruzelli, F. Helfferich, L. Liberti, J. Millar and R. Passino, *React. Polym.*, *7*:1 (1987).

45. A. I. Kalinitchev, T. D. Semenovskaya, E. V. Kolotinskaya, A. Ya. Pronin and K. V. Chmutov, *J. Inorg. Nucl. Chem.*, *43*:787 (1981).

46. A. I. Kalinitchev, E. V. Kolotinskaya and T. D. Semenovskaya, *J. Chromatogr.*, *17*:243 (1982).

47. A. I. Kalinitchev, E. V. Kolotinskaya and T. D. Semenovskaya, *Theor. Osn. Khim. Technol.*, *17*:313 (1983).

48. A. I. Kalinitchev, E. V. Kolotinskaya and T. D. Semenovskaya, *Zh. Fiz. Khim.*, *58*:2807 (1984).

49. A. I. Kalinitchev, E. V. Kolotinskaya and T. D. Semenovskaya, *React. Polym.*, *7*:123 (1988).

50. A. I. Kalinitchev, T. D. Semenovskaya and E. V. Kolotinskaya, *Proc. Int. Symp. on Metal Speciation, Separation and Recovery*, Chicago, 1986, pp. IV-39–IV-62.

51. N. N. Matorina et al., *J. Chromatogr.*, *365*:89 (1986).

52. L. Liberti and G. Schmuckler, *Desalination 27*:253 (1978).

53. L. Liberti, D.Petruzelli, G. Boghetich and R. Passino, *React. Polym.*, *2*:111 (1984).

54. L. Liberti, D. Petruzelli and R. Passino, *Desalination*, *48*:55 (1983).

55. T. Kataoka, H. Yoshida and Y. Ozasa, *Chem. Eng. Schi.*, *32*:1237 (1977).

56. T. Kataoka, H. Yoshida and S. Ikeda, *J. Chem. Eng. Jpn.*, *11*:156 (1978).

57. F. Helfferich, L. Liberti, D. Petruzelli and R. Passino, *Isr. J. Chem.*, *26*:3 (1985).

58. J. Plicka, K. Stamberg, J. Cabicar and V. Spevakova, *React. Polym.*, *7*:141 (1988).

59. G. Kraaijeveld and H. Wessselingh, in *Ion Exchangers* (K. Dorfner, ed.), de Gruyter, Berlin, 1991, 1. 3.

60. A. E. Tchalich, A. E. Aliev and A. Rubtsov, *Electronic Zond Microanalyses in Investigations of Polymers*, Nauka, Moscow, 1990. (Russian)

61. T. Gudmen, in *Advances in Heat Transfer*, Vol. 1 (T. F. Irvine and J. P. Harnett, eds.), Academic Press., New York, 1964.

62. A. I. Kalinitchev *Theor. Osn. Khim. Technol.*, *12*:673 (1978).

63. P. P. Zolotarev and L. V. Radushkevitch, *Zh. Fiz. Khim.*, *1*:244 (1970).

64. A. I. Kalinitchev and E. V. Kolotinskaya, *Tehor. Osn.Khim. Technol.*, in press.

65. R. M. Nicoud and D. Schweich, in *Proc. Int. Conf. on Ion Exchange IEX-86*, Balatonsheplak, Hungary, 1986.

66. A. I. Kalinitchev, E. V. Kolotinskaya and T. D. Semenovskaya, *React. Polym.*, *3*:235 (1985).

67. D. Petruzelli, L. Liberti, R. Passino, F. Helfferich and Y. L. Hwang, *React. Polym.*, 5:219 (1987).

68. T. D. Semenovskaya, A. I. Kalinitchev and E. V. Kolotinskaya, in *Ion Exchange Technology* (M. Streat and D. Naden, eds.), Ellis Horwood, Chichester, 1984, p. 257.

69. T. D. Semenovskaya, V. T. Avgul, A. Ja. Pronin and K. V. Chmutov, *Zh. Fiz. Khim.*, *53*:1850 (1979).

70. F. Helfferich, *J. Phys. Chem.*, *67*:1157 (1963); *69*:1178 (1963).

71. V. Ya. Filimonov, T. D. Semenovskaya and A. I. Kalinitchev, *Zh. Fiz. Khim.*, *66*:1949 (1992).

72. T. D. Semenovskaya, L. V. Shepetjuk and A. I. Kalinitchev, *Zh. Fiz. Khim.*, *65*:795 (1991).

73. A. I. Kalinitchev, in *New Developments in Ion Exchange* (M. Abe et al., eds.), Kodansha-Elsevier; Amsterdam and Tokyo, 1991, pp. 7–12.

5

Equilibrium Analysis of Complexation in Ion Exchangers Using Spectroscopic and Distribution Methods

Hirohiko Waki

Kyushu University, Hakozaki, Higashi-ku, Fukuoka, Japan

I. INTRODUCTION

For the knowledgeable study of ionic interactions in an ion-exchanger gel phase, it should be helpful to compare the features of its internal solution with those of the simple electrolyte solution in equilibrium with it. The ion exchanger contains an organic network with a fixed ionic group (acid or base, weak or strong; chelate) repeated throughout its three-dimensional structure [1–2]. The weak or strong acid (base) defines the properties of the ion exchanger. The concentration of this fixed group is quite sizable since the matrix defined by the organic network resists the entry of solvent from the dilute aqueous electrolyte solution in equilibrium with it. This leads to a lower water activity in the exchanger phase and a higher pressure exerted on the exchanger phase.

Such lowering of water activity, the higher pressure, and the organic nature of the charged sites may lead to lower hydration of counterions than one encounters in the simple electrolyte in equilibrium with the exchanger. Furthermore, there may be a lowering in dielectric constant from $\varepsilon = 78$ due to the organic component of the ion-exchanger phase. These two factors would enhance the interaction between the fixed sites and counterion in the exchanger phase. This would be expected to lead to stronger complexation of a particular counterion by the particular functionality repeated in the exchanger phase than by the same functionality appearing together with the same counterion in simple electrolyte solution.

Steric factors in the ion-exchanger phase present a complexation depressing effect by restricting rotational and translational motion. This reduces the number of collisions leading to association. The rate of the reverse reaction, hydrolytic dissociation of associated species is, however, not necessarily decreased because of lowered or less ordered hydration of the species in the ion-exchanger phase. In any case, the kinetic aspects seem to exercise a net negative effect on the complexation tendencies. The increase in ion interaction due to the lower medium dielectric constant more than compensates for the above. As a consequence, the lowering in effective collision frequency may become important only when very weak complexes are involved.

With the weak carboxylic acid ensemble repeated throughout the cross-linked ion exchanger, e.g., Amberlite IRC-50, Sephadex CM-25, Sephadex CM-50, serving as the metal ion complexation site [3–4] it is found that the accessibility of these sites for complexation is low. Most metal ions are observed to form only the unidentate species. Because the nonideality of bound and free states is the same they cancel in the mass action expression for the unidentate complexation reaction. The small value of the nonideality term which remains uncancelled explains the scarcity of multidentate species [4]. Correction for the Donnan-based enrichment of the metal ion and its nonideality leads to resolution of an intrinsic formation constant that agrees well with the constant published for the unidentate complex of the weak acid that resembles closely the weak acid ensemble repeated in the three-dimensional ion-exchanger matrix [2, 4]. Since this molecule is repeated often enough in the cross-linked ion exchanger to make them statistically equivalent, such a resemblance in stability constants is to be expected as long as the effect of other factors (osmotic pressure, medium dielectric) are small enough to neglect.

The resemblance between metal ion complexation of the repeating chelate moiety constituting the reactive sites of a chelating resin, Dowex A-1, and the simple iminodiacetic acid molecule itself is also reasonably consistent with the above estimate of the situation [5].

There is another, more straightforward path to the comparison of the stability of complexes formed in ion exchangers with the stability of the same complexes formed in solution. For example, by employing a strong-acid or strong base exchanger for this purpose, use, as before, of the repeating functionality of the resin itself as the ligand is avoided. With the strong base exchanger the counterion, X^-, becomes the potential ligand. Unlike the earlier resin systems considered this potential ligand is free and relatively unrestricted in its motion. At equilibrium the activity of X⁻ in the resin phase designated by subscript R is related to its activity in the solution phase by a Donnan potential, DP, term.

$$a_{X_R^-} = a_{X^-}(DP) \tag{1}$$

When mixtures of MX_n and HX are equilibrated with the resin in its X⁻ ion form all diffusible species, eg, M^{n+}, X^-, $\Sigma MX_i^{(n-i)}$ and H^+ bear a similar relationship:

$$a_{M_R^{n+}} = a_{M^{n+}}(DP)^{-n} \tag{2}$$

$$\Sigma a_{MX_{i_R}^{n-i}} = \Sigma a_{MX_i}(DP)^{n-1} \tag{3}$$

$$a_{H_R^+} = a_{H^+}(DP)^{-1} \tag{4}$$

and $a_{X_{i_R}} = a_{X_i}(DP)$ as before. It is of special interest to note that at equilibrium $\beta_{MX_i} = \beta_{MX_{i_R}}$. This is demonstrable in the following manner. Consider the species MX_3^- to be formed in solution in the manner shown for this purpose

$$M^{+2} + 3X^- \Leftrightarrow MX_3^-; \tag{5}$$

then

$$\beta_{MX_3^-} = a_{MX_3^-}/(a_{M^{+2}})(a_{X^-})^3 \tag{6}$$

$$\beta_{MX_{3_R}^-} = \frac{a_{MX_{3_R}^-}}{(a_{M_R^{+2}})(a_{X_R^-})^3} = \frac{(a_{MX_3^-})(DP)}{(a_{M^{+2}})(DP)^{-2}(a_{X^-})^3(DP)^3} = \frac{a_{MX_3^-}}{(a_{M^{+2}})(a_{X^-})^3} \tag{7}$$

so that

$$\beta_{MX_{3_R}^-} = \beta_{MX_3^-} \tag{8}$$

With the strong acid exchanger, X⁻ ion, the coion in this instance, can once again function as the potential ligand through its presence in the aqueous phase as HX together with MXn. In this instance the species entering the exchanger phase are H^+, M^{n+}, $\Sigma_i MX_i^{n-i}$, and X^-. At equilibrium the following disposition of species between resin and solution occurs

$$a_{H_R^+} = a_{H^+}(DP) \tag{9}$$

$$a_{M_R^{n+}} = a_{M^{n+}}(DP)^n \tag{10}$$

$$a_{MX_{i_R}^{(n-i)}} = a_{MX_i^{(n-i)}}(DP)^{(n-i)} \tag{11}$$

$$a_{X_R^-} = a_X(DP)^{-1} \tag{12}$$

and once again $\beta_{MX_i^{(n-i)+}} = \beta_{MX_i^{(n-i)+}}$

Let n = 2, i = 1 to demonstrate this:

$$\beta_{MX_R^+}^{int} = \frac{a_{MX^+}(DP)}{(a_{M^{+2}})(DP)^2(a_{X^-})(DP)^{-1}} \tag{13}$$

and

$$\beta_{MX_R^+}^{int} = \beta_{MX^+}^{int} \tag{8}$$

Examination of complexation in such systems through spectroscopic means seems appropriate. The author believes that with such an approach a comparison of complexation properties in the solution and exchanger phases would be most useful and informative. All spectral peaks common to both phases were expected to permit a meaningful comparison of nonideality terms in the resin and solution phases to show the influence of dielectric change* in the exchanger phase introduced by the organic content of the ion-exchanger three-dimensional matrix.

In addition, it had been shown by Kraus and coworkers [6,7] that the use of ligands in this manner for the separation of trace level concentrations of metal ions led to remarkably effective separations of difficulty separable elements.

The use of spectroscopic methods to facilitate the most informative analysis of the factors involved in such selective complexation of metal ion by strong acid and strong base ion exchangers has provided an important direction of research for the author and his colleagues.

It has provided measurements that show how the difference in resin phase environment from that provided by the solution phase, in equilibrium with it, i.e., the ion exchanger matrix itself, can lead to behavior unobservable in the solution phase. Since 1960 (8,9) the measurement of absorption spectra in the ion-exchanger has been studied to make them as usefully employable as they are in solution. To achieve this objective it was determined that a true spectrum of the resin sorbed species was obtainable only through the elimination of the huge light-scattering background from the solid particle layer.

---

*Editor's note: It is believed that lowering of the dielectric of the gel phase medium would require finite dissolution of the organic matrix. Since this does not occur there is no reason to expect lowering of the dielectric of the gel phase medium.

In the course of these studies both the anion and cation exchanger were observed to yield spectral peaks not observable in the solution phase, thereby pointing to the ion exchanger matrix as the source of the differences between the two phases [10–12]. These extra peaks could only be attributed to special enhancement in the exchanger phase of one or more complexed metal ion species, $\Sigma_{MX_i^{(n-i)}}$, formed in the system. These species, greatly concentrated in the resin phase, are apparently attracted by the organic portions of the ion-exchanger matrix.* Various methods, based on metal ion distribution studies, were developed to attempt elucidation of the nature and composition of the complexed metal species bound to the exchanger matrix [11–13]. The formation constant of these complexes was sought by attempting to measure spectroscopically two or more components encountered in the resin phase-enhanced complexation reactions [14,15].

It is the above aspects of the author's research over a number of years that have influenced the emphasis of this chapter on the spectroscopic approaches developed to measure the influence of free and labile ligand, in solution or resin, on its metal ion complexation properties in the ion-exchanger phase.

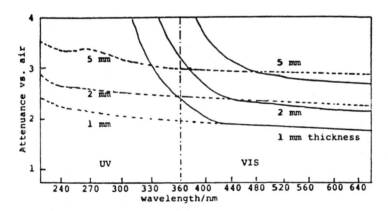

Figure 1 Background spectra of chloride form ion exchangers at different particle-layer thickness: ———— Dowex 1-X2 (100-200 mesh); – – – – – QAE-Sephadex A-25. (From Ref. 9.)

---

*Editor's note: The resin matrix serves as a membrane permeable to diffusible components. At equilibrium their chemical potentials are equal in separate phases. The fact that one or more species are observable in only the gel phase suggests that these species are immobilized by their interaction with the charged surface of the organic gel matrix.

II. ADAPTABILITY OF ABSORPTION SPECTROMETRY FOR ANALYSIS OF THE ION-EXCHANGER GEL PHASE

The use of absorption spectrometry to facilitate the analysis of complex formation between metal ions and anionic ligands in a particular ion exchanger is interfered with by the sizable light-scattering properties of the resin particles. Figure 1 shows the background scatter encountered with different ion-exchanger particle layers [16]. The very sizable attenuation is almost entirely ascribable to light scattering by solid beads, except in the ultraviolet region where absorption becomes increasingly important. The attenuation depends greatly on the material and the particle size. Ion exchangers of higher degrees of cross-linking and particles of smaller size, in general, lead to higher attenuation. The attenuation of a 1- to 2-mm-thick layer usually reaches 2–3 absorbance units. Its value is not directly proportional to layer thickness as in an ordinary absorbance measurement. Nonaromatic ion exchangers such as the Sephadex gels introduce somewhat lower scatter and offer the possibility for use even in the near ultraviolet region [17]. Correction for such attenuation has to be performed to permit meaningful resolution of the absorption spectra attributable solely to the species sorbed by the resin.

When attenuation is high, instrumental stray light as well as detector sensibility may interfere with the absorbance measurement. However, since the stray light is also scattered by the particle layer before entering the detector, only the detector sensibility becomes important.

When light-absorbing species are bound to the ion exchanger their absorbance is added to the high attenuation background. To facilitate correction for this background, an ion exchanger layer in the same condition, except for the presence of sorbed species, is employed to provide a measurement of this background. However, this corrective maneuver is still inadequate, leaving residual attenuation unaccounted for. The reason for this is that the optical path to the detector from the sample cell has been lengthened by the insertion of a mirror in an ordinary double-beam-type spectrophotometer. This leads to a greater loss of the scattered light from the sample cell than from the reference cell. To obtain the correct absorbance, an additional subtraction of the residual attenuation background is needed. Such a correction is easily made on the recording chart.

The absorption spectrum of Fe(II)-phenanthroline complex in a cation-exchanger resin obtained in this way, however, almost duplicates the pattern obtained in the aqueous solution (Fig. 2). Even though there is a possibility that the light scattering from a sample layer containing light-absorbing species is weaker than light scattering from the reference layer, Beer's law has been found to hold for the sorbed component in

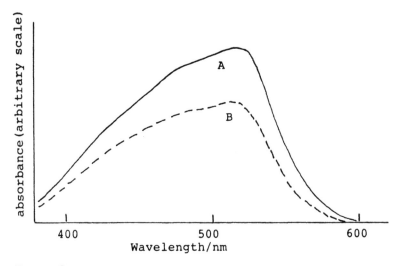

Figure 2 Comparison of ion-exchanger phase and solution phase spectral patterns obtained from the same species, the Fe(II)-ortho-phenanthroline complex (from Ref. 29): A: Fe(phen)$_3^{2+}$ sorbed in Dowex 50W-X2; B: Fe(phen)$_3^{2+}$ in aqueous solution.

most systems [18,19]. The spectral treatment described above can be considered to be satisfactory on this basis.

III.   STRONG COMPLEXATION IN ANION-EXCHANGER PHASE

One of the most remarkable and useful properties of fully dissociated strong-base anion exchangers is their unique tendency to interact selectively with metal ions that are complexed by the exchanger counterions. When metal ion is added to the electrolyte solution equilibrated with the exchanger a portion of the absorption spectrum resolved for the exchanger phase is not matched by the absorption spectrum of the solution phase even when care has been taken to keep the concentration level of counterion in the two phases the same. Advantage of this property has been taken for the development of a number of anion-exchange facilitated separations of metals [6,7,20].

A typical example of such different behavior is given in Fig. 3. The absorption spectra of the anion-exchange resin, AG-1-X8, and the 4 M hydrochloric acid solution of Co(II) used to prepare the resin sample for such a comparison are presented in this figure. It is immediately apparent that species exhibiting tetrahedral symmetry, e.g., CoCl$_3^{1-}$, or CoCl$_4^{2-}$, pre-

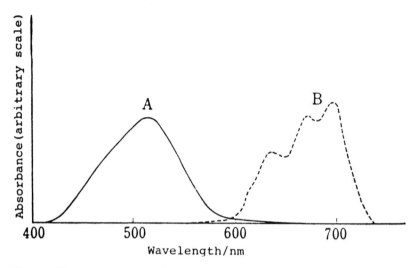

Figure 3 Comparison of the spectra for the cobalt complexes in the solution and the anion-exchanger phase in equilibrium: A: 4 M HCl solution of Co(II); B: AG 1-X8 equilibrated with solution A. (From Ref. 15.)

vail in the resin phase while octahedral symmetry prevails for the Co^{+2} and $CoCl^+$ species of the solution phase [11].

The fact that the selective uptake of metal ion by the ion exchanger is associated with highly coordinated species undetectable in the solution, even when ligand concentrations in the two phases have been kept similar, has to be attributed to either the immobilization of these species by their interaction with the organic component of the exchanger matrix or to sizable reduction of activity coefficients of ion species in the resin phase because of the lowering of the dielectric constant of the resin phase media.

Their absence in the solution phase is consistent with both explanations. If species immobilization is involved only the diffusible components of the system can be observable in both phases. The absence of the highly coordinated species in the solution phase can then be attributed to the immobilization phenomenon projected as shown.

$$Co_R^{+2} + 3Cl_R^- \rightarrow CoCl_{3_R}^-$$

$$CoCl_{3_R}^- + H_R^+ \rightarrow HCoCL_{3_R}$$

The several peaks in the absorption spectrum obtained with the AG 1-X8 resin equilibrated in a 4 M HCl solution of Co(II) (Fig 3) are consistent with the above expectation that both dissociated and undissociated species may be involved.

If nonideality enhancement in the ion exchanger phase is responsible the presence of the more highly ligand-coordinated species, the aqueous phase can be expected to be reduced to nondetectable concentration levels.

In order to examine more completely the above phenomenon, experiments have been carried out with both an anion and cation exchanger [14]. The resin phase chloride ion concentration levels were provided by equilibrating the AG 1-X4 with 1.84 M HCl and the Dowex 50 W-X4 with 5.4 M HCl, respectively. To obtain the resin phase absorption spectra for study, both resins were equilibrated with a 2.64 M HCl solution containing cobalt. The absorption spectra are essentially duplicated in the two exchangers (Fig 4). Once again the sizable presence of $CoCl_3^-$ or $CoCl_4^=$

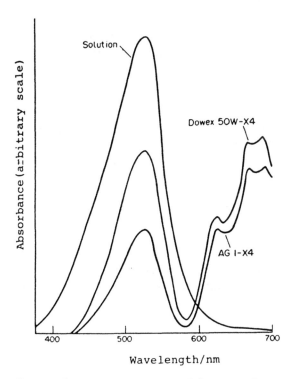

Figure 4 Absorption spectra of the ion-exchanger and solution phases for the cobalt(II)-chloride system at the same 2.64 M chloride concentration. (From Ref. 17.)

and $HCoCl_3$ or H_2CoCl_4 in the resin phase spectra is not observable in the solution absorption spectra. In this instance, however, the spectral properties of the solution phase attributable to Co^{+2} or $CoCl^+$ is detectable in the resin phase as well. The fact that the presence of Co^{+2} or $CoCl^+$ in the anion exchanger is less than it is in the cation exchanger is consistent with Donnan potential based expectations (see Equations 2 and 9). The duplication of the $HCoCl_3$ or H_2CoCl_4* and $CoCl_3^-$ or $CoCl_4^-$ spectra in both resin phases while it continues to be undetectable in both solution phases is also consistent with its earlier assignment to the combined presence of the undissociated and dissociated ensembles considered to be concentrated in the course of their interaction with the organic component of the resin matrices either by immobilization or by marked reduction of activity coefficients.[†]

Additional examples of ion-exchanger phase absorption spectra attributable to metal complexation in the resin phase are considered in subsequent sections of this chapter.

IV. HETEROGENEOUS TWO-PHASE DISTRIBUTION ANALYSIS OF COMPLEXATION IN ANION EXCHANGERS

Since 1961 [20], determination of the composition of complex species present in an anion-exchanger phase has been sought through the analysis of equilibrium distribution measurements between resins and solutions. The distribution ratio of a metal ion, M^{m+}, defined as the ratio of total concentrations of the metal in the two phases of a system containing a monovalent anion as ligand, L^-, can be represented by

$$D = \frac{[\Sigma M]_R}{[\Sigma M]} = \frac{[M]_R + [ML]_R + [ML_2]_R + \cdots}{[M] + [ML] + [ML_2] + \cdots} \qquad (14)$$

where the subscript R is used, as before, to identify the particular quantity with the ion-exchanger phase; the ionic charge of the various species is omitted. Introducing the overall stability constant, β_i, for ML_i the ith complex in the two phases as shown

$$\beta_i = \frac{[ML_i]}{[M][L]^i} \qquad (15)$$

---

*The possibility that $HCoCl_3$ or H_2CoCl_4 is present is introduced by the editor.
[†]It is difficult to accept this rationalization by the author; the lowering of the gel phase dielectric is deemed unlikely by the editor (see editor's note on p. 200).

$$\beta_{i_R} = \frac{[ML_i]_R}{[M]_R [L]_R^i} \tag{16}$$

By using the Gibbs-Donnan relationship

$$[M]_R [L]_R^m = G[M][L]^m \tag{17}$$

where

$$G = y_M y_L^m / y_{M_R} y_{L_R}^m \tag{18}$$

and y corresponds to the activity coefficient of M^{m+} and L^- in the two phases. Eqn. (14) is transformed to Eq. (19).

$$D = G \frac{[L]^m}{[L]_R^m} \cdot \frac{1 + \sum_i^R \beta_i [L]_R^i}{1 + \sum_i \beta_i [L]^i} \tag{19}$$

Differentiation of Eq. (19) then yields

$$\frac{d \log D}{d \log[L]_R} = \frac{d \log G}{d \log[L]_R} + m - \bar{n} \frac{d \log[L]}{d \log[L]_R} + \bar{n}_R - m \tag{20}$$

where $\bar{n}_R$ and $\bar{n}$ are respectively the average ligand number of complex species in the ion-exchanger and solution phases at equilibrium

$$\bar{n}_R \equiv \frac{\sum_i i[ML_i]_R}{\sum_i [ML_i]_R} \tag{21}$$

$$\bar{n} \equiv \frac{\sum_i i[ML_i]}{\sum_i [ML_i]} \tag{22}$$

In systems where the ionic strength of both phases is kept constant, G may be nearly constant. Furthermore, when the change of ligand concentration in the equilibrated solution is very much smaller than in the anion-exchanger phase, Eq. (20) can be simplified to

$$\frac{d \log(D/[L]^m)}{d \log[L]_R} = \bar{n}_R - m \tag{23}$$

or to

$$\frac{d \log D}{d \log[L]_R} = \bar{n}_R - m \tag{24}$$

With D and $[L]_R$ values experimentally accessible, log D is plotted against $\log[L]_R$ to determine $\bar{n}_R$ from the slope. In this case, of course, the assumption that $d\log[L]/d\log[L]_R$ is very small must be shown to be reasonably valid before simplified Eqs. (23) or (24) are applied. The reliability of this treatment has been confirmed for the chloro complexes of cobalt(II) and iron(III) [10]. Even if the chloro-complexed species under examination with this treatment exist as H_2CoCl_4 and $HFeCl_4$* together with $CoCl_4^=$ and $FeCl_4^-$, the fact that such species as $CoCl_4^=$ and $FeCl_4^-$ are presumed to exist in the development of Eqns. [23] and [24] implies that their use does not need to be abandoned. As long as the concentration of H^+ ion is kept constant in the course of the D measurements, the resolution of $\bar{n}_R$ is similarly valid.

For the analysis of these systems, as described, the employment of perchlorate ion as indifferent electrolyte proved most convenient because it assured sizable transfer of chloride ion from the exchanger to the solution at equilibrium. This is ascribable to the much greater selectivity of the ion exchanger for perchlorate ion. As complexation of metal ion occurs the change of chloride ion concentration in the exchanger is much larger than it is in the solution to satisfy the requirement that $\Delta L/\Delta L_R$ be small. As seen in Table 1, the great difference in the chloride ion concentration change for the two phases should make it possible to assume the $d\log[L]/d\log[L]_R$ term in Eq. (20) is small enough to permit use of Eq. (23) or (24) for the analysis.

Examples of this analysis mode are presented in Fig. 5. The slope of the log D versus $\log[\bar{L}]_R$ plots for the cobalt(II)–8 M HCl and the Fe(III)–6 M HCl systems indicate the formation of $CoCl_4^{-2}$ and $FeCl_4^-$, respectively, as the predominant complexes in the Dowex 1-X8 phase. These results are believed to prove the reliability of this approach, the presence of these complexes in concentrated hydrochloric acid being well documented. The chloride complex of uranium(IV), UCl_5^-, whose unique absorption spectrum is shown in Fig. 6 [11], in the HCl, $HClO_4$ (I = 10 M)

TABLE 1 Respective Chloride Concentrations in the Anion-Exchanger AG-X8 Phase and the Solution Phase in Equilibrium with It

(a) HCl + HClO$_4$ = 6 M					
$[Cl^-]_R$ mmol/g resin	1.00	2.05	3.60	4.34	5.36
$[Cl^-]$ M	5.86	5.85	5.84	5.85	5.86
(b) HCl + HClO$_4$ = 10 M					
$[Cl^-]_R$ mmol/g resin	2.17	4.16	6.75	8.10	
$[Cl^-]$ M	9.24	9.49	9.81	9.87	

*See first note on p. 206.

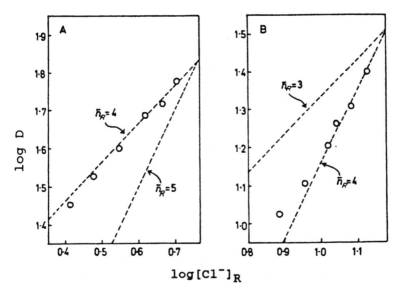

Figure 5 The slope analysis tests with standard complexes, $FeCl_4^-$ and $CoCl_4^{2-}$, present in anion-exchanger phase: A: Fe (III) HCl + $HClO_4$ (I = 6); B: Co (II) HCl + $HClO_4$ (I = 8); O experimental plots; – – – theoretical slope. (From Ref. 10.)

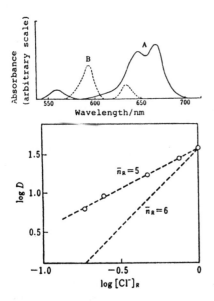

Figure 6 Absorption spectra of uranium(IV)-chloride complexes in solution and the anion exchanger in equilibrium with this solution (from Ref. 11) coordination of the resin phase complex (from Ref. 29): A: 10 M HCl solution of U(IV); B: AG 1-X8 equilibrated with solution A; O experimental plot; – – – theoretical slope; solutions for slope analysis: HCl + $HClO_4$ (I = 10).

system, was found to be the main species in the same anion-exchange resin when it was equilibrated with 10 M HCl. Whether it is a combination of the undissociated protonated species and the negatively charged anionic species is not determinable with this approach.

Even in systems where the complexes formed are weak, the ligand number of the complexed species found in the anion-exchanger phase is high. In spite of their low stability nitrate complexes of mercury(II), silver, and utanyl ions in nitrate solutions, e.g., $Hg(NO_3)_3^-$, $Hg(NO_3)_4^{2-}$ [13], $Ag(NO_3)_2^-$ [21], $UO_2(NO_3)_3^-$, and $UO_2(NO_3)_4^{2-}$ [22], are found as new species in Dowex 1X8 when equilibrated with nitrate solutions.

In water-organic solvent mixtures, even higher complexes may be formed in the anion exchanger phase. The presence of $Ag(NO_3)_3^{2-}$ [23], $Be(NO_3)_4^{2-}$ [24], $Mg(NO_3)_4^{2-}$, and $Ca(NO_3)_5^{4-}$ [25], etc., has been reported in the anion-exchange resin phase upon equilibration with solutions containing a high percentage of alcohol. For example, a reaction such as

$$Mg(NO_3)_3^- + NO_3^- = Mg(NO_3)_4^{2-}$$

is conceivable in the lower dielectric constant medium, provided by the charged surface of the gel phase,* since the attraction between the ligand anion and the central metal cation can be shown to overcome the repulsion between the anions in electrostatic energy calculations employing an appropriate geometrical model. Even though these complexes are not spectroscopically detectable, their presence in the anion-exchanger phase should be deducible with the distribution analysis whose applicability has been confirmed spectroscopically.

V. THREE-PHASE DISTRIBUTION ANALYSIS FOR COMPLEXATION IN ANION EXCHANGERS

When the ligand ion is not selectively removed from the anion exchanger by the anion of the supporting electrolyte used to define the constant ionic strength of the system, $d \log[L]/d \log[L]_R$ may become too large to permit the use of Eqs. (23) or (24) for species analysis. Direct application of the original Eq. (20) may be difficult, as well, since the values of $\bar{n}_R$ are often obscured. In this case, the three-phase distribution technique using anion and cation exchangers at the same time may be convenient [12]. The presumption of the absence of highly coordinated species in the cation exchanger because of their negative charge, that has to be employed with this approach, is a potential source of some error. In this

---

*Surface sorption concept introduced by the editor.

method the same solution is employed for the equilibration with anion and cation exchangers, though the equilibration of three phases in a single batch is not always necessary. After determining the quantities of metal sorbed by the two ion exchangers at equilibrium, the coordination number of the complex species present in the anion exchanger can be determined as follows: the distribution ratios for anion exchanger (AR) and cation exchanger (CR) are defined by

$$D_{AR} = \frac{[M]_{AR} + [ML]_{AR} + [ML_2]_{AR} + \cdots}{[M] + [ML] + [ML_2] + \cdots} \tag{25}$$

$$D_{CR} = \frac{[M]_{CR}}{[M] + [ML] + [ML_2] + \cdots} \tag{26}$$

For Eq. (26) only free metal ion is assumed to enter the cation exchanger phase. This assumption is acceptable only when Donnan invasion of negatively charged species is sufficiently low. With the two distribution ratios made available from a common equilibrating solution their ratio is examined to provide the following alternate path to species identification:

$$\frac{D_{AR}}{D_{CR}} = \frac{[M]_{AR}}{[M]_{CR}}\left(1 + \sum_{I}^{AR} (\beta_I)[L]_{AR}\right) \tag{27}$$

Employment of the Donnan relationship in the presence of the supporting electrolyte, B^+A^-, used to provide a constant ionic strength, leads to the derivation of the above equation and those that follow:

$$\frac{[M]_{AR}}{[M]_{CR}} = G'\left(\frac{[B][L]}{[B]_{CR}[L]_{AR}}\right)^m \tag{28}$$

where

$$G' = \frac{CRy_M}{ARy_M}\left(\frac{y_B \cdot y_L}{CRy_B \cdot ARy_L}\right)^m \tag{29}$$

Since $[B]/[B]_{CR}$ is constant, Eq. (27) can be rewritten in the form

$$\frac{D_{AR}}{D_{CR}[L]^m} = \frac{1 + \Sigma_i^{AR}\beta_i[L]_{AR}^i}{[L]_{AR}^m} \times \text{const.} \tag{30}$$

Differentiation of the logarithm of this equation gives

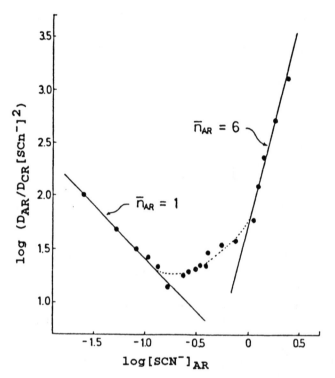

Figure 7 Coordination of nickel(II)-thiocyanate complexes forms in anion-exchange resin phase by three phase distribution method: Resin: Dowex 1-X4 (50–100 mesh, SCN⁻ form), Dowex 50W-X4 (50–100 mesh, Na⁺ form); Solution: Ni(II)–NaSCN–NaClO₄ (I=1.0); ● experimental plot; ——— theoretical slope. (From Ref. 12.)

$$\frac{d \log(D_{AR}/D_{CR}[L]^m)}{d \log[L]_{AR}} = \bar{n}_{AR} - m \tag{31}$$

and the average ligand number of complex species in the anion exchanger can be determined from the slope of a plot of $\log(D_{AR}/D_{CR} \cdot [L]^m)$ versus $\log[L]_{AR}$.

An example of the use of this approach is shown [12] for the nickel-thiocyanate, Dowex-50W-X4, and Dowex 1-X4 system (Fig. 7). In the lowest SCN⁻ concentration range, the slope is –1 and NiSCN⁺ is identified; in the middle SCN⁻ concentration range the slope increases from negative to positive and finally with further increase of SCN⁻ concentration it reaches +4 to signal the predominance of the fully coordinated Ni(SCN)$_6^{4-}$ complex in the anion-exchanger phase. The absorption spectrum of this last complex is shown in Fig. 8. It has not yet been observed in concen-

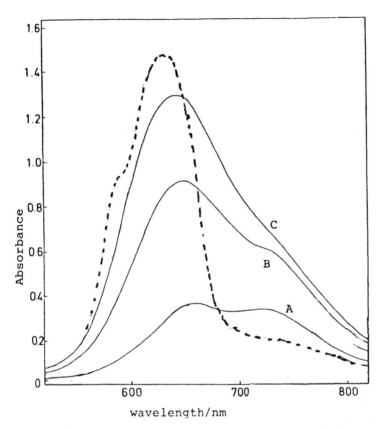

Figure 8 Absorption spectra of anion exchanger and solution containing nickel(II)-thiocyanate complexes: solution spectrum (10 mm cell); A: 0.01 M NH$_4$ SCN + Ni(SCN)$_2$; B: 4 M NH$_4$SCN + Ni(SCN)$_2$; C: 8 M NH$_4$SCN + Ni(SCN)$_2$; – – – – Dowex 1-X4 (50-100 mesh, SCN form) spectrum obtained after equilibration with solution B. (From Ref. 12.)

trated thiocyanate solutions. The same type of analysis has also been carried out with the copper(II)-bromide ion-exchanger systems [12].

VI. COMPLEXATION AT HIGH LOADING IN ANION-EXCHANGER PHASE

In spite of uncertainties with respect to the assessment of ion activity coefficients in ion-exchanger phases, it has been experimentally demonstrated that use of distribution measurements in Eq. (20) promotes valid identification of the monomeric complex species absorbed by the anion exchanger phase. A different distribution approach has been attempted at high metal loading where the formation of polynuclear complexes is expected to complicate the ion species assessments. It has been reported, for example, that the absorption spectrum of an anion-exchanger phase sometimes varies with increasing load of the metal to support this estimate. The fact that an anion-exchanger sorbing both metal ion and ligand anion in excess may be considered an effective medium for forming polynuclear complexes seems quite reasonable. Figure 9 shows the change in

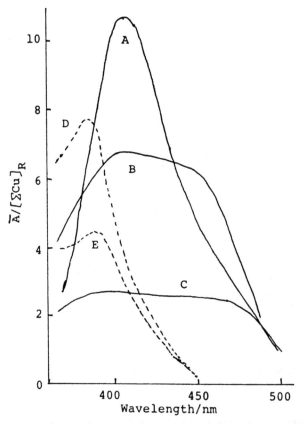

Figure 9 Absorption spectra for copper(II) retained in anion-exchange resin at different loads: ——— resin phase spectrum ($\overline{A}$: absorbance); resin: AG 1-X8 (100–200 mesh, Cl⁻ form); solution: $CuCl_2$ + 6 M HCl; cell thickness: 0.5 mm; load: A, 0.02; B, 0.063; C, 0.26 mmol Cu/g resin ($[\Sigma Cu]_R$); – – – solution spectrum for copper (in arbitrary scale) D: 11.5 M; E: 6 M HCl. (From Ref. 10.)

the absorption spectrum of the resin phase when Dowex 1-X8 is equilibrated with hydrochloric acid solutions containing copper at different
concentration levels. At high copper loading, spectral peak broadening
and peak expansion to higher wavelengths appear, the spectral pattern
being different from that at low loading and also from that observed with
any of the aqueous solutions [10].

In order to explain this change in spectral behavior, Waki et al. have
proposed the following method for studying the polymerization phenomenon presumed to prevail at high loading in the anion-exchanger [10]. By
assuming the polynuclear complex formed from M^{m+} and L^- to be $M_pL^{i-}_{mp+i}$,
the only species in the anion exchanger, and by assigning only monomeric complexes to the solution in equilibrium with the exchanger, the
following equation can be derived:

$$D = [\Sigma M]^{p-1} \left\{ [\Sigma L]_R - \left(m + \frac{i}{p} \right)[\Sigma M]_R \right\}^i \times \text{const.} \tag{32}$$

In this equation i and p are unknown, while $[\Sigma M]$, $[\Sigma M]_R$, and $[\Sigma L]_R$ are
measurable. In determining i and p with such an equation, the use of a
slope-fitting approach, where the slope for an experimental curve of log
D versus $\log[\Sigma M]_R$ is compared with slopes of reference curves constructed from assumed values for p and i, and measured values of $[\Sigma L]_R$,
may be most convenient.

In testing the applicability of this graphical approach for identifying
polynuclear species, chlorocomplexes of iron(III) and cobalt(II), whose
compositions are known, were employed. The variation of log D with
$\log[\Sigma M]_R$ in the iron and AG 1–4 M HCl system, and in the cobalt and AG
1–8 M HCl system, corresponded to the curve constructed using p = 1,
i = 1 ($FeCl_4^-$) and p = 1, i = 2 ($CoCl_4^{2-}$), respectively (Fig. 10).

Even though a valid test of the applicability of this approach is not
possible because independent confirmation of polynuclear complex formation at equilibrium has not been made, the success achieved with standard monomeric species by the use of such an approach is believed to
support the validity of the proposed method at high loading.

The results from the above analysis of the copper-chloride system
in question are shown in Fig. 11. As seen in the figure, the copper-
chloride complex at high loading in the anion-exchanger phase appears
to change from monomeric $CuCl_3^-$ to dimeric $Cu_2Cl_7^{3-}$ with increase in the
degree of cross-linking in Dowex 1 resin. The binuclear species which may
have the structure

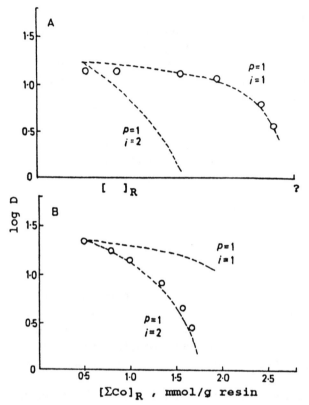

Figure 10 The slope-fitting analysis tests with standard complexes, FeCl$_4^-$ and CoCl$_4^{2-}$, present in anion-exchanger phase: A: FeCl$_3$ + 4 M HCl; B: CoCl$_2$ + 8 M HCl; ○ experimental plots; – – – theoretical slope. (From Ref. 10.)

$$\begin{bmatrix} & Cl & & Cl & \\ & | & & | & \\ Cl- & Cu & -Cl- & Cu & -Cl \\ & | & & | & \\ & Cl & & Cl & \end{bmatrix}^{3-}$$

has not been reported in ordinary aqueous chloride solutions. It appears reasonable, on the basis of the above, to conclude that such resin phase promotion of polymerization is possible in other L$^-$ complexed metal ion, anion-exchanger systems. Just as before, one can attribute the absence of detectable quantities of the corresponding polymeric species in the solution phase to either their immobilization by the ion-exchanger matrix or to their depressed activity coefficient values.*

*See Editor's note on pp. 200 and 206.

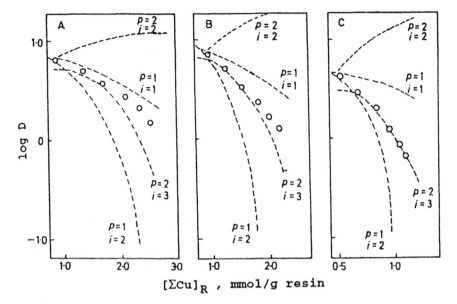

Figure 11 Slope-fitting analysis for estimate of the composition of copper(II)-chloride complexes in the anion-exchanger phase at high loading: resin (50–100 mesh, Cl$^-$ form): A Dowex 1-X1, B AG 1-X8; C Dowex 1-X16; solution: CuCl$_2$ + 6 M HCl. (From Ref. 10.)

VII. COMPLEX FORMATION OF METAL IONS IN CHELATE-FORMING RESINS

In the foregoing sections the anion-exchanger phase has been described as a complexation enhancing medium for metal ions and ligand anions entering from electrolyte solutions equilibrated with it. It has also been of interest to the author of this chapter to determine whether the conclusions gained in earlier studies of the direct complexation of metal ions by the fixed ligand anions, such as the repeating iminodiacetate groups of a chelating resin or the repeating monocarboxylate groups of a weak-acid-type cation exchanger, are consistent with their spectral properties. This author believes that the ligand to ligand distances in the chelating resin are sufficiently small to permit interligand chelation. The coexistence of ML, ML$_2$, or even ML$_3$ is considered possible from a statistical point of view even though earlier studies (1968) of the complexation of Ni^{+2}, Co^{+2}, Zn^{+2} and Cu^{+2} by the Dowex A-1 chelating resin [5] have indicated that this is not the case. When compared with the complexation pattern developed in parallel studies of the binding of these elements by benzyl iminodiacetic acid (BIDA), the functionality repeated throughout the three-dimensional matrix of the Dowex A-1 gel, it was observed that only the first of the two chelated species (MBIDA and M(BIDA)$_2$) that occur simultaneously with the simple chelating molecule in solution was encountered with the ion exchanger. The respective stability constants, $pK_{M(BIDA)}$ resolved for the singly-coordinated species in these parallel studies were in reasonably good agreement as well [5,26].

Fig. 12 shows the absorption spectrum of an iminodiacetate (IDA) chelating resin, Dowex A-1, reacted with nickel(II) ions. For comparison the spectra of Ni(IDA) and Ni(IDA)$_2^{-2}$ species in aqueous solutions are also given in the figure. The absorption band of nickel coordinated to the resin iminodiacetate groups appears in a position between those for the Ni(IDA) and Ni(IDA)$_2$ complexes measured in solution. The fact that Ni(IDA)$_2$ may also have been formed, but in smaller quantities than Ni(IDA) on the basis of the peak position being closer to that observed for Ni(IDA), is consistent with the author's expectations. The fact that the sole presence of Ni(IDA)$_2$ did not prevail as it would have with the simple iminodiacetic acid molecule in solution at the concentration level encountered in the chelating resin, however, continues to support the concept of restricted ligand acceptability in the crosslinked exchanger [2–4].

In the case of cobalt(II), the spectral band, A, of the A-1 resin-containing cobalt complex is broader than expected. This broadening could be attributed to a composite of peaks due to Co(IDA)$_2$, D, and Co $\cdot$ nH$_2$O^{+2}, also presented in Fig 13. Since the presence of Co(IDA)$_2$ is less likely than the presence of Co(IDA) in the resin phase, the spectrum observed has been attributed to distortion introduced by the coexistence of octahedral and tetrahedral configurations of the complex that depends on the localized steric environments provided by the resin network. The observation that the spectral pattern for a thicker sample layer (2 mm) is considerably different from that for a thin layer (1 mm) (Fig. 13) is very unusual. In the thick layer sample, the absorbance at lower wavelengths (500 nm) decreased, while it increased at higher wavelengths (600 nm). This unusual behavior has been tentatively attributed to an internal photoreaction by which the intermediate complexes with incomplete ligand numbers(II) between octahedral(I) and tetrahedral(III) conformations are excited to an ordinary bridged tetrahedral structure(III) by the irradiation at approximately 500 nm. Such a transformation may be almost instantaneous and reversible, occurring only at the moment of irradiation. In a thick layer, the irradiation becomes locally more intensive at the surface, part of the particles being diverted by multiple reflections. This leads to more effective photochemical transformation and a decrease in unstable complexes (~500 nm). Multiple reflections in the thicker layer, on the

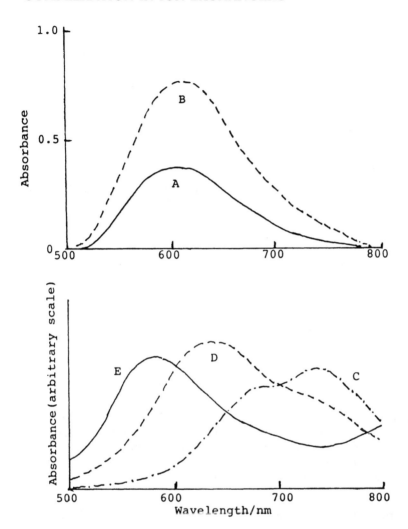

Figure 12 Comparison of absorption spectra for nickel(II) in the IDA chelating A-1 resin phase and in the solution: A: Muromac A-1 (100–200 mesh, Na$^+$ form) spectrum at 26% Ni loading and external pH 6.1 with 1-mm cell thickness; B: same as A with 2-mm cell thickness; C: solution spectrum for Ni^{2+} aqua ion; D: solution spectrum for Ni(IDA) complex; E: solution spectrum for Ni(IDA)$_2^{2-}$ complex.

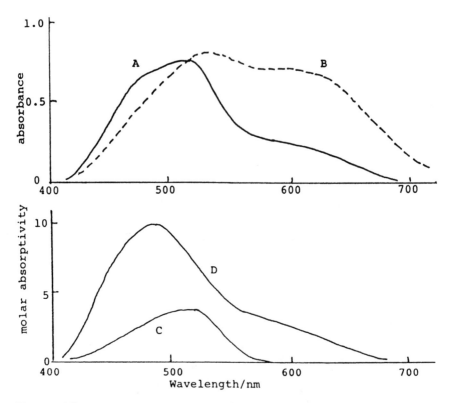

Figure 13 Comparison of absorption spectra for cobalt(II) in the IDA chelating resin A-1 phase and in the solution: A: Muromac A-1 (100–200 mesh, sodium
form) spectrum at 27% Co loading and external pH 6.0 with 1-mm cell thickness;
B: same as A with 2-mm cell thickness; C: solution spectrum for Co^{2+} aqua ion;
D: solution spectrum for $Co(IDA)_2^{2-}$ complex.

other hand, may produce an increase in the apparent absorptivity (~600
nm) of the stable bridged complex, through the extended light path
affected.

For such coordination behavior the distance between fixed ligand groups and steric disturbance of the coordinated atom are important factors. Thus "Chelex" resin of the lowest crosslinking or monocarboxylate "CCR-2" resins promote a spectral pattern similar to that of ordinary solutions and do not exhibit unique spectral behavior by irradiation, because of the long distance between two adjacent fixed ligands in the Chelex resin and the simple ligand structure in the (CCR-2) resin, respectively.

VIII. EVALUATION OF STABILITY CONSTANTS OF METAL COMPLEXES IN THE ION-EXCHANGER PHASE USING A SPECTROPHOTOMETRIC METHOD

In order to examine more closely than before the difference in complexibility between the ion-exchanger phase and ordinary solution, the stability constants of metal complexes in the ion-exchanger phase have been evaluated directly by using spectroscopic methods such as absorption spectrophotometry [14] and NMR spectrometry [15]. As mentioned earlier, the stability constants of metal complex with the fixed ligand of an ion exchanger had been shown to be comparable in magnitude to those with the corresponding monomeric ligand in ordinary solution [2–5]. However, it was felt by the author of this chapter that a more knowledgable assessment of ion-exchanger matrix effect on the internal solution of the ion exchanger than existed was desirable. It was decided to investigate complex formation between a metal ion and free ligand anion from the external solution to eliminate direct involvement of the matrix as the fixed ligand source. By such elimination of the direct involvement of the exchanger matrix contribution of the organic component of the resin matrix to the dielectric properties of the internal solution medium was expected to become available.

The use of a strong acid type cation exchanger without involving the matrix site as the fixed ligand was felt to be suitable for this purpose. Since the complexes involved in such systems are usually restricted to one or two species of low ligand number, spectral analysis was expected to be rather simple.

To facilitate the program it was decided to study the cobalt(II)-thiocyanate complex system which was as accessible to direct assay in the exchanger phase as it was in the solution phase by absorption spectrophotometry in the visible region. Both Co(II) and $Co(SCN)^+$ exhibit sizably different spectral absorption properties in the visible region. Assuming that the presence of higher complexes than one to one in the cation exchanger is negligible, the net absorbance $\overline{A}$ for the Co^{2+}–$CoSCN^+$ mixture, after cancellation of background attenuation, is represented by

$$\overline{A}/\overline{l} = \overline{\varepsilon}_0[\overline{Co^{2+}}] + \overline{\varepsilon}[\overline{CoSCN^+}] \tag{33}$$

where the bar refers to the cation-exchanger phase, $\overline{l}$ is light path length, and the ε's correspond to the molar absorptivity of the corresponding species. The apparent molar absorptivity $\overline{\varepsilon}$ for cobalt becomes

$$\overline{\varepsilon} \equiv \frac{\overline{A}}{\overline{C}_M\overline{l}} = \frac{\overline{\varepsilon}_0 + \overline{\varepsilon}_1\overline{K}_1[\overline{SCN^-}]}{1 + \overline{K}_1[\overline{SCN^-}]} \tag{34}$$

where $\overline{C}$ is the total concentration of cobalt in the ion-exchanger phase, and $\overline{K}_1$ the stability constant of the $CoSCN^+$ complex.

$$\overline{K}_1 = \frac{[\overline{CoSCN^+}]}{[\overline{Co^{2+}}][\overline{SCN^-}]} \tag{35}$$

Equation (34) is transformed to

$$\overline{\phi} \equiv \frac{\overline{\varepsilon} - \overline{\varepsilon}_0}{[\overline{SCN^-}]} = \overline{K}_1\overline{\varepsilon}_1 - \overline{K}_1\overline{\varepsilon} \tag{36}$$

The thiocyanate concentration in the ion-exchanger phase is provided by the Donnan relation

$$[\overline{SCN^-}] = G \cdot \frac{[Na^+]}{[\overline{NA^+}]} \cdot [SCN^-] \approx k[SCN^-] \tag{37}$$

Since bulk electrolyte in the solution phase is adjusted to yield an ionic strength equal to the ionic strength of the gel phase in equilibrium with it the $[Na^+]$ is essentially constant over the concentration range of SCN^- employed in the experimental program. The activity coefficient term, G, is as a consequence nearly constant as well and $[\overline{SCN^-}]$ is calculable through the known quantities k and $[SCN^-]$. Finally, $\overline{K}_1$ can be determined from the slope of the $\overline{\phi}, \overline{\varepsilon}$ experimental plot.

Such an analysis of data compiled with different cation exchangers is presented in Fig. 14 [14]. For comparison results of analysis of complexes in the solution phase by the conventional method are also shown in the same figure. The stability constant values resolved in this manner are listed in Table 2. The results that were obtained indicate that the K_1 values in the cation-exchanger phase tend to be larger by about a factor of 2 than those in the corresponding solution. The K_1 value measured in the ion exchanger with the highest degree of cross-linking, however, was

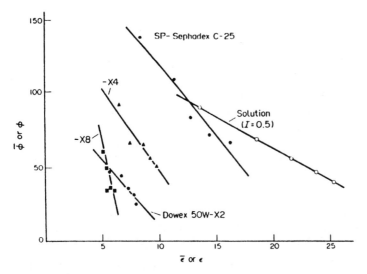

Figure 14 Determination of stability constants for the CoSCN$^+$ complex in cation-exchanger Dowex 50W phase and in the aqueous solution at 25°C with the spectrophotometric method. (From Ref. 14.)

raised another factor of 3 while measuring only 10% higher in the solution phase at the same ionic strength.

The fact that the concentration terms employed in calculating the K_1 values associated with the resin phase are not the same as those employed in the solution phase may be responsible for at least a part of the discrepancy observed between them. In computing K_1, values concentrations in the resin phase were based on 1 ml of resin whereas K_1 values in the

TABLE 2 Stability Constants of the CoSCN$^+$ Complex in Cation-Exchanger Phases and the Corresponding Aqueous Solutions, Obtained by Spectrophotometric Method

Ion exchanger	$\bar{I}$	$\bar{K}_1$	I	K_1
SP-Sephadex C-25	0.42	9.6	0.5	5.3
Dowex 50W-X2	1.7	8.9	1.7	5.3
Dowex 50W-X4	2.0	11	2.0	5.2
Dowex 50W-X8	3.4	28	3.4	6.0

$\bar{I}$ = ionic strength of ion-exchanger phase.
I = the corresponding solution ionic strength.
$\bar{K}_1$ = stability constant for ion-exchanger phase.
K_1 = stability constant for solution.
Source: From Ref. 14.

solution phase were based on 1 ml of solution. Since the resin phase solution volume in 1 ml of resin is a fraction of this quantity and becomes a smaller fraction as the degree of cross-linking increases and since the dimensions of $\overline{K}_1$ should be (millimoles × ml gel phase solution)$^{-1}$, just as they are for the solution phase for a more meaningful comparison the numbers obtained are increasingly larger with increasing cross-linking than they would have been if compared on an equivalent basis. In addition activity coefficients, instead of remaining constant, as presumed, would have increased with the resultant increase in ionic strength as the degree of cross-linking rose from 2% to 8% to tend to level the $\overline{K}_1$ values even further.*

IX. EVALUATION OF STABILITY CONSTANTS OF METAL COMPLEXES IN THE ION-EXCHANGER PHASE USING NMR SPECTROSCOPIC METHOD

Another approach to the assessment of ion speciation in the resin and solution phases with equal facility for the determination of formation constants of complexes in these two environments is provided by NMR spectroscopy. The method can provide sharp, well-separated signals for each successive complex species formed, the peak area being proportional to the atomic concentration of the element being measured, irrespective of its complex form. For a completely labile system, a well-defined chemical shift change is often observed with successive complex formation. These features make this analytical procedure well suited for study of metal ion complexation by the ion-exchanger phase.

Oxoanion complexes of aluminum are suitable candidates for study using NMR spectroscopy, because of their moderate ligand exchange rate [27]. Figure 15 is presented to show a ^{31}P NMR spectrum of the aluminum-phosphinate complex system. The ^{31}P NMR signal for free phosphinate anion, $PH_2O_2^-$, in the cation-exchanger phase is separated from its signal for this species in the equilibrated solution as well as from the signal for the $AlPH_2O_2^{2+}$ complex in the cation-exchanger phase [28]. With the availability of data, the concentrations of $PH_2O_2^-$ and $AlPH_2O_2^{2+}$ are calculable directly from their corresponding peak areas using a $PH_2O_2^-$ calibration curve. The other quantity, the free-aluminum-ion concentration $[\overline{Al}^{3+}]$ can be calculated by subtracting $[AlPH_2O_2^{2+}]$ from the total aluminum concentration in the cation-exchanger phase which is chemically analyzable. The stability constant is then accessible with Eq. (38):

$$\overline{K}_1 = \frac{[\overline{AlPH_2O_2^{2+}}]}{[\overline{Al}^{3+}][\overline{PH_2O_2^{-1}}]} \tag{38}$$

---

*Editor's assessment of discrepancy between K_1 and $\overline{K}_1$ values.

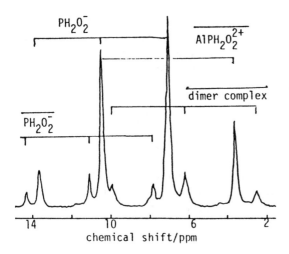

Figure 15 ^{31}P NMR spectrum for aluminum-phosphinate complexes retained in cation-exchanger Muromac 50W-X4 equilibrated with 0.016 M Al $(NO_3)_3$–0.04 M $NaPH_2O_2$ solution (pH 3). Overbar refers to the resin phase. (From Ref. 28).

Here the concentration in the solid phase is expressed as mmol/total volume of solid phase (mL) thereby including the organic skeleton as before. The $\overline{K}_1$ values so obtained are given in Table 3, together with the corresponding K_1 values in the solution phase. In this instance, the resin phase $\overline{K}_1$ values are larger by only 20% to 30%.

In the case of much more labile complex systems, the NMR spectrum yields a single peak at the position which corresponds to the weighted average of the chemical shift values characterizing each successive complex and their individual abundances at equilibrium. The cadmium-phosphinate system is such a labile complex system. It has been studied by monitoring the shift of the single peak formed, the coordination be-

TABLE 3 Stability Constants of the $AlPH_2O_2^{2+}$ Complex in Cation-Exchanger Phases and the Corresponding Aqueous Solution Phases, Obtained by NMR Method

Ion exchanger	$\overline{K}_1$	$\overline{I} = I$	K_1
Muromac 50W-X4	150	2.3	110
Muromac 50W-X8	190	3.4	160

Symbols are the same as in Table 2.
Source: From Ref. 29.

havior being considered simple enough to use this approach for such assessment of the NMR data [15]. Since the complexes are fairly weak, only the one-to-one complex was assumed to form at low phosphinate concentrations. On this basis the ^{31}P chemical shift observed for the cation exchanger phase was expressed as

$$\bar{\delta}_{obs} = \frac{\bar{\delta}_0[\overline{PH_2O_2}] + \bar{\delta}_1[\overline{CdPH_2O_2^+}]}{[\overline{PH_2O_2^-}] + [\overline{CdPH_2O_2^+}]} = \frac{\bar{\delta}_0 + \bar{\delta}_1\bar{K}_1[\overline{Cd^{2+}}]}{1 + \bar{K}_1[\overline{Cd^{2+}}]} \qquad (39)$$

where $\bar{\delta}_0$ and $\bar{\delta}_1$ are the characteristic chemical shifts for $PH_2O_2^-$ and $CdPH_2O_2^+$ in the ion-exchanger phase, respectively. For the aqueous solution, a similar equation without bars was employed. Equation (39) transforms into

$$\frac{\bar{\delta}_{obs} - \bar{\delta}_0}{[\overline{Cd^{2+}}]} = -\bar{K}_1\bar{\delta}_{obs} + \bar{\delta}_1\bar{K}_1 \qquad (40)$$

The stability constant of the $CdPH_2O_2^+$ complex, $\bar{K}_1$, is then determinable from the slope of the $(\bar{\delta}_{obs} - \bar{\delta}_0)/[\overline{Cd^{2+}}]$ versus $\bar{\delta}_{obs}$ plot. Since the free cadmium ion concentration in the cation-exchanger phase is initially unknown, the total cadmium concentration was first substituted for $[\overline{Cd^{2+}}]$ in Eq. (40) to obtain an approximate value for $\bar{K}_1$. The cadmium ion concentration calculated using the approximate $\bar{K}_1$, value was then approached by means of successive approximations. The final plots are presented in Fig. 16 together with the plots resolved for the solution phase in equilibrium with the resin phase in the three ionic strength experiments conducted. The stability constants obtained are compared in Table 4 [15]. The values for the cation-exchanger phase were in all cases

TABLE 4 Stability Constants of the $CdPH_2O_2^+$ Complex in Cation-Exchanger Phases and the Corresponding Aqueous Solution Phases, Obtained by NMR Method

Ion exchanger	$\bar{K}_1$	$\bar{I} = I$	K_1
AG 50W-X2	1.0	1.2	6.8
AG 50W-X4	1.1	2.0	6.2
AG 50W-X8	2.0	2.9	7.4

Symbols are the same as in Table 2.
Source: From Ref. 15.

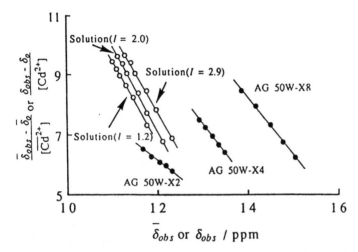

Figure 16 Determination of stability constants for the $CdPH_2O_2^+$ complex in cation–exchanger Muromac AG 50W, phases and in the aqueous solution phases with NMR method. (From Ref. 15).

considerably lower than those for the corresponding solutions, a result contrary to the one obtained in the various systems described earlier.

This stability reversal may be attributable to ion interaction depressing factors promoted by restricted ionic motion in the exchanger phase and the confinement of ligand and complex by the charged gel matrix geometry. The restricted motion of the polarized PH2O2– ligand appears to be sufficient to produce a net negative effect on the complexation reaction

X. CONCLUSIONS

It is believed that the application of spectroscopic methods as described have been useful for providing a better understanding of the various factors influencing ionic interactions in the anion-exchanger phase. Of special importance has been the identification of highly ligand coordinated metal species in anion exchangers while absent in the equivalent solution phases equilibrated with them. New insight with respect to the role played by the organic component of anion exchangers on the enhancement of metal ion separations affected through the use of the anion exchanger counterions as the complexing ligand in effecting these separations has been gained.

REFERENCES

1. F. Helfferich, *Ion Exchange*, McGraw-Hill, New York, 1962.
2. J. A. Marinsky, in *Ion Exchange and Solvent Extraction* (J. A. Marinsky and Y. Marcus, eds.), Dekker, New York, Vol 11, Chap. 5 1993 237–334.
3. J. A. Marinsky, T. Miyajima, E. Högfeldt and M. Muhammed, React. Polym., *11*: 279 (1989).
4. J. A. Marinsky, *J. Phys. Chem., 86*: 3318 (1982).
5. C. Eger, J. A. Marinsky and W. M. Anspach, *J. Inorg. Nucl. Chem., 30*: 1911 (1968).
6. K. A. Kraus and G. E. Moore, *J. Am. Chem. Soc., 75*: 1460 (1953).
7. F. Nelson, R. A. Day, Jr., and K. A. Kraus, *J. Inorg. Nucl. Chem., 15*: 140 (1960).
8. J. L. Ryan, *J. Phys. Chem., 64*: 1375 (1960).
9. C. Heitner-Wirguin and R. Cohen, *J. Phys. Chem., 71*: 2556 (1967).
10. H. Waki, S. Takahashi and S. Ohashi, *J. Inorg. Nucl. Chem., 35*: 1259 (1973).
11. H. Waki, *Kagaku no Ryoiki, 27*: 65 (1973).
12. K. Yoshimura, H Waki and S. Ohashi, *J. Inorg. Nucl. Chem., 39*: 1697 (1977).
13. H. Waki, *Bull. Chem. Soc. Jpn., 34*: 829 (1961).
14. H. Waki and Y. Miyazaki, *Polyhedron, 7*: 859 (1989).
15. Y. Miyazaki and H. Waki, *Polyhedron, 11*: 3031 (1992).
16. H. Waki, in "Ion Exchange Technology," Proc. of IEX '84, (D. Naden and M. Streat eds.), Ellis Horwood, Chichester, England 1984, p 595.
17. H. Waki and J. Korkisch, *Talanta, 30*: 95 (1983).
18. K. Yoshimura, H. Waki and S. Ohashi, *Talanta, 23*: 449 (1976).
19. K. Yoshimura and H. Waki, *Talanta, 32*: 345 (1985).
20. J. Yoshimura and H. Waki, *Bull. Chem. Soc. Jpn.: 35*, 416 (1962).
21. H. Waki, *Bull. Chem. Soc. Jpn.*: 1842 (1961).
22. J. Yoshimura, H. Waki and S. Tashiro, *Bull. Chem. Soc. Jpn., 35*: 412 (1962).
23. J. S. Fritz and H. Waki, *J. Inorg. Nucl. Chem., 26*: 865 (1964).
24. K. Kurokawa, *J. Chem. Soc. Jpn., 89*: 1076 (1966).
25. H. Waki and J. S. Fritz, *J. Inorg. Nucl. Chem., 28*: 577 (1966).
26. L. G. Sillén and A. E. Martell (Eds.), *Stability Constants of Metal-Ion Complexes,* Special Publication No. 17, London, 1964.
27. Q. Feng and H. Waki, *Polyhedron, 10*: 659 (1991).
28. H. Waki and G. Miyazaki in "New Developments in Ion Exchange", Proc. ICIE '91 (M. Abe, T. Kataoka and T. Suzuki eds., Kodanska and Elsevier, Tokyo, 1991 p. 35.
29. H. Waki in *Chelate Chemistry* Vol 3, (K. Ueno, ed.) Nankoko, Tokyo, 1977.

6

Ion-Exchange Kinetics in Heterogeneous Systems

K. Bunzl

*GSF-Forschungszentrum für Umwelt und Gesundheit,
Institut für Strahlenschutz, Neuherberg, Germany*

I. INTRODUCTION

A detailed knowledge of the kinetics of ion exchange is important not only for economic employment of ion exchange materials in the laboratory and in industry but also for a better understanding of these processes in natural systems, as e.g., in the soil or in biological membranes. For this reason numerous theoretical and experimental investigations on this subject are available. For excellent review articles see e.g., [1–10].

Usually, ion-exchanger materials are employed, which are homogeneous with respect to quantities which affect the rate of ion exchange (as, e.g., the particle size), are selected for such investigations to avoid complicating the derivation and validation of rate equations unnecessarily. Nevertheless, for many inorganic synthetic ion exchangers and especially for natural ion exchangers in the soil (clay minerals, humic substances, sesquioxides) inhomogeneous mixtures of ion exchangers (at least with respect to particle size) are invariably present.

In this presentation we, therefore, investigate the kinetics of ion exchange in such mixtures for the case where diffusion of the ions across a hydrostatic boundary layer (Nernst film) surrounding the particles is the rate controlling step (*film diffusion*). In well-stirred systems, liquid-phase mass transfer will usually be favored by a low concentration of the external solution, a high ion-exchange capacity, and a small particle size [1].

Several methods are available to ascertain whether film diffusion is rate determining for a given system (e.g., interruption test, determination of the Helfferich number) but will not be discussed here in more detail.

Two experimental procedures employed to obtain kinetic data for ion-exchange processes are discussed next: (i) The widely used batch technique, where a given amount of ion exchanger is stirred together with an electrolyte solution in a thermostated vessel. In this case the rate of ion exchange is followed, e.g., by measuring continuously pH or electrical conductivity; by withdrawing, after predetermined times, small samples of the solution for analysis; or by interrupting, after a given time, the experiment for analysis of the particles or the solution. (ii) The stirred flow cell (also called continuous flow-stirred reactor), which is used successfully for investigating the sorption kinetics of ions by natural ion exchangers [11–14]. Here, the ion exchanger particles are stirred in a closed, thermostated cell into which electrolyte solution is pumped continuously at a given rate. Again, the concentration of the ions in solution can be measured continuously by an immersed electrode. Alternatively, the effluent solution is collected by a fraction collector for analysis. Because the system is well stirred, the concentration of the outflowing solution and the cell solution are equal. In this way the rate of ion exchange can be monitored conveniently also for ions which are impossible to analyze in situ.

The theory presented here for the prediction of heterogeneous ion-exchange kinetics in these two systems is compared, as far as possible, with experiments. For a better understanding of the essential differences between the kinetics of heterogeneous and homogeneous systems, the rate equations for homogeneous systems are also discussed briefly. Because the differential equation used to describe the rate of film-diffusion controlled ion exchange is of fundamental importance for all subsequent considerations (irrespective of whether homogeneous or heterogeneous systems are considered), its derivation and underlying assumptions are also outlined.

As we will see later, both the batch procedure and the stirred-flow method promote advantages or suffer disadvantages when used for determining the kinetic properties of inhomogeneous mixtures of ion exchangers. In particular, we will see that predictions for the rate of ion exchange based only on measurements of concentration changes in the solution phase can be quite misleading.

II. THE RATE EQUATION FOR FILM DIFFUSION

The derivation of the rate equation used subsequently for film diffusion controlled ion-exchange processes in *homogeneous* systems is outlined briefly below, since it is needed later for initiation of the description of heterogeneous systems. For simplicity we consider here only the exchange of two monovalent ions A and B by a cation exchanger. The anion in solution is also monovalent. For more complicated cases see [15,16]. If ions A and B and the common anion are identified with indices 1, 2, and 3, respectively, their corresponding flows in the solvent fixed (SF) reference frame according to the theory of irreversible thermodynamics can be written as

$$J_i = -\sum_{k=1}^{3} l_{ij} \text{ grad } \bar{\mu}_j \qquad (i = 1,2,3) \tag{1}$$

where l_{ij} are the solvent fixed ionic transport coefficients and $\bar{\mu}$ is

$$\bar{\mu} = \mu_i + z_i F \varphi \tag{2}$$

Here μ denotes the chemical potential, z the signed valencies of the ions, F the Faraday number and φ the electrical potential. The flow of solvent is zero in the SF frame. The electrical current density has to vanish, i.e.,

$$I = \sum_{k=1}^{3} z_k F J_k = 0 \tag{3}$$

Substituting Eqs. (2) and (1) into Eq. (3) yields

$$\text{grad } \varphi = -\frac{\sum_{k,l=1}^{3} z_k l_{kl} \text{ grad } \mu_l}{F \sum_{k,l=1}^{3} z_k l_{kl} z_l} \tag{4}$$

Since no electrical field is applied, φ is the diffusion potential originating from the different mobilities of the ions in the film. The ionic flows are obtained by substituting Eq. (4) in Eq. (1) as shown

$$J_i = -\frac{\sum_{j,k,l=1}^{3} l_{ij} l_{kl} [z_l \text{ grad } \mu_j - z_j \text{ grad } \mu_l]}{\sum_{k,l=1}^{3} z_k l_{lk} z_l} \tag{5}$$

For an 1:1 electrolyte the measurable chemical potentials μ_{13} and μ_{23} are equated to their inaccessible single ion chemical potentials μ_i as shown

$$\mu_{13} = \mu_1 + \mu_3; \qquad \mu_{23} = \mu_2 + \mu_3 \tag{6}$$

The solvent fixed flows J_{i3} of the two electrolytes can then be written as

$$J_{i3} = J_i = \frac{-\sum_{j=1}^{2} [\sum_{k,l=1}^{3} z_k z_l (l_{ij} l_{kl} - l_{il} l_{kj})] \text{ grad } \mu_{j3}}{\sum_{k,l=1}^{3} z_k l_{kl} z_l} \qquad (i = 1,2) \tag{7}$$

or

$$J_{i3} = -\sum_{j=1}^{2} L_{ij} \, \text{grad} \, \mu_{j3} \qquad (i = 1,2) \tag{8}$$

Equation (8) can be identified with Fick's law, namely

$$-J_{13} = \overline{D}_{11} \, \text{grad} \, c_1 + \overline{D}_{12} \, \text{grad} \, c_2$$
$$-J_{23} = \overline{D}_{21} \, \text{grad} \, c_1 + \overline{D}_{22} \, \text{grad} \, c_2 \tag{9}$$

where $\overline{D}_{ij}$ corresponds to the four diffusion coefficients in the SF frame and are functions of l_{ij} the solvent fixed ionic transport coefficients, and c_1 and c_2 are the concentrations of the electrolytes in mole electrolyte/L. The unit of $\overline{D}_{ij}$ is L cm^{-1} s^{-1} and it is related to the usual D_{ij} units of cm^{-2} s^{-1} by $\overline{D}_{ij} = D_{ij}/1000$. If we assume that the structure of the water in the film is the same as that of bulk water, the D_{ij} values are identical to those determined by measurements in the corresponding ternary electrolyte solution containing the cations A, B and the common anion at the same concentration as in the film. Unfortunately, very few diffusion measurements are available for ternary electrolyte solutions [17,18].

Since electroneutrality in the ion exchanger has to be conserved we have

$$z_1 J_1 = -z_2 J_2; \qquad \text{or} \qquad z_1 J_{13} = -z_2 J_{23} \tag{10}$$

This condition, connected with those defined by Eq. (3), results in a vanishing flow of the co-ions J_3, although the corresponding concentration gradient grad μ_3 does not vanish. This, however, is not contradictory since, as shown by Eq. (1) and Eq. (2), J_3 is not only determined by grad μ_3 but consists of several terms, the sum of which only has to vanish. Combination of Eqs. (9) and Eq. (10) yields

$$-J_{13} = \overline{D}_A \, \text{grad} \, c_1, \qquad -J_{23} = \overline{D}_B \, \text{grad} \, c_2 \tag{11}$$

where

$$\overline{D}_A = \frac{\overline{D}_{11}\overline{D}_{22} - \overline{D}_{12}\overline{D}_{21}}{\overline{D}_{12} + \overline{D}_{22}}$$

$$\overline{D}_B = \frac{\overline{D}_{11}\overline{D}_{22} - \overline{D}_{21}\overline{D}_{12}}{\overline{D}_{11} + \overline{D}_{21}} \tag{12}$$

Again, $\overline{D}_A$ in L cm^{-1} s^{-1} is related to D_A in cm^2 s^{-1} by $\overline{D}_A = D_A/1000$, and analogously $\overline{D}_B = D_B/1000$. The sum of the counterions A and B in the ion exchanger has to remain constant in the ion exchanger throughout the experiment, i.e.,

$$C_1(t) + C_2(t) = C \tag{13}$$

where C_1 and C_2 are the concentrations of the counterions A and B in the ion exchanger in mole ion per liter, and C is the constant volume ion-exchange capacity. In accordance with the Nernst concept of film diffusion we assume that the concentrations vary linearly within the film of thickness δ from the respective c_1 and c_2 concentrations in solution to their respective $\overline{c}_1$ and $\overline{c}_2$ concentrations at the surface of the ion exchanger, i.e.,

$$J_1 = -\frac{\overline{D}_A(c_1 - \overline{c}_1)}{\delta}; \qquad J_2 = -\frac{\overline{D}_B(c_2 - \overline{c}_2)}{\delta} \tag{14}$$

If film diffusion is rate determining, the equilibrium distribution of the ions at the surface of the ion exchanger will be established more rapidly than the diffusion processes proceed in the film. This equilibrium will be characterized by the separation factor α_B^A (equivalent basis)

$$\alpha_B^A = \frac{C_1\overline{c}_2}{C_2\overline{c}_1} = \frac{X_{A,\infty}x_{B,\infty}}{X_{B,\infty}x_{A,\infty}} \tag{15}$$

where $X_A = C_1/C$ and $X_B = C_2/C$ are the equivalent fractions of the ions A and B in the ion exchanger, x_A and x_B the corresponding values in the solution, and the subscript ∞ denotes the equilibrium value ($t \to \infty$). Here it is assumed that (i) α_B^A at the surface of the particle does not differ from the corresponding quantity inside the particle, and (ii) that α_B^A does not depend on the ionic composition X_A of the ion exchanger. The first assumption is true for homogeneous ion exchangers only, but not necessarily for heterogeneous particles (see Sec. B). The second assumption usually applies if the ion exchanger is only converted to a rather small extent during the kinetic experiment. If this is not the case (e.g., complete conversion of the ion exchanger from A to the B form), the dependence of α_B^A on the ionic composition of the ion exchanger has to be measured separately for consideration in the evaluation of kinetic experiments (see later).

The coupling of diffusion and the ion-exchange process is obtained by the material balance

$$\frac{dC_i}{1000\,dt} = -\frac{\overline{F}}{V}J_i \qquad (i = 1,2) \tag{16}$$

where $\bar{V}$ is the volume in cm^3 and $\bar{F}$ the surface area in cm^2 of the ion-exchanger particles. If spherical particles of radius r are used, $\bar{F}/\bar{V} = 3/r$. The factor 1000 appears in Eq. (16) since the unit of C_i is mole L^{-1}. Equations (14) now become

$$\dot{C}_1 = \frac{\bar{F}D_A}{\bar{V}\delta}(c_1 - \bar{c}_1)$$

$$\dot{C}_2 = \frac{\bar{F}D_B}{\bar{V}\delta}(c_2 - \bar{c}_2)$$

(17)

Differentiation of Eq. (13) with respect to t and combination with Eq. (17), Eq. (15) and Eq. (13) finally yields the following differential equation for the measurable rate of uptake of ion A by the ion exchanger

$$\frac{dX_A}{dt} = R\frac{\alpha_B^A c_A(1-X_A) - c_B X_A}{\alpha_B^A(D_B/D_A(1-X_A) + X_A}$$

(18)

where R, the rate coefficient, is given by

$$R = \frac{\bar{F}D_B}{\bar{V}C\delta} \qquad \text{(for arbitrary particles)}$$

$$R = \frac{3D_B}{rC\delta} \qquad \text{(for spherical particles)}$$

(19)

As shown above, the interdiffusion coefficients D_A and D_B are related quantitatively to the D_{ij} set in a corresponding ternary electrolyte solution, and can thus be determined by independent measurements. Equation (18) has been used successfully to describe the kinetics of film-diffusion-controlled ion exchange in homogeneous systems for various initial and boundary conditions [15, 19–23]. It has also been shown for the exchange K$^+$/Li$^+$, where the set of four D_{ij} quantities is known for the corresponding ternary solution H$_2$O–KCl–LiCl [17], that the ratio $D_B/D_A = D_{Li}/D_K$, calculated with Eq. (12) to be 0.746 agrees, within experimental error, with the corresponding value (0.75) obtained by using the observed ion-exchange kinetic data in Eq. (18) [19]. Because Eq. (18) seems to describe film-diffusion-controlled ion exchange in homogeneous systems quite well, it is subsequently used as a starting point for considerations of heterogeneous mixtures of ion exchangers.

III. BATCH SYSTEMS

A. Homogeneous systems

The kinetics of ion exchange in homogeneous systems is not discussed here in detail. However, because some of the results observed for homogeneous systems are to be considered later, the solution of Eq. (18) is reported now for this eventuality. The system for this purpose consists of V (L) solution volume (well stirred), to which a given amount of ion exchanger Q (equiv) is added. If the mass of the ion exchanger is m (kg), $Q = mC'$, where C' is the ion-exchange capacity in equiv/kg. The initial ionic composition of the ion exchanger at $t = 0$ is characterized by the equivalent fraction $X_{A,0}$ of the ions A in the ion exchanger. In the solution (total concentration c_{tot} (equiv/L)), the ions B are initially present at a concentration $c_{B,0}$ (equiv/L), i.e., the initial equivalent fraction of B in solution $x_{B,0} = c_{B,0}/c_{tot}$. Because in a batch process all counterions leaving the ion exchanger have to appear in solution, the boundary condition is

$$c_B(t) = c_{B,0} + \frac{Q[X_A(t) - X_{A,0}]}{V} \tag{20}$$

$$c_A(t) = c_{tot} - c_B(t) \tag{21}$$

Integration of Eq. (18) then yields for the time dependence of X_A as a function of t:

$$Rt = \frac{E}{2A} \ln \frac{AX_A^2 + BX_A + D}{AX_{A,0}^2 + BX_{A,0} + D} \tag{22}$$
$$+ \frac{FA - EB/2}{2AS} \ln \frac{[S - B/2 - AX_A][S + B/2 + AX_{A,0}]}{[S - B/2 - AX_{A,0}][S + B/2 + AX_A]}$$

where $A = \dfrac{(\alpha_B^A - 1)Q}{V}$

$B = \dfrac{QX_{A,0}(1 - \alpha_B^A)}{V} - \alpha_B^A\left(c_{tot} - c_{B,0} + \dfrac{Q}{V}\right) - c_{B,0}$

$D = \alpha_B^A\left(c_{tot} - c_{B,0} + \dfrac{QX_{A,0}}{V}\right)$

$E = 1 - \alpha_B^A D_B/D_A$

$F = \alpha_B^A D_B/D_A$

$S = [B^2/4 - DA]^{1/2}$

The concentrations of ion B in solution as a function of t are subsequently obtained with Eq. (20) and Eq. (21). The equilibrium ionic composition of the ion exchanger at $t \rightarrow \infty$ becomes

$$X_{A,\infty} = \frac{[b^2 + ML]^{1/2} - b}{2M} \qquad (23)$$

where
$$b = \frac{[\alpha_B^A - X_{A,0}(1 - \alpha_B^A)]Q}{V} + \alpha_B^A(c_{tot} - c_{B,0}) + c_{B,0}$$

$$M = \frac{Q(1 - \alpha_B^A)}{V}$$

$$L = 4\alpha_B^A\left(c_{tot} - c_{B,0} + \frac{QX_{A,0}}{V}\right)$$

The equilibrium concentrations $c_{B,\infty}$ and $c_{A,\infty}$ in the solution can subsequently be obtained from Eq. (20) and Eq. (21) by substituting $X_{A,\infty}$ for $X_A(t)$.

Conventionally, the rates of ion exchange are presented in terms of the fractional attainment U(t) of equilibrium. This quantity is defined for the ion-exchanger phase as

$$U(t)_{exchanger} = \frac{[X_A(t) - X_{A,0}]}{[X_{A,\infty} - X_{A,0}]} \qquad (24)$$

and for the solution

$$U(t)_{sol} = \frac{[c_B(t) - c_{B,0}]}{[c_{B,\infty} - c_{B,0}]} \qquad (25)$$

Thus, for a homogeneous system we show through use of Eq. (20) that $U(t)_{exchanger} = U(t)_{sol}$. It is, therefore, sufficient to determine the concentration changes in the solution as a function of time to obtain the rates for the ion exchanger phase as well. As we will see later, this is no longer the case for heterogeneous mixtures of ion exchangers.

B. Heterogeneous Systems

Mixtures of ion-exchanger particles which differ in their kinetic properties are considered next to examine the consequences of such heterogeneity. Eqs. (18) and (19) show that for a given pair of counterions the quantities which are material specific and which affect the ion-exchange kinetics are the particle size (e.g., radius r), the ion-exchange capacity C (both influencing the rate coefficient R) and the separation factor α_B^A. Derivation of the theoretical framework for the ion-exchange kinetics of such mixtures that follows subsequently uses experimental results for Cs/ H exchange by polydisperse mixtures of resin ion exchangers for illustration.

To describe the kinetic behavior of a heterogeneous mixture of ion-exchanger particles we consider it to consist of n different fractions of particles, each of which has a given rate coefficient R_i (i = 1, 2, . . . , n), e.g., as a result of different particle sizes and/or ion-exchange capacities. In addition, the selectivity of each fraction for the counterions in solution (as characterized by $\alpha_{B,i}^A$) may also be different. Heterogeneity of the separation factor within the particles (e.g., presence of different functional groups) can be tolerated as long as their spatial distribution is uniform. If this is not the case (e.g., if the separation factor at the particle surface differs from that inside), the value of α_B^A obtained from an equilibrium experiment which yields only the mean $\alpha_{B,i}^A$ of the whole particle of a given fraction i cannot be used in the rate equation. The separation factor at the particle surface has to be substituted. This follows from Eq. (15), which shows that in film-diffusion-controlled ion-exchange processes the separation factor at the particle surface is the quantity to be used.

Let each ion-exchanger fraction in the solution be present in Q_i (in equiv) amounts. If the mass of each fraction is m_i (kg), $Q_i = m_i C_i'$, where C' is again the ion exchange capacity in equiv/kg. For simplicity and in agreement with the experiments described later, the following initial condition at t = 0 is assumed:

$X_{A,0} = 0$ (no A ions in the ion exchanger)

$c_{B,0} = 0$, $c_{A,0} = c_{tot}$ (only A ions in solution)

A material balance requires that all ions leaving the ion exchanger must appear in solution,

$$c_B(t) = \frac{1}{V} \sum_{i=1}^{n} Q_i X_{A,i}(t) \tag{26}$$

Elimination of c_A and c_B from Eq. (18) with help of Eqs. (21) and (26) yields the system of n differential equations, that describe the rate of ion exchange in each fraction i of the heterogeneous mixture as

$$\frac{dX_{A,i}}{dt} = \frac{R_i}{V} \frac{\alpha_{B,i}^A (1 - X_{A,i})[c_{tot}V - \sum_{i=1}^n Q_i X_{A,i}] - X_{A,i}[\sum_{i=1}^n Q_i X_{A,i}]}{\alpha_{B,i}^A (1 - X_{A,i})(D_B/D_A) + X_{A,i}}$$

(27)

$$i = 1, 2, \ldots, n$$

These n nonlinear differential equations have to be solved simultaneously and numerically to obtain the rate of ion exchange of each fraction in the mixture. Once all X_i are known as a function of time, the concentration c_B and c_A of the counterions in solution as a function of time can be calculated from Eqs. (26) and (21).

Equilibrium Values. In many kinetic investigations of ion exchange the rates are not given in terms of $X_A(t)$ or $c_A(t)$ but rather by relating these values to the corresponding values attained at equilibrium. This fractional attainment U(t) of equilibrium of each fraction i of the ion exchanger and the solution, respectively, is given for the above initial conditions as

$$U_i(t) = \frac{X_{A,i}(t)}{(X_{A,i,\infty})}, \qquad U_{sol}(t) = \frac{c_B(t)}{c_{B,\infty}}$$

(28)

U(t) thus ranges from 0 at $t = 0$ to 1 at $t \to \infty$ (equilibrium). The n equilibrium values $X_{A,i,\infty}$ can be obtained with Eq. (27) by letting $dX_{A,i}/dt = 0$; at equilibrium all rates have to vanish. The resulting n equations for the n unknown $X_{A,i,\infty}$ values have to be solved numerically with a suitable algorithm. Alternatively, it is possible to obtain the $X_{A,i,\infty}$ values by using Eq. (27) to calculate the $X_{A,i}$ values from $t = 0$ to a time beyond which *all* $X_{A,i}$ values become practically constant. The equilibrium concentration of the ions in solution can subsequently be obtained as

$$c_{B,\infty} = \frac{1}{V} \sum_{i=1}^n X_{A,i,\infty} Q_i, \qquad c_{A,\infty} = c_{tot} - c_{B,\infty}$$

(29)

The time dependence of the fractional attainment U of each fraction of a mixture which may be heterogeneous with respect to particle size, the ion-exchange capacity, and the separation factor is accessible with Eq. (27) and Eq. (28). Subsequently, the fractional attainment of the ionic composition of the solution as a function of time is calculated with Eqs. (26), (28), and (29). For the sake of clarity the effect of heterogeneities arising from particles size, ion-exchange capacity and the separation factor are treated separately next.

1. Heterogeneity of Particle Size

Mixtures of ion-exchanger particles which differ in their particle size are usually called polydisperse. In studies of ion-exchange kinetics polydispersivity is usually deliberately avoided by carefully sieving the material, or, if this is difficult to achieve, by using an average particle radius for the calculations. As we will see, however, this can be a rather poor approximation of the situation and in special cases lead to completely misleading predictions for the system.

To demonstrate the above, let us first consider a mixture consisting of five fractions of spherical ion-exchanger particles which differ only in their radius r. These five fractions were obtained by wet-sieving a strong-acid ion-exchange resin (AG-50W-X8; Bio-Rad Laboratories) in the H^+-form as (1) r = 0.0025–0.005, r_1 = 0.00375 cm; (2) r = 0.005–0.001 cm, r_2 = 0.0075 cm; (3) r = 0.010–0.020 cm, r_3 = 0.015 cm; (4) r = 0.020–.040 cm, r_4 = 0.030 cm; (5) r = 0.040–0.055 cm, r_5 = 0.0475 cm, where r_i denotes the mean radius of each fraction. The ion-exchange capacity of this material was 4.21 mequiv/g air dry. To determine first the ion exchange kinetics of each fraction *separately* (homogeneous system), 1 g of air dry ion exchanger (corresponding to Q = 4.21 mequiv) of a given radius was suspended in a thermostated reaction cell (25 ± 0.05°C) filled with (V = 300 mL) deionized water. While stirring at 280 rpm the exchange of Cs^+ for H^+ was initiated by adding 2 mL CsCl (1 M) as rapidly as possible with a piston pipette. The rate of ion exchange was measured by an immersed electrical conductivity probe, connected to a conductivity meter and a recorder (details can be found in [24]). The five rate curves obtained in this way are shown in Fig. 1. They are expressed in equivalent fractions, X_{Cs}, of Cs ion in the resin as a function of time. Because each of the five experiments represents a homogeneous system, the rate curves for the ions in solution are the same as for the particles (after proper normalization to the equilibria attained) and are not presented. Before examining the utility of the theory given above for assessment of the rate curves it is necessary first to verify that with the experimental conditions affected, film diffusion had been rate determining. The Helfferich number He [1] resolved for this purpose yielded values between ≅130 (smallest particle size) and 10 (largest particle size). Because these numbers are >1, it can be concluded that film diffusion was indeed rate controlling. The observed rate curves may now be used to evaluate with the help of Eq. (22) the rate coefficient, R, for each fraction and the value for D_H/D_{Cs} (which is the same in all fractions). With the separation factor α_H^{Cs} = 3.16, accessible from the equilibria attained (average value of all five fractions) and the total solution concentration also determined (c_{tot} = 0.00662 M), curve fitting yielded the following values (in cm^3 s^{-1} $mequiv^{-1}$): R_1 = 7.5, R_2 = 3.4, R_3 = 1.4, R_4 = 0.58, and R_5 = 0.415. D_H/D_{Cs} = 1.5. Inserting these values in Eq. (22) yielded the solid lines shown in Fig. 1. There is good agreement with the experimental values

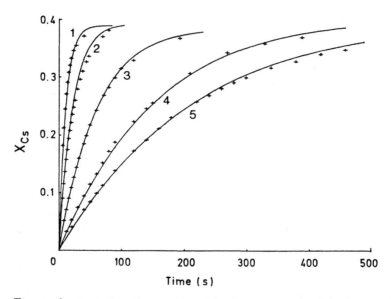

Figure 1 Equivalent fraction X_{Cs} of the Cs ions in each of the five size fractions of a resin cation exchanger as a function of time, when placed *individually* into the CsCl solution in their H form (monodisperse system). Amount of each fraction Q = 4.21 mequiv. Total concentration 0.00662 M. (+++) experimental values. Solid line: calculated curve based on rate coefficients R_i of each size fraction resolved with Eq. (22). (Reprinted with permission from Ref. 24, Copyright 1978 American Chemical Society.

over the whole range. A plot of R versus 1/r for the five fractions yields a straight line. This also demonstrates that film diffusion was the rate-determining step (see Eq. (19) in this study, because for particle diffusion the rate is proportional to r^{-2} [1]).

After having characterized the kinetic and equilibrium properties of the individual fractions of the ion exchanger, it should now be possible to describe quantitatively the kinetics of the Cs^+/H^+ exchange in an arbitrary mixture of the resin fractions without further adjustment of parameters. The experiment consisted in this case of stirring *simultaneously* 0.2 g air dry ion exchanger in the H^+ form of each of the above size fractions (corresponding to Q_i = 0.842 mequiv; i = 1–5) in the solution, while the initial and boundary conditions were the same as for the individual fractions described above (initially only Cs ions in solution; c_{tot} = 0.00662 M, V = 302 mL; α_H^{Cs} = 3.16). The experimentally observed decrease of Cs ions in solution as a function of time (again determined by an immersed conductivity probe) is shown in Fig. 2 (top). To calculate this rate curve, Eq.

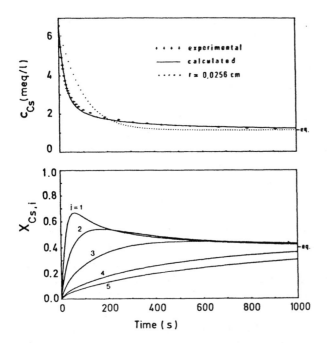

Figure 2 Equivalent fraction $X_{Cs,i}$ of the Cs ions in each of the five size fractions of the cation exchanger (bottom) and concentration c_{Cs} of the Cs ions in solution (top) as a function of time, when placed *simultaneously* into the CsCl solution in their H form (polydisperse systems). Amount of each fraction Q_i = 4.21 mequiv. Total concentration 0.00662 M. Solid line: calculated with Eqs. (30), (26), and (21) using the rate coefficients obtained for the individual size fractions (see Fig. 1). (+++) experimental values. Dotted line: rate curve calculated for a monodisperse system with a particles radius taken to be the median radius of the polydisperse mixture. (Reprinted with permission from Ref. 24, Copyright 1978 American Chemical Society.)

(27) when applied to the case where the separation factors of all fractions are constant now has to be used.

$$\frac{dX_{A,i}}{dt} = \frac{R_i}{V} \frac{\alpha_B^A (1 - X_{A,i})[c_{tot}V - \sum_{i=1}^{n} Q_i X_{A,i}] - X_{A,i}[\sum_{i=1}^{n} Q_i X_{A,i}]}{\alpha_B^A (1 - X_{A,i})(D_B/D_A) + X_{A,i}}$$

$$i = 1, 2, ..., n \tag{30}$$

If this set of five differential equations is solved simultaneously (e.g., numerically with the Runge-Kutta method), using the above values for R_1 and R_5 and for D_H/D_{Cs}, the rate curves for the uptake of Cs ions by each size fraction in the mixture is obtained as a function of time. These rates, characterized by the equivalent fraction X_{Cs} of Cs ions in each size fraction as a function of time are shown in Fig. 2 (bottom). They illustrate clearly how the various size fractions compete for Cs ions in solution. Fractions 1 and 2, with the smaller particles, react initially much faster (large rate coefficient) than the larger particles and even overshoot their final equilibrium value. As a result, they have to release, after some time, part of the sorbed Cs ions for the larger but slower reacting particles (fractions 4 and 5). During such redistribution of Cs ions between the particles (which nevertheless occurs via film diffusion and not by contact ion exchange), the Cs concentration in solution changes only very little, because all ions released by the faster reacting fractions are taken up by the slower ones.

After calculating the rate curves in Fig. 2 (bottom), the overall rate curve for the composite resin sample has been calculated with Eq. (26) and Eq. (21). This rate curve is shown as a solid line in Fig. 2 (top) together with the measured rate curve. As can be seen, good agreement is obtained between the calculated and the experimental curves, even though only the parameters of the experiments involving the individual fractions were used, and no other parameters could be adjusted.

It is of interest to learn to what extent the kinetics of this polydisperse system deviates from a monodisperse system with a particle radius r' which corresponds to the *median* particle radius with respect to the Q_i of the polydisperse mixture. In the present case, i.e., for a rectangular distribution of the fractions according to their weight, we obtain $r' = (0.00375 + 0.0475)/2 = 0.0256$ cm. The corresponding value for R', obtained from interpolation of the $1/r$ versus R plot mentioned above is $R' = 0.746$ cm^3 s^{-1} mequiv^{-1}. With this value and Eqs. (22), (20) and (21) the rate curve for the solution can be calculated (see Fig. 2, top, dotted line). It illustrates that even for the solution phase a polydisperse system is only rather poorly approximated by a monodisperse one. Compared to the monodisperse system the polydisperse system reacts much faster initially, but slower toward the end of the ion-exchange process.

In the above experiment involving five size fractions the rate curves for Cs uptake by each ion-exchanger fraction was not determined experimentally. This formidable task would have required several interruptions of the kinetic experiment after given periods of time, followed by reproducible separations of the particles by sieving into the five size fractions and subsequent analysis of each fraction. For a two component mixture of particles, however, the above sequence of operations becomes practicable and experimental results are available for such systems (see Sec. B.4, where anomalous ion-exchange kinetics is discussed). Another way to avoid the complexity described is facilitated as described next. If for each size fraction only one single bead with a given radius is used, the separation of the mixture will always be uncomplicated. In this way and by using radioactive labelling with ^{134}Cs the kinetics of the exchange Cs^+-H^+ and Cs^+-Cs^+ (isotopic exchange) in such a "mixture," involving three particles with different diameters can be investigated [25]. For various initial conditions a quantitative agreement with the above theory was always obtained.

The rate curves shown in Fig. 2 for the five size fractions indicate that while the concentration of the Cs ions in solution is after 1000 s almost at equilibrium, this is not quite yet the case for the size fractions 4 and 5 of the ion exchanger. As will be seen later, this effect is enhanced extraordinarily if there is an excess of ion-exchanger material (in equivalents) in the system as compared to the total amount of ions A initially in solution (sorption of traces, exhaustion of the solution). In this case rather anomalous kinetic effects will be observable. They will be treated separately in more detail in Sec. B.4.

2. Heterogeneity of Ion-Exchange Capacity

Equations (19) and (27) reveal that ion exchange capacity, C, has an effect on the rate of ion exchange only through the rate coefficient R. Because the radius, r, of the particles and capacity, C, both appear in the denominator of Eq. (19), the effect of C is essentially the same as that of r discussed above. To obtain, for instance, the same values for rate coefficients and thus rate curves as in the earlier example (Fig. 2), let $r = r_3 = 0.015$ cm = constant for *all* particles, and make $C_3 = 2.9$ mequiv/mL ion exchanger. Because in this case $C_3 \cdot R_3 = 2.91(1.4) = 4.06 \text{ s}^{-1}$, the same values for R_i as before are obtained when $C_i = 4.06/R_i$. This determines the ion-exchange capacities of the five fractions (in mequiv/mL): $C_1 = 0.54$; $C_2 = 1.29$; $C_3 = 2.90$; $C_4 = 7.0$; $C_5 = 9.78$. Such a mixture is expected to yield the rate curves as presented in Fig. 2. The particles with the·smaller ion exchange capacities (fractions 1 and 2) are now expected to overshoot their eventual equilibrium values and subsequently release part of the already sorbed ions in favor of the fractions with the larger capacities. *Experimental* verification of the above prediction is presently not available and will be difficult to achieve, because it would require a separation of particles which differ only in their ion-exchange capacities but exhibit the same diameter.

3. Heterogeneity of the Separation Factor

In case the ion exchanger is heterogeneous only with respect to the separation factor, the rate coefficients of all fractions are identical but the $\alpha^A_{B,i}$ in Eq. (27) are expected to be different. No experimental investigation of the kinetics of such a system is available, but the effect of such heterogeneity on its behavior is predictable with Eq. (27). To examine this aspect let us use experimental conditions similar to those used for the previous experiments (exchange Cs^+/H^+; five fractions, $V = 302$ mL; $c_{tot} = 0.00662$ M; $D_H/D_{Cs} = 1.5$; $c_{H,0} = 0$; $X_{Cs,0} = 0$; $Q_i = 0.842$ mequiv for i = 1–5). Let the rate coefficients for all fractions be 1.4 cm^3 s^{-1} $mequiv^{-1}$. Assign the following separation factors to the five fractions: $\alpha_{H,1}^{Cs} = 0.1$; $\alpha_{H,2}^{Cs} = 3$; $\alpha_{H,3}^{Cs} = 10$; $\alpha_{H,4}^{Cs} = 30$; $\alpha_{H,5}^{Cs} = 100$. In contrast to the previous experiments discussed, the equivalent fraction of Cs in the ion-exchanger particles at equilibrium, $X_{Cs,\infty}$, is now different for each of the five fractions. They are obtained with Eq. (27) by computations using the $X_{Cs,i}$ value at t = 0 up to the time when $X_{Cs,i}$ reaches a constant value. The values for the present case are $X_{Cs,1,\infty} = 0.007567$; $X_{Cs,2,\infty} = 0.1862$; $X_{Cs,3,\infty} = 0.4326$; $X_{Cs,4,\infty} = 0.6958$; $X_{Cs,5,\infty} = 0.8840$. The value for the solution phase, obtained subsequently with Eq. (29), is $c_{H,\infty} = 0.0061511$M. The fractional attainment U(t) for each of the five fractions and the solution, which has always to approach 1.0 for t → ∞ can then be calculated with Eqs. (27), (28) and (29). These rate curves are presented in Fig. 3. The result is qualitatively similar to the earlier examples (Fig. 2), where the ion exchangers were heterogeneous only with respect to their rate coefficients. Fractions 1, 2, and 3, which have the smaller separation factors, initially overshoot their equilibrium uptake of Cs^+ and subsequently have to release some of the

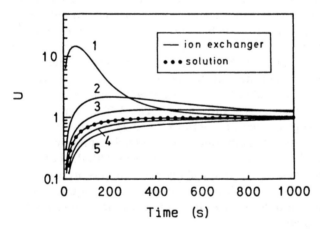

Figure 3 Fractional attainment U(t) of the equilibrium for a heterogeneous mixture five fractions with identical particle diameters but with different separation factors $\alpha_{B,i}^A$ ranging from 0.1 to 100. Amount Q_i and rate coefficient of each fraction 0.842 mequiv and 1.4 cm^3 s^{-1} $mequiv^{-1}$, respectively. Total solution concentration 0.00662 M. Rate curves calculated with Eqs. (27), (28), and (29).

Cs ions to particle fractions 4 and 5, which have the higher separation factors and can thus take up more Cs ion. The solution again reaches its equilibrium somewhat faster than the ion-exchanger fractions.

4. Anomalous Kinetic Effects

As mentioned above, calculations of the kinetics in heterogeneous mixtures of ion exchangers indicate that the concentration of ions in solution approach equilibrium faster than the individual fractions of ion-exchanger particles. This results from the fact that during the redistribution phase the faster reacting particles release previously sorbed ions which are subsequently taken up by the slower reacting particles. As a result the concentration of ions in solution changes only slightly during this time period and can simulate an equilibrium, which is not yet reached by the ion-exchanger particles. This effect can become extremely striking if the excess, E, of ion exchanger in the system is large compared to the total amount of ions initially in solution. In our notation this means

$$E = \frac{\sum_{i=1}^{n} Q_1}{c_{A,0} V} \gg 1 \tag{31}$$

Unless α_B^A is very small, the concentration of A in solution becomes very low during equilibration (exhaustion of the solution), and the redistribution of the ions between the particles occurs extremely sluggishly as a consequence. This situation, where the solution reaches its equilibrium very much faster than the particles will be referred to as anomalous ion-exchange kinetics from this point.

For an ion exchanger that is heterogeneous only with respect to particle size, with the above condition Eq. (30) can be simplified to yield

$$\frac{dX_{A,i}}{dt} = \frac{R_i^*}{V} \frac{\alpha_B^A (1 - X_{A,i})[c_{tot} V - \sum_{i=1}^{n} Q_i X_{A,i}] - X_{A,i}[\sum_{i=1}^{n} Q_i X_{A,i}]}{\alpha_B^A (1 - X_{A,i})} \tag{32}$$

$$i = 1, 2, \dots, n$$

where $R_i^* = R_i D_A / D_B$. The advantage provided by Eq. (32) is that in this case one has only to know R_i^* for each size fraction i rather than R_i and D_B / D_A. These R_i^* values can be obtained from kinetic measurements of the individual size fractions. The equilibrium value for the uptake of A by each ion-exchanger fraction can be obtained in this case (only polydispersivity) from Eq. (23); in the case where $\alpha_B^A = 1$ (isotopic exchange) the equilibrium value for the uptake of A by each ion-exchanger fraction is accessible with Eq. (33)

$$X_{A,\infty} \approx \frac{c_{tot}V}{c_{tot}V + \sum_{i=1}^{n} Q_i} \tag{33}$$

Situations with an excess of ion exchanger occur if, for example, a given ionic species has to be removed from the solution by a batch operation. An excess of particles with ion exchange properties is usually also present when the sorption of trace ions by soils or soil components (clay minerals, oxides, humic substances) is investigated. Especially in this latter case the particles will be invariably polydisperse. This anomalous kinetic behavior will, of course, only be observed experimentally if the concentration of ions A in the solution *and* in the various fractions of the ion-exchanger particles in the mixture are measured continuously.

To demonstrate this effect experimentally, the exchange of Cs^+/Cs^+ (isotopic exchange) and Cs^+/H^+ in a mixture two different size fractions (0.04 and 0.08 cm in diameter) of a sulfonated polystyrene-type cation exchanger (AG-50W-X8; Bio-Rad Laboratories) has been investigated in well-stirred batch experiments (V = 50 mL; 21 ± 0.5°C). For further details see [26]. The total concentration of Cs ions initially in solution and labelled with ^{137}Cs was always $c_{Cs,0} = c_{tot} = 0.0001$ M. The amount of each size fraction of ion exchanger in the mixture was $Q_1 = Q_2 = 0.41$ mequiv so that the amount of ion exchanger (in mequiv) exceeded the total amount of Cs^+ in solution by a factor of 164 (E = 0.82/0.005 = 164). For the isotopic exchange experiment the ion exchanger was initially in unlabeled Cs form; for the ion exchange experiment it was initially in the H form. After predetermined time periods the ion-exchange process was interrupted and the ion exchanger separated from the solution. The ion-exchanger particles were separated in their two size fractions with help of a nylon screen, and the ^{137}Cs activity of each fraction and the solution was determined.

In Fig. 4 we show first the rate curves obtained when only *one* of the two size fractions is present in the solution. In this case the isotopic exchange is fairly rapid for both size fractions. After about 1000 s the equilibrium reaction is to more than 90% completed; the smaller particles, of course, react somewhat faster. The rate curves for the corresponding solution phase and the ion exchanger when presented in fractional attainment terms, U(t), are in this case (homogeneous mixture) identical and not given separately. To obtain the rate coefficients R* of the two size fractions from these rate curves with Eq. (34), use is made of Eq. (32) for the case n = 1, and of Eq. (24) for the initial conditions as described (isotopic exchange: $D_A/D_B = 1$; $\alpha_B^A = 1$; R* = R)

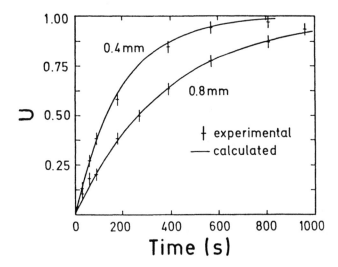

Figure 4 Fractional attainment U(t) of equilibrium for the isotopic exchange, Cs^+/Cs^+, by a resin cation exchanger (0.4- and 0.8-mm particles in diameter), when added *individually* to a CsCl solution (c = 0.0001 M). (+++) experimental values. Solid lines: Fit of the experimental values according to Eq. (34) to obtain the rate coefficients R_i of each size fraction. (Reprinted with permission from Ref. 26, Copyright 1991 American Chemical Society.)

$$U(t) = 1 - \exp[-R^*(c + \frac{Q}{V})t] \tag{34}$$

Because all quantities, with the exception of R*, are known, Eq. (34) can be used to evaluate R* from the rate curves in Fig. 4 by curve fitting. If we put $R_1^* = 0.70$ and $R_2^* = 0.32$ cm³ s⁻¹ mequiv⁻¹ (for the 0.04- and 0.08-cm fraction, respectively) a good fit between the experimental points and the calculated ones (solid lines) is obtained (Fig. 4).

If both size fractions of the ion exchanger are present *simultaneously* in the solution, the rate curves shown in Fig. 5 are obtained. They reveal that in this case the solution has reached equilibrium after about 600 s. After this time interval, however, the ion-exchanger particles have (with respect to their eventual equilibrium) sorbed either too much ¹³⁷Cs (0.04-cm fraction) or too little (0.08-cm fraction). They subsequently approached their equilibrium (U = 1) so slowly that 95% of the equilibrium value is reached only after 50,000 s. Equation (32) can now be applied for a comparison of the experimental results with the theoretically based predictions. For this purpose first Eqs. (33) and (29) are used first

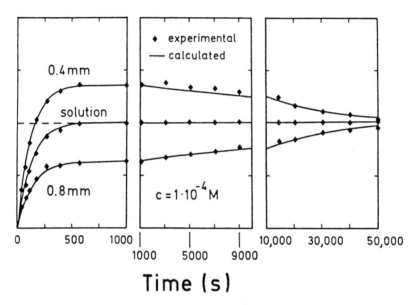

Figure 5 Fractional attainment U(t) of equilibrium for the isotopic exchange, Cs^+/Cs^+, when ion-exchanger particles of 0.4 and 0.8 mm in diamter are added *simultaneously* to a CsCl solution (c = 0.001 M). (+++) experimental values. Solid lines: Calculated by using Eqs.(32), (26), and (28) and the rate coefficients obtained with the individual size fractions (see Fig. 4). (Reprinted with permission from Ref. 26, Copyright 1991 American Chemical Society.)

to obtain the equilibrium values $X_{A,\infty}$ and $c_{B,\infty}$. Next, the two differential equations (32) are solved simultaneously to obtain $X_{A,1}(t)$ and $X_{A,2}(t)$. This is possible because all quantities in Eq. (32) are known (the values for R_1^* and R_2^* are from the kinetic measurements obtained with the individual size fraction samples; see above). $U_{sol}(t)$ is then obtained with Eqs. (26) and (28), and $U_1(t)$ and $U_2(t)$ with Eq. (28). The resulting rate curves are shown in Fig. 5 as solid lines. The agreement between the experimental and calculated results is quite good, even though parameter could not be adjusted.

   To investigate the rates of ion exchange Cs^+/H^+, the resin beads in the H^+ form were added to the CsCl solution (again 0.0001 M), labeled with ^{137}Cs. When the two size fraction samples were placed *individually* in the solution (V = 50 mL) for separate study, the rate curves shown in Fig. 6 were obtained. The quantities Q of ion exchanger used were 0.34 and 0.30 mequiv for the 0.04- and 0.08-cm fraction, respectively. The separation factor (see Eq. (15)) obtained for both size fractions was α_H^{Cs} = 3.5 for the equilibrium uptake of Cs by the exchanger. To obtain the values of R* for the individual size fractions from these rate curves, the rate law, Eq. (35), derived from Eq. (32) for the limiting case $X_{1,A} \ll 1$ (large excess of ion exchanger) was used:

$$U(t) = 1 - \exp\left(\frac{-R^*Qt}{V}\right) \tag{35}$$

All quantities in this equation, except R*, are known. The value of R* had to be obtained from the experimental rate curve by curve fitting. By letting $R_1^* = 0.67$ and $R_2^* = 0.33$ cm^3 s^{-1} mequiv^{-1}, a good fit was obtained for each size fraction (Fig. 6). These values are slightly different from those observed above for the isotopic exchange of Cs$^+$, because different values for D_B/D_A are involved. Besides, in contrast to the isotopic exchange experiments, the ion exchanger is now in the H$^+$ form, and this results in a larger surface of the particles that is attributable to greater swelling of the particles.

If both size fractions are *simultaneously* in the solution, the rate curves given in Fig. 7 are observed. In this case $Q_1 = 0.34$ and $Q_2 = 0.30$ mequiv in the H$^+$ form (0.04- and 0.08-cm size fraction, respectively) were added to a 0.0001 M CsCl solution. Due to the large excess of ion ex-

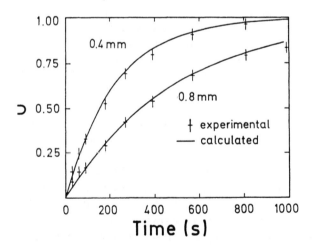

Figure 6 Fractional attainment U(t) of equilibrium for the exchange, Cs$^+$/H$^+$, by a resin cation exchanger (0.4- and 0.8-mm particles in diameter), when added *individually* to a CsCl solution (c = 0.0001 M). (+++) experimental values. Solid lines: calculated curve based on rate coefficients R_i obtained for the two size fractions with Eq. (35). (Reprinted with permission from Ref. 26, Copyright 1991 American Chemical Society.)

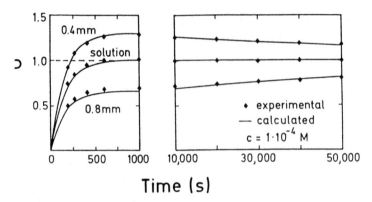

Figure 7 Fractional attainment $U(t)$ of the equilibrium for the exchange Cs^+/H^+, when ion-exchanger particles of 0.4 and 0.8 mm in diameter are added *simultaneously* to a CsCl solution (c = 0.0001M). (+++) experimental values. Solid lines: Calculated using Eqs.(32), (26), and (28) and the rate coefficients obtained from the individual size fractions, see Fig. 6. (Reprinted with permission from Ref. 26, Copyright 1991 American Chemical Society.)

changer, almost complete removal of the Cs-ions in solution ($c_{Cs,\infty}$ = 2.2 $\times$ 10^{-7} M) occurs and this leads once again to anomalous kinetic behavior. At the time when the solution reached its equilibrium ($U = 1$ after about 700 s), the small particles are still far removed from equilibrium, the 0.04-cm size fraction at $U_1 = 1.3$, and the 0.08-cm fraction at $U_2 = 0.65$. The subsequent attainment of their equilibrium is so slow that even after 50,000 s $U_1 = 1.16$ and $U_2 = 0.81$. Only about 95% of the equilibrium composition is reached by the particles even after 150,000 s (ca. 42 h). Again we can compare the experimental rate curves with the theoretical predictions. First we obtain from Eq. (23) and (29) for the equilibrium uptake by both fractions $X_{Cs,\infty} = 0.0078$ and $c_{H,\infty} = 9.98 \times 10^{-5}$ M. Because the rate coefficients R_1^* and R_2^* have already been determined (see above) all quantities needed next to solve the two differential equations (32) simultaneously are known, the $X_{Cs,1}(t)$ and $X_{Cs,2}(t)$ values become numerically accessible. Finally $U_{sol}(t)$, $U_1(t)$ and $U_2(t)$ are calculated with Eqs. (26) and (28). The resulting rate curves are shown in Fig. 6 as solid lines. Again the agreement between calculated and observed rates is quite good, even though no parameter were adjusted. An additional experiment involving a different ratio of the above two ion exchangers in the mixture [26], showed similarly good agreement between experimental and calculated rate curves.

Finally, to illustrate the extent of anomalous ion-exchange kinetics, a polydisperse mixture, the particle size of which is given by a normal distribution, is considered. To reveal especially the effect of the excess E of an ion exchanger in the system, this quantity is varied from 5 to 500. No experimental results, however, are available at present for such a system and they must be synthesized. Consider for this purpose, that the frequency distribution of the radii of the ion-exchanger particles is characterized by 11 fractions, which vary from $r_1 = 0.005$ cm to $r_{11} = 0.105$ cm in 0.01-cm steps. If it is assumed that they are distributed according to the standard normal frequency function ($\bar{r} = 0.055$ cm, $\sigma = 0.02$ cm) and if the total amount of ion exchanger in the system is $Q_{total} = 5$ mequiv, 11 values of Q_i illustrated in Fig. 8 are obtained. Equation (30) rather than Eq. (32) is used for the calculation of the rate curves because the condition $E \gg 1$ is not always satisfied. The corresponding 11 rate coefficients R_i required to solve Eq. (30) can be obtained with Eq. (19), if the radii of the 11 fractions given in Fig. 8 are used and $D_B/\delta C$ is assigned a value of 0.007 cm^4 s^{-1} mequiv^{-1}, which is typical for Cs$^+$/H$^+$ exchange in conventional ion-exchange resins (e.g., AG-50-WX-8, Bio-Rad Laboratories, see above). The resulting 11 values for R_i (in cm^3 s^{-1} mequiv^{-1}) then become 4.20, 1.40, 0.84, 0.60, 0.47, 0.38, 0.32, 0.28, 0.25, 0.22, and 0.20. For the remaining parameters left in Eq. (30) D_B/D_A is assigned a value of 1.5 (as above), $V = 100$ mL, and $\alpha_B^A = 2$. Initially, the A ions are only in the solution and not in the ion exchanger.

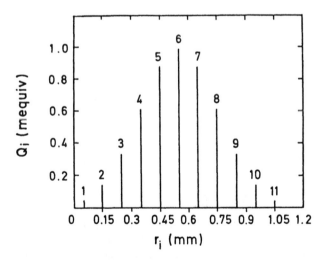

Figure 8 Frequency distribution of the radius of the ion-exchanger particles in the mixture. Each of the 11 fractions is present in the amount Q_i. Total amount of ion exchanger $\Sigma Q_i = 5$ mequiv. (From Ref. 27).

To project first the rate of ion exchange if each of the 11 size fractions (5 mequiv each) were present *individually* in the above solution (c_{tot} = 0.001 M), U(t) is calculated for each fraction by using Eq. (22). The rate curves obtained are illustrated in Fig. 9. They show that even the largest particles (fraction 11) reach 95% of their equilibrium state already after 500 s. Increasing or decreasing c_{tot} by a factor of 10 barely changes the rate curves in Fig. 9.

If the 11 fractions are *simultaneously* in the solution in the Q_i amounts given above, the rate curves are quite different and the effect of the concentration c_{tot} (or rather the excess E) is large. To reveal the characteristics of anomalous ion exchange kinetics, c_{tot} is varied in the following manner: c_{tot} = 0.01, 0.001 and 0.0001 M. These values correspond to E, excess of ion exchanger in the three systems, of E = 50; 50 and 500, respectively (see Eq. (31)). The numerical solution of the set of 11 differential equations (Eq.(30)) then yields the equivalent fraction, $X_{A,i}$, of ion A in each fraction as a function of time. The equilibrium values of $X_{A,i,\infty}$ are identical for a given value of c_{tot}, and are obtained from Eq. (23) for Q_{total} = 5 mequiv and the above three values of c_{tot}, 0.1812, 0.01992 and

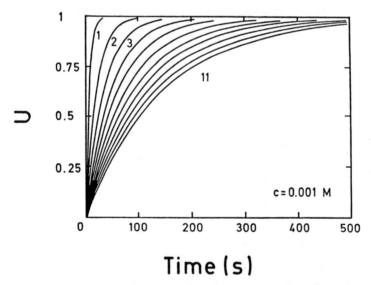

Figure 9 Fractional attainment U(t) of equilibrium in a batch system when each of the 11 size fractions given in Fig. 8 is added *individually* (5 mequiv each, $X_{A,0}$ = 0) to the solution (c_{tot} = $c_{A,0}$ = 0.001 M). The particle radius increases from 0.05 mm (fraction 1) to 1.05 mm (fraction 11) in steps of 0.1 mm. The ions in solution and in the ion-exchanger particles reach their equilibrium along the same rate curve. (From Ref. 27.)

0.002010, respectively. The resulting equilibrium concentrations in solution $c_{A,\infty}$ are obtained subsequently as 9.9602×10^{-4}, 1.0058×10^{-5}, and 1.006×10^{-7} M, respectively. From Eqs. (24) and (25) the fractional attainment of ion A in each fraction and in the solution is finally obtained. These values are graphically presented in Fig. 10 as a function of time. When $c_{tot} = 0.01$ M or the excess E = 5 (Fig. 10a), it is observed that fractions 1–5 with the smaller particle radii (larger rate coefficients R) overshoot their eventual equilibrium uptake values considerably and have to release part of the ions taken up initially in favor of fractions 7–11 with the larger, but slower reacting particles. Fractions 5 and 6 reach equilibrium at about the same rate as the solution. However, while the solution reaches 95% of its equilibrium value (U = 0.95) after about 200 s, the fractions with the large particles, and, to a smaller extent, those with the very small particles are still far removed from equilibrium. Thus, fraction 1 has approached to within 5% of its equilibrium value after 750 s, fraction 3 after 1180 s, and fraction 11 only after 2700 s. This effect becomes more noticeable, when the concentration of the solution is decreased to $c_{tot} = 0.001$ M (excess E = 50; Fig. 10b) or to $c_{tot} = 0.0001$ M (E = 500; Fig. 10c). In this latter case the solution reaches 95% of its equilibrium value again after about 200 s, while fractions 1, 3, and 11 approach within 5% of their equilibrium values only after 80,000 s, 140,000 s, and 320,000 s, respectively.

Qualitatively similar results are obtained, if the radius (or rate coefficient R) of all particles is kept constant, but the separation factor α_B^A of each fraction is different. These rate curves can be found in [27], where the values of α_B^A for the 11 fractions used, were between 0.1 and 100.

Numerous computer simulations, such as those given above, have been performed, using arbitrary values for Q_i, R_i, and $\alpha_{B,i}^A$ of the mixture. In every case it was found, that whenever a considerable excess of ion exchanger is present in the system, a very long lasting redistribution period is observed, during which the initially faster-reacting particles release some of the already sorbed counterions in favor of the initially slower-reacting particles. However, the system balances itself always in such a way that concentration of the counterions in the solution phase remains constant at its equilibrium value during this period.

IV. STIRRED-FLOW CELL

The rates of ion exchange for heterogeneous mixtures of ion exchanger may also be investigated in a stirred-flow cell. For a better understanding of the characteristics of the stirred-flow method, however, the results obtained for homogeneous systems are discussed briefly first [28]. This

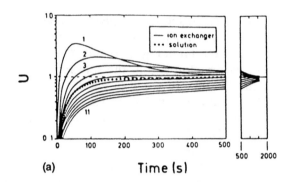

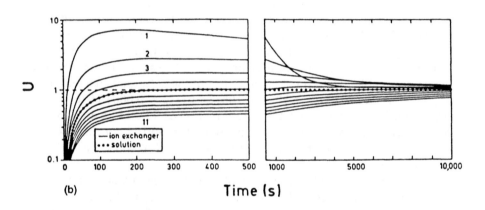

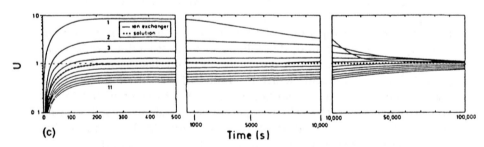

Figure 10 Fractional attainment U(t) of equilibrium in a batch system containing the 11 size fractions (1 to 11) of an ion exchanger *simultaneously* (all $X_{A,0}$ = 0). The amount Q_i and the corresponding particle radius of each fraction are given in Fig. 8. The initial concentration of the counterions A in solution $c_{A,0}$ = c_{tot} are (a) 0.01 M, (b) 0.001 M and (c) 0.0001 M. Equilibrium for each fraction and the solution is attained when U = 1. (From Ref. 27.)

theory developed is then extended to heterogeneous systems. Again, film diffusion is assumed to be the rate-controlling step.

A. Homogeneous Systems

Let us consider a thermostated cell containing Q equivalent of ion exchanger, dispersed in V mL deionized water by continuous stirring. (If C' is the ion-exchange capacity (equiv/kg) and m (kg) its weight, then $Q = mC'$). Initially (time t = 0) the ion exchanger contains the two monovalent counterions A and B in equivalent fractions $X_{A,0}$ and $X_{B,0}$, where $X_{A,0} + X_{B,0} = 1$. To this suspension an electroyle solution, also containing ions A and B, is added via an inlet opening at a constant flow rate of J $(L\ s^{-1})$, while the effluent solution is collected in a fraction collector for analysis. To avoid the loss of ion-exchanger particles from the cell, the outlet is covered with a fine nylon screen.

   If, for simplicity, only the case of equivalent ion exchange between monovalent counterions is considered, the total amount q_B of ions B (in equiv) released by the ion exchanger is related to the corresponding change of the equivalent fraction X_A in the ion exchanger as shown:

$$q_B(t) = Q[X_A(t) - X_{A,0}] \qquad (36)$$

The change in the concentration of ions B in the cell as a result of the ion-exchange process is given by the following material balance:

$$V\left(\frac{dc_B}{dt}\right) = Jc_{B,in} - Jc_{B,out}(t) + \frac{Qd[X_A(t) - X_{A,0}]}{dt} \qquad (37)$$

where c_B, $c_{B,in}$ and $c_{B,out}$ are the concentrations of ion B in the cell, in the influent and in the effluent (in mol/L), respectively. The last term in Eq. (37) represents the amount of ion B released by the ion exchanger as a result of its uptake of ion A from the influent solution. If the ion exchanger is completely saturated with ion B ($X_{A,0} = 0$) initially, and the influent solution contains only ion A ($c_{B,in} = 0$), the ion exchanger is converted completely to the A form. Also to be considered is the case where $X_{A,0}$ and $c_{B,in}$ are not zero. Here, the ion exchanger is converted only partially to the A form. If the equivalent fraction of ion B in the influent solution is expressed as $x_{B,in}$:

$$c_{B,in} = x_{B,in}c_{in,tot} \qquad (38)$$

where $c_{in,tot}$ is the total and constant concentration of A and B in the influent solution (mol/L). Because the solution in the cell is always well mixed $c_{B,out} = c_B$ and one obtains for Eq. (37).

$$\frac{Qd[X_A(t) - X_{A,0}]}{dt} + Jx_{B,in}c_{in,tot} - \frac{Vdc_B(t)}{dt} - Jc_B(t) = 0 \qquad (39)$$

Equation (39) can be integrated to yield the time dependence of c_B, i.e., the quantity which is directly measurable in the effluent as a function of time

$$Vc_B(t) = Q[X_A(t) - X_{A,0}] - J\int_0^t c_B(t)\,dt + Jx_{B,in}c_{in,tot}t \qquad (40)$$

Because it is assumed that film diffusion is rate controlling, the rate of ion exchange, i.e., $X_A(t)$ in Eq. (40) is given by Eq. (18) and the rate coefficient R by Eq. (19) with α_B^A is again the equivalent separation factor. Because the ion exchanger is also in equilibrium with the influent solution for $t \to \infty$, α_B^A is given by

$$\alpha_B^A = \frac{X_{A,\infty}x_{B,in}}{(1 - X_{A,\infty})(1 - x_{B,in})} \qquad (41)$$

Thus, with exception of the case where $x_{B,in} = 0$ (total conversion of the ion exchanger to the A form), Eq. (41) can be used to obtain α from the known value of $x_{B,in}$ and the measured value of $X_{A,\infty}$. Because an ion exchanger exchanges counterions in equivalent amounts, the total equivalent solution concentration c_{tot} (equiv/L) in the cell, given by Eq. (21),

$$c(t)_{tot} = c_A(t) + c_B(t) \qquad (42)$$

does not change as a result of the ion-exchange process. Therefore, the time dependence of $c(t)_{tot}$ is independent of the presence of an ion exchanger in the stirred-flow cell and can be obtained from the mass balance in the absence of ion exchanger:

$$J[c_{in,tot} - c_{tot}] = \frac{Vdc_{tot}}{dt} \qquad (43)$$

After integration

$$c(t)_{tot} = c_{in,tot}\left[1 - \exp\left(\frac{-Jt}{V}\right)\right] \qquad (44)$$

Here $c_{tot} = 0$ at $t = 0$, because the cell is filled initially only with pure water. By using Eqs. (44) and (42) c_A is eliminated in Eq. (18) to obtain

$$\frac{dX_A}{dt} = R\frac{\alpha_B^A(1-X_A)\{c_{in,tot}[1-\exp(-Jt/V)]-c_B(t)\}-X_A c_B(t)}{\alpha_B^A(D_B/D_A)(1-X_A)+X_A} \qquad (45)$$

Equation (45) and Eq. (40) both contain the unknown quantities $X_A(t)$ and $c_B(t)$. However, because $c_B(t)$ cannot be solved with Eq. (40) so that it can be inserted in Eq. (45), the two equations have to be solved simultaneously and numerically by an iteration process (e.g., Runge-Kutta method). The integral term in Eq. (40) is also evaluated numerically. At the end of each infinitesimal time interval, Δt, the resulting value of c_B is used to calculate $c_A(t)$ with Eqs. (42) and (44).

The conversion of a cation exchanger (AG-50W-X8, wet-sieved to 0.04 cm in diameter) from the H^+ form to the Cs^+ form (forward reaction) and back to the H^+ form (reverse reaction) is an *experimental* example selected for consideration next. Besides these two total conversions the rates for partial conversions of the ion exchanger in four steps from the H^+ form to the Cs^+ form, and back in the same steps to the H^+ form have also been measured. The cell, filled initially with deionized water (V = 20 mL) and Q = 0.781 mequiv of ion exchanger was stirred at 730 rpm. For the four partial forward reactions the initial ionic compositions of the ion exchanger were $X_{Cs,0}$ = 0, 0.34, 0.57, and 0.81. The corresponding compositions of the influent solution were $X_{Cs,in}$ = 0.1, 0.25, 0.55, and 1.0. For each experiment $c_{in,tot}$ = 0.005 M). The ionic compositions reached by the ion exchanger at equilibrium were for these experiments: $X_{Cs,\infty}$ = 0.34, 0.57, 0.81 and 1.0. In this way the initial ionic composition of the ion exchanger was always the equilibrium ionic composition of the preceding experiment. The reverse reactions were investigated in the same way. For total conversions of the ion exchanger the influent solution was either 0.005 M CsCl or HCl. The flow rate, J, was always 10 mL min^{-1} = 0.1666 mL s^{-1}. The Cs was labelled with radioactive ^{137}Cs. The effluent was collected by a fraction collector in 10-mL portions and analyzed for Cs by gamma counting. For the total conversions the concentration of H ions in the effluent was determined with a glass electrode. The amount of Cs taken up by the ion exchanger as a function of time was obtainable as the difference between the quantity of Cs pumped into the cell and the observed quantity of Cs in solution (in the cell as well as in the collected fractions). For details see [28]. The separation factor was obtained from the equilibria attained at the end of each experiment.

As an example of the observed rate curves the rate of the first partial forward reaction (X_{Cs} from 0 to 0.34, see Fig. 11a) and for the total conversion (X_{Cs} from 0 to 1; see Fig. 11b) are shown. The rates are given for the ion exchanger and for the solution phase in terms of the fractional attainment, U, of equilibrium defined as

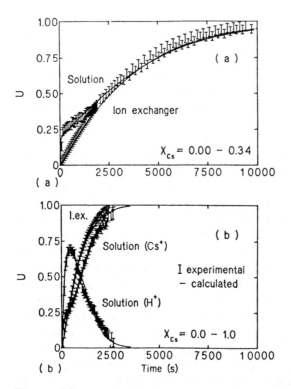

Figure 11 Fractional attainment U(t) of equilibrium of Cs^+ in the effluent solution and for its uptake by the ion exchanger initially in the H^+ form. Initial and equilibrium fraction of Cs^+ in the ion exchanger (a) $X_{Cs,0} = 0$; $X_{Cs,\infty} = 0.34$. (b) $X_{Cs,0} = 0$; $X_{Cs,\infty} = 1$. The fraction $x_{Cs,in}$ of Cs^+ in the inflowing solution for experiment (a) 0.1; and for (b) 1.0. In the latter case the release of H ions by the ion exchanger was also determined. Flow rate J = 10 mL/min. Total concentration in the influent $c_{in,tot} = 0.005$ M. The calculated curves (solid lines) shown were obtained with Eqs. (45) and (40). (From Ref. 28.)

$$U_{exch}(t) = \frac{X_{Cs}(t) - X_{Cs,0}}{X_{Cs,\infty} - X_{Cs,0}} \qquad (46a)$$

$$U_{sol}(t) = \frac{c_{Cs}(t)}{c_{Cs,in}} \qquad (46b)$$

All rates for the forward and reverse reactions are summarized in Fig. 12, where, characterized by the corresponding half-times $t_{1/2}$ observed with the ion-exchanger particles, they are plotted as a function of the initial

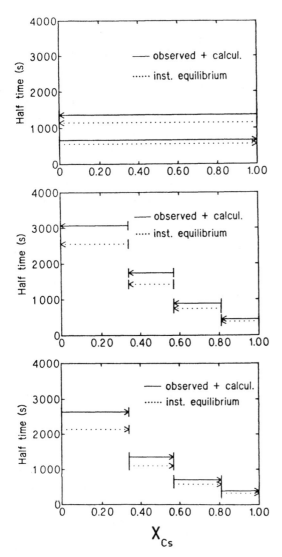

Figure 12   Half-times for the uptake of Cs^+ by the ion exchanger obtained from experiments like those shown in Fig. 11. The beginning of each arrow denotes the initial fraction, $X_{Cs,0}$, the end of each arrow denotes the equilibrium fraction, $X_{Cs,\infty}$ of Cs^+ in the ion exchanger for each experiment. Solid lines: experimental and calculated results. Dotted lines: half-times calculated by assuming that the equilibrium between the ion exchanger and the solution is always attained instantaneously in the cell. Bottom: forward reactions (uptake of Cs ions), successive partial conversions; Middle: reverse reactions (release of Cs ions), successive partial conversions. Top: total conversion in one step (forward and reverse reaction).

equivalent fraction of Cs^+ in the ion exchanger. This figure reveals that if the ion exchanger is converted in steps from the H^+ form to the Cs^+ form, the rate of Cs ion uptake increases considerably, the more the ion exchanger is already saturated with Cs ions initially. If the ion exchanger is converted in one step from the H^+ to the Cs^+ form, the rate of ion exchange exhibits an intermediate value. A comparison of the reverse reactions (release of Cs ions) shows that these rates are always somewhat slower than those for the corresponding forward reactions.

The observed and calculated rate curves, the latter calculated with Eqs. (45) and (40), are plotted in Fig. 11 to permit comparison. In Eqs. (45) and (40), where ion A = Cs^+ and ion B = H^+, the quantities, V, J, Q, $c_{tot,in}$ $X_{Cs,0}$ and $x_{H,in}$ are known (see above). The separation factor, α_H^{Cs}, obtained from the equilibrium intake of Cs at the end of each experiment, depended somewhat on X_{Cs}, the fraction of Cs in the ion-exchanger phase. For X_{Cs} = 0.34, 0.57, and 0.81, α_H^{Cs} was found to be 4.7, 4.0, and 3.5, respectively. The value of α_H^{Cs} at X_{Cs} = 1, determined to be equal to 3.0, was obtained by extrapolation of the above values. The value for D_H/D_{Cs}, independent of the particle size and the ion-exchanger properties, was assigned the value of 1.5, extracted from batch experiments described earlier (Sec. III.B.). The value of the rate coefficient, R, depends on the rate of stirring and has to be obtained by curve fitting of a single ion-exchange experiment in the stirred-flow cell. Because R depends on the particle diameter (see Eq. (7)), slightly different values of R have to be expected for the C^+ and the H^+ form of the ion exchanger due to swelling effects [29]. However, because the diameters of the ion-exchanger particles used here differ for these two ionic forms by less than 3% (as determined under the microscope), this effect is neglected. When a value of R = 1.3 cm^3 s^{-1} mequiv^{-1} is used, the calculated rate curves for the forward reaction, shown in Fig. 11 as solid lines, are obtained. They are in good agreement with the experimental rate curves. An equally good fit was obtained when the same value for R was used for the other partial forward reactions [29]. Thus, even though all forward reactions exhibited quite different rates (see Fig. 12), they could be described with one single value of R.

For calculation of the rates of the reverse reactions (Fig. 12 middle) it is only necessary to exchange the A and B indices in Eqs. (35)–(45), because the release of Cs ion can be equated to the uptake of H ion. This means that for the assignment of parameters in Eqs. (45) and (40) α_B^A = α_{Cs}^H = $1/\alpha_H^{Cs}$. The values of α_{Cs}^H used for the four reverse reactions are thus the inverse values of those given above for α_H^{Cs}. Similarly we have now for $D_B/D_A = D_{Cs}/D_H$ = 1/1.5. If we exchange the indices A and B in Eq. (19), the rate coefficient R of the reverse reaction is obtained.

$$\frac{R_{forward}}{R_{reverse}} = \frac{D_B}{D_A} \quad\quad\quad (47)$$

In the present case this yields $R = 1.3/1.5 = 0.77 \text{ cm}^3 \text{ s}^{-1} \text{ mequiv}^{-1}$ for the reverse reaction. Now, the values for $X_{A,0}$ and $x_{B,in}$ become $X_{H,0} = 1 - X_{Cs,0}$ and $x_{Cs,in} = 1 - x_{H,in}$ and no parameters remain to be adjusted for the reverse reactions. Comparison of the calculated and observed rate curves showed the agreement between them to be good [29]. Within experimental error (circa $\pm$ 4%), the observed and calculated half times shown in Fig. 12 were identical and are not given separately.

Equations (40) and (45) can be used again to disclose why the half-time of ion exchange for partial conversions depends so strongly on the initial ionic composition $X_{A,0}$ of the ion exchanger (Fig. 12). If $t_{1/2}$ is calculated for successive ion-exchange processes with increasing values of $X_{A,0}$ and with $D_B/D_A = \alpha_B^A \to 1$ (no selectivity, isotopic exchange), it is discovered that $t_{1/2}$ does *not* depend on $X_{A,0}$. If we let $\alpha_B^A \to 1$ and $D_B/D_A > 1$, the values for $t_{1/2}$ decrease when $X_{A,0}$ increases. This effect is easy to understand, because at very low equivalent fractions of ion B (here H^+) in the Nernst film (i.e., high values of ion A (here Cs^+)) the rate of diffusion is controlled by the more dilute ion [1,2] (here H^+). In the present case, however (experimental values as described, $D_B/D_A = 1.5$), this effect is rather small. If we compare, e.g., the two reactions starting at $X_{A,0} = 0$ and 0.81 (see Fig. 12) and assume hypothetically $\alpha_B^A \to 1$, the calculated half-times become 900 s and 840 s, respectively. The experimentally observed much larger difference for these two half-times (factor 7), thus can only be due to the effect of the separation factor α. The calculations do indeed show that if $\alpha_B^A > 1$, the half-time will increase strongly for reactions starting at low values of $X_{A,0}$ and will decrease for high values of $X_{A,0}$. If $\alpha_B^A < 1$, the opposite behavior is observed.

It is of interest also to consider experimental conditions where the exchange equilibrium in the cell is always attained instantaneously (*instantaneous equilibrium*) for elucidation of the strong effect of selectivity on the kinetic behavior of exchangers in a flow cell. If the ion exchanger attains its equilibrium with the influent solution in the cell at any time interval instantaneously, the resulting half-time for the (partial or total) conversion of the ion exchanger for a given flow, J, is minimal (though not zero). The value of J has, of course, to be sufficiently high to avoid this situation when evaluating the rate of ion exchange in a stirred-flow experiment. To calculate the rate curves in the limit of instantaneous equilibrium the stirred-flow process is divided into a series of successive, differentially small batch experiments in which equilibrium is always attained. A computer algorithm is then employed to calculate for each small time interval Δt: (1) the amount of solute (volume $J\ \Delta t$) entering the cell, (2) the equilibrium uptake and release of ions by the ion exchanger and the subsequent change of the concentration of the ions in the cell, and (3) the amount of solute leaving the cell with the outflow. The equilibrium uptake of ion A by the ion exchanger in each batch process (step 2) is calculated with Eq. (23). The concentrations, $c_{A,\infty}$ and $c_{B,\infty}$, in the cell at the end of each time interval Δt are calculable with the values obtained for $X_{A,\infty}$. The subscript 0 in Eq. (23) identifies initial values. These are obtained always in a DO LOOP from the equilibrium values for the ions in solution and in the ion exchanger attained in the preceding differentially small batch experiment.

If the half-times for the successive partial conversions of the ion exchanger are calculated in this way for the experiments described above assuming instantaneous equilibrium the values for $t_{1/2,\text{inst}}$ shown in Fig. 12 as dotted lines are resolved. These values are always sufficiently smaller than the observed values to demonstrate that the selected flow, J, in the experiments was adequately high. More interesting, however, is the observation that $t_{1/2,\text{inst}}$ decreases also substantially with increasing values of $X_{Cs,0}$.

If $t_{1/2,\text{inst}}$ is calculated for an ion-exchange reaction with zero selectivity ($\alpha \rightarrow 1$), it is found that $t_{1/2,\text{inst}}$ does not depend on $X_{A,0}$ or $x_{B,\text{in}}$, just as was shown above for stirred-flow experiments, where equilibrium is not attained instantaneously. Obviously, the characteristic dependence of $t_{1/2}$ on $X_{Cs,0}$ and $x_{B,\text{in}}$, as illustrated in Fig. 12, is, to a large extent, not the result of diffusion of the ions across a Nernst film but rather of the fact that the ion-exchange isotherm is nonlinear, i.e., that the ion exchanger is selective for a given ion. Only when $\alpha_B^A \approx 1$, can the effect of the diffusion coefficients, D_A and D_B, modify this behavior substantially. One might suspect that the similar kinetic behavior of $t_{1/2}$ for the instantaneous equilibrium condition and for our experimental conditions would disappear if a higher flow rate, J, or an ion exchanger which reacts slower had been selected. Model calculations show, however, that this is not the case.

If the flow rate in the cell is too high compared to the reaction rates of the ion exchanger, the difference in the concentration of the ions A in the influent and the effluent solutions becomes too small to use for an accurate evaluation of the rate parameters of the particles. In this case, it is, however, possible to obtain the rate parameters by interrupting the experiment after predetermined time intervals and subsequent analysis of the ionic composition of the ion-exchanger particles. If, in the limit, the flow rate is high enough, however, to permit the concentration in the influent and the effluent to be considered identical, i.e., $c_{B,\text{in}} = c_B = c_{B,\text{out}}$

constant and $c_{A,in} = c_A = c_{A,out}$ = constant (corresponding to an infinite volume of a batch experiment), an analytical solution for the rate ion exchange is obtained with Eq. (48) [19]:

$$\ln(1 - U) + \frac{\kappa(X_A - X_{A-0})}{X_{A,\infty}} = -\tau t \tag{48}$$

where U is defined by Eq. (24), and

$$\tau = \frac{c_{tot} R \alpha_B^A}{[X_{A,\infty}(1 - \alpha_B^A) + \alpha_B^A][X_{A,\infty} + \alpha_B^A D_B(1 - X_{A,\infty})/D_A]}$$

and

$$\kappa = \frac{1 - \alpha_B^A D_B/D_A}{1 - \alpha_B^A D_B/D_A + D_B \alpha_B^A/(D_A X_{A,\infty})}$$

The value for $X_{A,\infty}$ can be obtained with Eq. (41). This method has been used earlier to evaluate the rate coefficient and the D_B/D_A values for a resin ion exchanger [19].

Because the rates of ion exchange in a stirred-flow experiment depend so strongly on the separation factor and only to a smaller extent on the actual kinetic properties of the particles, it is conceivable that kinetic behavior qualitatively similar to that described here would also prevail under conditions where particle diffusion rather than film diffusion (as considered here) would be rate determining. In this case the similarity should become especially striking if the separation factor were far from unity. For a rigorous theoretical treatment of this situation, however, more information on the accessibility and concentrations of the free mobile ions in the exchanger phase and its explicit dependence on the separation factor are needed. The Donnan-based model developed recently by J. Marinsky [30–33] should also provide a valuable basis for theoretical mass-transfer investigations in synthetic and natural organic ion exchangers. If the diffusion processes within the particles cannot be defined precisely (as, e.g., in soils), the approach of Aharoni and Sparks [34] to heterogeneous diffusion should facilitate characterization of these systems.

B. Heterogeneous Systems

The kinetic behavior of a well-defined heterogeneous mixture of ion exchangers in a stirred-flow cell has obviously not yet been investigated, neither theoretically nor experimentally. To remedy this situation an attempt is made to extend the theory for homogeneous ion exchangers to heterogeneous mixtures by using model calculations to predict the kinetic behavior of such a system. Again only the exchange of monovalent ions is considered and film diffusion is assumed to be rate controlling to assure clarity of the treatment.

The heterogeneous mixture of ion exchangers consists again of n different fractions present in amounts Q_i mequiv, which differ in their rate coefficients and separation factors. The initial ionic composition of each fraction is the same as before and is characterized by the equivalent fraction $X_{A,0}$ of the counterions A. If this mixture is initially stirred in V mL of pure water, Eqs. (42) and (44), given before for the homogeneous case, are still valid. Equation (39), however, has to be modified, because in a heterogeneous mixture all n fractions will release counterions B as a result of the uptake of ions A. This yields

$$\sum_{i=1}^{n} Q_i \frac{d[X_{A,i} - X_{A,0}]}{dt} + Jx_{B,in}c_{in,tot} - V\frac{dc_B(t)}{dt} - Jc_B(t) = 0 \qquad (49)$$

After integration we obtain Eq. (50) for the time dependence of c_B, the quantity which is directly measurable in the effluent as function of time

$$Vc_B(t) = \sum_{i=1}^{n} Q_i[X_{A,i}(t) - X_{A,0}] - J\int_0^t c_B(t)\, dt + Jx_{B,in}c_{in,tot}t \qquad (50)$$

The rate of ion exchange, i.e., $X_A(t)$ is given for each fraction by Eqs. (18) and (19). Because each fraction has its own rate coefficient R_i and separation factor $\alpha_{B,i}^A$, Eq. (45) is replaced by Eq. (51) which consists of a set of n differential equations:

$$\frac{dX_{A,i}}{dt} = R_i \frac{\alpha_{B,i}^A(1 - X_{A,i})\{c_{in,tot}[1 - \exp(-Jt/V)] - c_B(t)\} - X_{A,i}c_B(t)}{\alpha_{B,i}^A(D_B/D_A)(1 - X_{A,i}) + X_{A,i}} \qquad (51)$$

$$i = 1, 2, \ldots, n$$

These n equations and Eq. (50) provide $n+1$ equations for evaluation of the unknown quantities $X_{A,i}(t)$ and $c_B(t)$. However, because Eq. (50) cannot lead to solution of $c_B(t)$ for insertion in Eq. (51), the n differential equations and Eq. (50) have to be solved simultaneously by an iteration process. The integral term in Eq. (49) is also evaluated numerically. The concentration $c_A(t)$ of the ions A in solution is always obtained with Eqs. (42) and (44). The fractional attainment U for each fraction i of the ion exchanger and for the ions A in solution are obtained as

$$U_{i,exch}(t) = \frac{[X_{A,i}(t) - X_{A,0}]}{[X_{A,i,\infty} - X_{A,0}]} \tag{52a}$$

$$U_{sol}(t) = \frac{c_A(t)}{c_{A,in}} \tag{52b}$$

Model calculations. For illustration let us consider the Cs^+/H^+ exchange in a cation exchanger, consisting of five fractions (amount of each fraction $Q_i = 0.2$ mequiv, $i = 1$–5) which are present initially in the H^+ form ($X_{Cs,0} = 0$) and stirred in ($V = 20$ mL) water. The ion-exchanger particles differ only in their diameters (polydisperse mixture), i.e., the rate coefficient, R_i, of each fraction is different, but the separation factors are identical for each fraction. In the present example the values given before in Sec. III.B.1, i.e., $R_1 = 7.5$, $R_2 = 3.4$, $R_3 = 1.4$, $R_4 = 0.58$, $R_5 = 0.415$ cm^3 s^{-1} mequiv^{-1}. For the separation factor and the diffusion coefficient $\alpha_H^{Cs} = 3.16$ and $D_H/D_{Cs} = 1.5$ as before. The influent solution CsCl at the same concentration selected for use in Sec. III.B.1, i.e., $c_{in,tot} = 0.00662M$; $x_{Cs,in} = 1$. In this instance all fractions of the ion exchanger are eventually converted in the cell from the pure H^+ form to the pure Cs^+ form ($X_{Cs,i,\infty} = 1$ for $i = 1$–5). The flux rate, J, is fixed at 1.5 mL s^{-1}. The resulting rate curves calculated with Eqs. (50), (51), and (52) are shown for the five fractions of the ion exchanger and the counterions in solution in Fig. 13 (bottom). The concentration of the H^+ ions released by the ion exchanger are plotted as $c_H/c_{in,tot}$.

In this case the five size fractions of the ion exchanger approach equilibrium with quite different rates. The range of the half-times extends from 45 s for the most rapid particles (fraction 1) to 394 s for the slowest reacting particles. The concentration of Cs ions in the cell and the effluent increase initially, i.e., for about 20 s rapidly, because during this time the pure water originally in the cell is replaced by the influent Cs solution and the ion-exchange process is initiated only slowly during this period. After this time the Cs ions are taken up by the particles more effectively and the Cs concentration approaches its equilibrium more slowly. The concentration of H ion released from the ion exchanger into the cell and effluent passes through a maximum after about 40s. In a computation similar to the one employed for the heterogeneous mixture considered in Sec. III.B.1, the rate curves for a homogeneous, i.e., monodisperse mixture with a particle radius which corresponds to the *median* of the above polydisperse mixture have been obtained for comparison. The rate coefficient is in this case $R = 0.746$ cm^3 s^{-1} mequiv^{-1} (see Sec,

III.B.1); the total amount of ion exchanger in the cell $Q = 5 \times 0.2 = 1.0$ mequiv. The resulting rate curves (obtained from Eqs. (40) to (45) are shown as dotted lines in Fig. 13 [top]). They illustrate that compared to the solution in a monodisperse mixture the concentration of Cs ions in the solution in contact with the polydisperse mixture initially increases more slowly, because the fraction of smaller particles take up Cs ions more rapidly than the particles of the monodisperse mixture. Later, however, opposite behavior is observed, because then the larger and slower reacting particles of the polydisperse mixture remain for Cs uptake. For the same reason the release of H ions is initially faster in the polydisperse mixture but slower later (see Fig. 13). Calculating the rate curves of a polydisperse mixture by assuming an average radius (or rate coefficient) for all particles, therefore, cannot result in satisfactory curve fits. On the other hand, if the measured rate curves are fit with a kinetic model for a homogeneous mixture and the deviations as shown in Fig. 13 are observed, one might conclude that the mixture was actually heterogeneous.

In a treatment similar to the one demonstrated with a homogeneous ion exchanger (see Sec. IV.A.1) the rate curves can be computed using once again the assumption that all fractions would attain their equilibrium in the cell instantaneously (*instantaneous equilibrium*). The calculated half time for the total conversion of the ion exchanger ($Q = 1$ mequiv) is in this case $t_{1/2,inst} = 60$ s. This is, however, no contradiction to the value of only 45 s calculated for the fastest fraction in the polydisperse mixture (see above), because this fraction is present in the mixture to an extent of only 0.2 mequiv. To check, therefore, whether the flow rate used above is sufficiently high to prevent always the fastest fraction in the mixture from not reaching its equilibrium state instantaneously, the rate curves of the polydisperse mixture were recalculated with the values given above, while the value of R_1 was increased by several orders of magnitude. It was then found that the half-time of the conversion of the first fraction approaches $t_{1/2} = 13$ s. Because this value is sufficiently smaller than 45 s, we can conclude that the above flow rate of $J = 1.5$ is high enough to prevent always the fastest fraction from its instantaneous equilibrium state.

To study the effect of flow rate on the kinetics of a polydisperse mixture in more detail we recalculated the rate curves shown in Fig. 13, but decreased J by factor of 10 to 0.15 mL s^{-1}. The rate curves resolved are presented in Fig. 14. If we compare the rate curves for the five particle size fractions with those obtained for the higher values of J (see Fig. 13) the corresponding rates for the conversion of the ion exchanger are of course much slower, but they do not differ from each other as much as they do at the higher flow rate. In particular, the rate curves for frac-

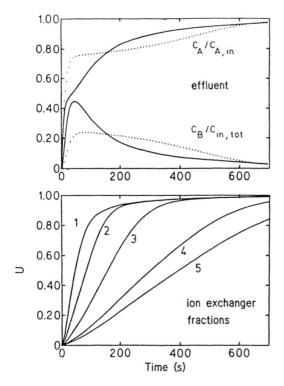

Figure 13 Fractional attainment of equilibrium, U(t), for five size fractions of a polydisperse cation exchanger, *simultaneously* in a stirred flow cell (bottom) during their conversion from the pure A form to the pure B form. For the effluent solution, the concentrations, $c_A/c_{A,in}$, for the ions A take up by the ion exchanger, and for the ions B released, $c_B/c_{in,tot}$, as a function of time are given (top). Because the total conversion of the ion exchanger is considered $c_{A,in}$ = $c_{in,tot}$. Flow rate: 1.5 mL/s. Solid lines: calculated with Eqs. (50), (51), and (52). Dotted lines: calculated for a monodisperse system which assumes the particle radius to be represented by the median radius of the polydisperse mixture.

tions 1 and 2 (the faster-reacting particles) tend to resemble each other more, obviously because the condition of instant equilibrium is nearly fulfilled in this instance. This is especially true for fraction 1. If the rate curves for this low flow rate are recalculated with the above values but the rate coefficient R_1 (simulation of instant equilibrium of fraction 1) is constantly increased, a half time of 300 s is eventually obtained. This value is already quite close to the corresponding value obtained for the nonequilibrium case (Fig. 14) where fraction 1 exhibits a half time of 370

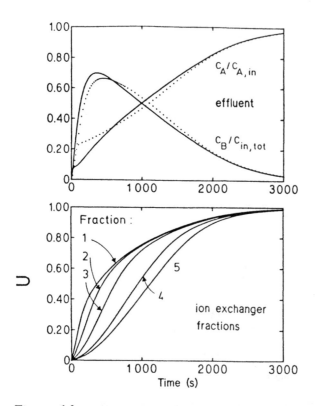

Figure 14   Same system as in Fig. 13, but flow rate $J = 0.15$ mL/s.

s. The fact that at this low flow rate the instant equilibrium condition is approached to a larger extent also explains why the rate curves calculated for the ions in solution for the case of a homogeneous mixture with a corresponding *median* particle radius (see dotted lines in Fig. 14; median rate coefficient again 0.746 cm^3 s^{-1} mequiv^{-1}) deviate much less from those of the polydisperse mixture that were observed for the higher flow rate (Fig. 13). It is obvious that in the limiting case, when the instant equilibrium condition is completely fulfilled, the rate curves observed for the ions in solution do not depend any longer on the rate coefficients of the individual particles, and the difference between a polydisperse and a monodisperse mixture disappears.

    The effect of a very high flow rate for the investigation of a polydisperse mixture is illustrated in Fig. 15, where $J = 15$ mL s^{-1} was selected for the above flow experiment. In this case the concentration of the Cs ions in the effluent solution is nearly the same as that in the influent and

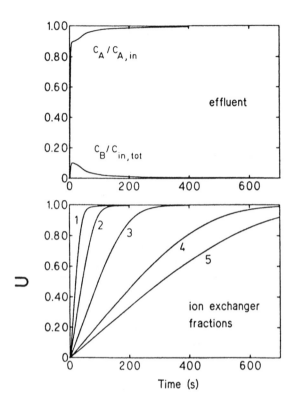

Figure 15 Same system as in Fig. 13, but flow rate $J = 15$ mL/s.

is, therefore, expected to yield very little information on the rates of ion exchange. (For this reason the rate curves for a homogenous mixture with a corresponding median particle radius is not given here). The rate curves of various ion-exchange fractions, however, differ from each other most substantially (Fig. 15). Interruption of the experiment after predetermined time periods, subsequent separation of the mixture into its individual fractions, and analysis of the ionic composition of each fraction, in this case, is expected to yield the rate coefficients of each fraction quite accurately. In this case competition of the various ion-exchanger fractions for the counterions in solution does not actually exist any longer because sufficient ions are always available in the influent. In the limit, each fraction will attain its equilibrium at the same rate that it would if the other fractions were not present in the cell. Therefore, this situation can be evaluated by applying Eq. (48) (i.e., the infinite volume solution), to each fraction in a heterogeneous mixture. Such an evaluation is, however, only possible if the heterogeneous mixture can be separated into several approximately homogeneous fractions.

V. CONCLUSIONS

The rates of ion exchange in heterogeneous mixtures of ion exchangers exhibit in the concentration range, where film diffusion is rate determining, several characteristics which can be predicted with the theory given here, and which have to be considered when experimental data are interpreted.

A. Batch Systems

In contact with a solution containing counterions A, some particles of a heterogeneous ion exchanger mixture will take up the A ions more rapidly than other particles (in response to different diameters, ion-exchange capacities or separation factors) and even overshoot, after some time, their eventual equilibrium uptake considerably. During the subsequent redistribution period, however, these particles have to release considerable amounts of these already sorbed counterions to the slower-reacting particles. Even though such redistribution of the ions also proceeds via film diffusion in the solution phase, the concentration of the counterions in solution changes only very little during this period, because the counterions released by the faster-reacting particles are taken up by the slower-reacting ones.

This effect is enhanced extraordinarily if there is an excess of ion exchanger (in equivalents) in the system as compared to the total amount of A counterions (also in equivalents) initially in solution. In this case the concentration of A in solution becomes very small at the beginning of the redistribution period (exhaustion of the solution). As a result we find, that even though the concentration of counterions in solution will still approach its equilibrium rather rapidly, the ion-exchanger particles will approach their equilibrium during the subsequent redistribution period very slowly (anomalous kinetics). The kinetic behavior of such a system is, therefore, not sufficiently characterized by only measuring the concentration of counterions in solution, or in the combined solid phase as a function of time, because these measurements would simulate, after a short time, an equilibrium which is not approached by the various fractions of the heterogeneous mixture. In this case it will be necessary to interrupt the experiment after predetermined time periods and analyze the ionic composition of the various individual fractions of the ion exchanger.

B. Stirred-Flow Cell

1. Homogeneous Mixture

If the rate of ion exchange is investigated in a stirred-flow cell, selection of a suitable flow rate is important, even for a homogeneous mixture. If the flow rate is too low compared to the rate of ion exchange, instant equilibrium is approached in the cell, and the course of counterion concentration with time in the effluent will not yield any information on the actual rates of ion exchange. Even in this latter case, however, the half-time for the (partial or total) conversion of the ion exchanger in the cell is not zero but attains only its minimal $t_{1/2,inst}$ value. If the flow rate is too high, the concentration of the counterions in the influent and the effluent solution will be nearly the same and cannot be used to evaluate the rates of ion exchange accurately.

The half-time of ion exchange, as observed in a stirred-flow cell, decreases even for a homogeneous ion exchanger if the initial fraction $X_{A,0}$ of the preferred ion A in the ion exchanger is increased. In addition, at a given value of $X_{A,0}$ the half-time also decreases if the ion exchanger is converted to a larger content. Only if the separation factor of the ion exchanger for the counterions is close to unity can the diffusion coefficients modify these observations. In this case the half-time decreases with increasing $X_{A,0}$ if $D_B/D_A > 1$, and increases if $D_B/D_A < 1$. To some extent this is also observed in batch experiments [20]. In a stirred-flow experiment, however, this strong dependence of the half-time on the initial ionic composition of the ion exchanger remains even for $t_{1/2,inst}$, i.e., for a case where the kinetic properties of the ion exchanger are no longer relevant. Theoretical considerations show that this behavior is, in both cases, essentially the result of the selectivity of the ion exchanger (i.e., nonlinearity of the ion-exchange isotherm). This demonstrates that the half-time of ion exchange, as observed in a stirred-flow cell, depends to a large extent on equilibrium properties rather than on the kinetic properties of the material investigated (e.g., rate coefficients and diffusion coefficients).

2. Heterogeneous Mixture

If the kinetics of a heterogeneous mixture of ion exchangers is investigated by the stirred-flow method, the selection of the appropriate flow rate becomes rather difficult. If the flow rate is adjusted to the rate coefficients of the slowest-reacting particles, the faster-reacting particles in the mixture will reach their equilibrium in the cell almost instantaneously. In this case the rate curves for all faster-reacting particles will almost coincide and the rate curve for the counterions in the effluent can also be fitted to that of a homogeneous mixture. To attempt to prove that the existence of kinetic homogeneity of an unknown mixture by fitting the rate curve of counterions in the effluent with a homogeneous model is thus not appropriate, because, at the flow rate selected, a large fraction of the particles may satisfy the instant equilibrium condition. If, on the other hand, a high flow rate is selected in order to adjust it to the fastest particles, the concentration of counterions in the effluent is hardly affected by the ion-exchange reactions of the slower-reacting particles and the rate curve for the counterions in the effluent will be primarily controlled by the faster-reacting particles.

As a consequence, to obtain meaningful information with respect to the kinetic parameters of an unknown heterogeneous mixture from the rate curve of the counterions in the effluent, experiments at various different flow rates have to be performed and the results compared with calculated rate curves obtained for the heterogeneous mixture with iteratively adjusted values for the rate parameters (R_i, D_2/D_1), for the quantities of Q_i of each fraction, and for the separation factors $\alpha_{B,i}^A$ of each fraction. Only if a set of such experiments yields consistent values for these quantities, should they be accepted with some confidence.

If the heterogeneous mixture can be separated into its various fractions, it is , of course, easier to determine the rate parameters and other quantities associated with the fractions of these isolated particles to predict the kinetic behavior of the mixture subsequently by applying the rate equations given above for heterogeneous systems. Separation of a heterogeneous mixture into individual fractions with respect to their kinetic properties might, however, be difficult to achieve, because the rate of ion exchange depends not only on the surface to volume ratio of the particles but also on the ion-exchange capacity and the separation factor. Fractionation by sieving is appropriate only if the mixture consists of different sized spherical particles of an otherwise homogeneous material. If a fractionation is not possible (e.g., for soils or soil components), the parameters R_i, Q_i and $\alpha_{B,i}^A$ of the various fractions may be obtained from a series of stirred-flow experiments at different flow rates. This is an advantage of the stirred-flow method in the case of heterogeneous mixtures, because in batch experiments a unique evaluation of the above parameters for each fraction from a single observed rate curve of the counterions in solution is hardly possible. In addition, anomalous kinetic effects, as observed for batch procedures under certain conditions (see above) are not present in stirred-flow experiments. This is due to the fact that in this case the competition of the various ion exchanger fractions for the ions in solution is only weak, because additional counterions are always flowing into the cell.

If the rate equations given above for heterogeneous mixtures are applied for the sorption of ions by soils or soil components, one has, of course, first to examine, whether film diffusion is in fact the rate-determining step. In several cases this was indeed observed to be the case [21,35,36,37]. If this is not the case, selection of the correct rate equation for the rate-controlling step can be quite difficult. As emphasized by Amacher [38], selection of an incorrect rate equation will produce rate coefficients which vary with initial concentration and time. The above considerations reveal that even when the correct rate equation is selected, the fact that in heterogeneous mixtures different particles will sorb the ions with different rates has also to be taken into account quantitatively (irrespective of the actual rate-controlling mechanism). Otherwise, the (mean) rate coefficient obtained from a stirred-flow experiment will depend on the flow rate selected, and not necessarily agree with the value obtained from batch methods.

REFERENCES

1. F. Helfferich, *Ion Exchange*, McGraw-Hill, New York, 1962.
2. F. Helfferich and Y. Hwang, in *Ion Exchangers* (K. Dorfner, ed.), Walter de Gruyter, New York, 1991, Chap. 6.2.
3. C.B. Amphlett, *Inorganic Ion Exchangers*, Elsevier, Amsterdam, 1964.
4. F. Helfferich, in *Ion Exchange* (J. A. Marinsky, ed.), Marcel Dekker, New York, 1966, Vol.1, Chap. 2.
5. F. Helfferich, in *Mass Transfer and Kinetics of Ion Exchange* (L. Liberti and F. Helfferich, Eds.), Martinus Nijhoff, The Hague, 1983, Chap. 5.
6. L. Liberti, *Mass Transfer and Kinetics of Ion Exchange* (L. Liberti and F. Helfferich, eds.), Martinus Nijhoff, The Hague, 1983, Chap. 6.
7. L. Liberti and R. Passino, in *Ion Exchange and Solvent Extraction* (J. Marinsky and Y. Marcus, eds.), Marcel Dekker, New York, 1985, Vol. 9, Chap. 3.
8. N.M. Brooke and L.V.C. Rees, *Trans. Faraday Soc.*, *65*: 2728 (1969).
9. I.V. C. Rees and A. Rao, *Trans. Faraday Soc.*, *62*: 2103 (1966).
10. D.L. Sparks, *Kinetics of Soil Chemical Processes*, Academic Press, San Diego, 1988.
11. A. Bar-Tal, D.L. Sparks, J.D. Pesek, and S. Feigenbaum, *Soil Sci. Soc. Am. J.*, *54*: 1273 (1990).
12. M.J. Eick, A. Bar-Tal, D.L. Sparks, and S. Feigenbaum, *Soil Sci. Soc. Am. J.*, *54*: 1278 (1990).
13. D.L. Sparks and P.M. Jardine, *Soil Sci. Soc. Am. J.*, *45*: 1094 (1981).
14. D.L. Sparks, L.W. Zelazny, and D.C. Martens, *Soil Sci. Soc. Am. J.*, *44*: 1205 (1980).
15. K. Bunzl, *Z. Phys. Chem. (Munich)*, *75*:118 (1971).
16. K. Bunzl, *J. Chromatogr.*, *102*: 169 (1974).
17. P.J. Dunlop, *J. Phys. Chem.*, *68*: 3062 (1964).
18. D. Miller, *J. Phys. Chem.*, *71*: 616 (1967).

19. K. Bunzl and G. Dickel, *Z. Naturforsch., 24a*: 109 (1969).
20. K. Bunzl, *J. Soil Sci., 25*: 343 (1974).
21. K. Bunzl, *J. Soil Sci., 25*: 517 (1974).
22. K. Bunzl, A. Wolf, and B. Sansoni, *Z. Pflanzenern. Bodenk., 4*: 475 (1976).
23. K. Bunzl, W. Schimmack, in *Rates of Soil Chemical Processes* (D.L. Sparks and D.L. Suarez, eds.) Soil Science Society of America, Madison, WI, 1991, Chap. 5.
24. K. Bunzl, *Anal. Chem., 50*: 258 (1978).
25. C. Wolfrum, K. Bunzl, G. Dickel, and G. Ertl, *Z. Phys. Chem. (Munich), 135*: 185 (1983).
26. K. Bunzl, *J. Phys. Chem., 95*: 1007 (1991).
27. K. Bunzl, *Radiochim. Acta, 58/59*: 461 (1992).
28. K. Bunzl, *J. Chem. Soc. Faraday Trans., 89*: 107 (1993).
29. G.E. Boyd and K. Bunzl, *J. Am. Chem. Soc., 96*: 2054 (1974).
30. J.A. Marinsky, in *Mass Transfer and Kinetics of Ion Exchange* (L. Liberti and F. Helfferich, eds.) Martinus Nijhoff, The Hague, 1983, Chap. 3.
31. J.A. Marinsky, *J. Phys. Chem., 89*: 5294 (1985).
32. J.A. Marinsky, R. Baldwin, and M.M. Reddy, *J. Phys. Chem., 89*: 5303 (1985).
33. J.A. Marinsky and M.M. Reddy, *J. Phys. Chem., 95*:10208 (1991).
34. C. Aharoni and D.L. Sparks, in *Rates of Soil Chemical Processes,* (D.L. Sparks and D.L. Suarez, eds.), Soil Science Society of America, Madison, WI, 1991, Chap. 1.
35. R. Salim and B.G. Cooksey, *Plant and Soil, 54*: 399 (1980).
36. R. A. Ogwada and D. L. Sparks, *Soil Sci. Soc. Am. J., 50*: 1162 (1986).
37. A. P. Jackman and K. T. Ng, *Water Res. Res., 22*: 1664 (1986).
38. M. C. Amacher, in *Rates of Soil Chemical Processes,* (D.L. Sparks and D.L. Suarez, eds.), Soil Science Society of America, Madison, WI, 1991, Chap. 2.

7

Evaluation of the Electrostatic Effect on Metal Ion-Binding Equilibria in Negatively Charged Polyion Systems

Tohru Miyajima

Kyushu University, Hakozaki, Higashi-ku, Fukuoka, Japan

I. INTRODUCTION

Counterion binding in polyion systems is of fundamental importance in various fields of science and has long been studied from both theoretical and practical points of view. However, in spite of recent rapid technological developments in synthetic polymer science and increasing needs for the deeper understanding of the delicate properties of naturally occurring polyelectrolytes, such as bipolyelectrolytes and humic substances, the analytical treatment of their binding equilibria still remains imperfect [1]. Due to the lack of a generalized treatment, the macroscopic binding equilibrium data compiled up to the present are not fully related to the reactions at the molecular level. Establishment of a unified analytical treatment of the ion-binding equilibria of the linear and cross-linked linear polyion systems is required. The objectives of the present review are (1) to present the experimental evidence for aspects common to the equilibria of ion binding in negatively charged polymer systems of various ionic groups, of various backbone structures, and of various dimensions, and (2) to propose a thermodynamic approach for relating the microscopic binding equilibria at the reaction site to the macroscopic binding data.

Several unique physicochemical properties of polyion systems originate from the combination of their polymeric and ionic character. In particular, an electrostatic effect attributable to counterions being concentrated around the polymer skeleton so as to reduce the electrorepulsive force among the fixed negative charges on the polymer backbone, is one of the characteristic properties in the binding equilibria of countercations to negatively charged polyions. Since the "polyelectrolyte effect" results from the accumulation of the electrostatic effect due to individual mono-meric ionic groups which constitute the polymer molecule, it is of inter-est to study how the ion-binding properties are dependent on the size of the polyion. In this review, therefore, we will treat not only polyion sys-tems but also simple and oligoion systems, in order to observe the tran-sition from simple ion to polyion. By such transition studies, a better understanding of the electrostatic effect inherent in the counterion-polyion binding equilibria, is expected to be facilitated.

A Gibbs-Donnan concept has been proposed by Marinsky and his co-workers [2–9] for the interpretation of counterion binding equilibria to linear and cross-linked linear polyions. Initially this approach was success-fully applied to describe the ion distribution equilibria in ion exchangers (polyion gels) [10–13]. More recently, the concept has been proven to be equally applicable for the description of water-soluble linear polyion-counterion binding systems [14]. In the case of ion exchangers (polyion gels) the boundary between the ion-exchanger phase and the bulk solu-tion phase is defined explicitly, and the volume of solvent uptake by the exchanger can be determined. This enables the calculation of the *aver-aged concentrations* of each chemical species, such as fixed ions, co-ions, and counterions in the exchanger phase. Contrary to the gel systems, *the effective polyelectrolyte phase volume of water-soluble linear polyions* is not available by any straightforward experimental procedure. Instead, estimates of the hypothetical polyelectrolyte phase volume have been shown to be extractable from comparison of polyion-counterion equilib-rium measurements with idealized projections of these equilibria [7, 15, 16]. This computational procedure, when applied to the polyion gel sys-tems, resolve polyelectrolyte phase volumes in excellent agreement with the experimentally based values in the case of rigidly cross-linked gels; however, it is also revealed that the effective polyelectrolyte phase volume is projected to be much smaller than the gel phase volume measured for highly swollen gels. Rather than attribute the result to failure of the Donnan model this inconsistency has been considered to be a conse-quence of failure of the method for the gel phase volume measurement. On the basis of this assessment it has been claimed that the equilibria of counterion and co-ion distribution between the polyelectrolyte phase and the bulk solution phase is described quantitatively by the Gibbs-Donnan equation and can be taken advantage of for the precise assessment of the hypothetical counterion-concentrating region around the polymer skeleton.

In order to visualize the essential role of the electrostatic effect on the polyion-binding equilibria, we have chosen to study their interaction with metal and H^+ ions at different salt concentration levels. Acidic dextrans (Fig. 1) and their gel analogs were chosen to serve as examples of polyion and polyion gel, because of their high hydrophilicity and the common backbone structures. By use of these ionic dextran samples, it has been possible to compare the computational procedures used to evaluate the electrostatic effect whether they are linear or cross-linked, weak or strong-acid polyions. In particular, one of the most important objectives of this project is to define the relationship between the electrostatic effect in the polyion systems and the structural parameters of the polyions [17], e.g., the linear charge separation of the linear polyions. Such treatment of the polyelectrolyte effect is in accord with Manning's theoretical approach [18,19] and comparison of the results obtained with Manning's prediction (two-variable theory [19]) was felt to be important. This objective has been reached in these systematic studies with both the carboxylated and sulfated dextrans studied. Five sections follow this introductory section. In Sec. II, both the standard and fundamental aspects encountered in various polyion systems are examined with the Gibbs-Donnan model. Sections III, IV, and V present detailed discussion of polyions of the respective ionic groups selected; the carboxylate (Sec. III), sulfate, sulfonate (Sec. IV), and phosphate groups (Sec. V). All the equilibrium data presented and discussed in this review were compiled at 298 K unless otherwise stated.

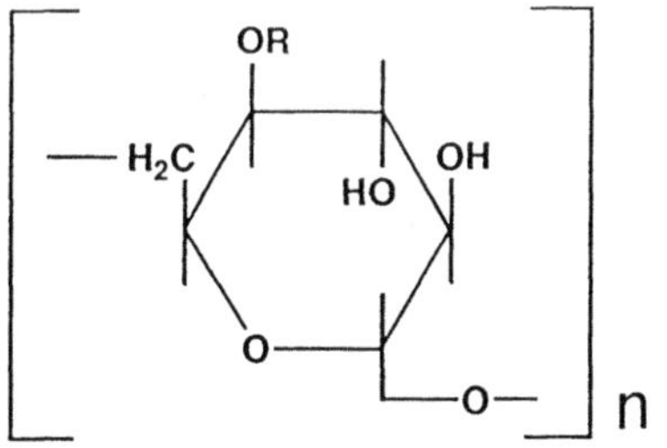

R : SO_3^- / Dextran sulfate (DxS)
CH_2COO^- / Carboxymethyldextran (CmDx)

Figure 1 Structures of ionic dextrans. The degree of substitution, DS, varies from 0 to 3.

II. FUNDAMENTALS

Polyelectrolyte properties arise from the eventual arrangement in response to mutual repulsions of similarly charged ionic groups (negatively charged groups in this review) attached to each other in the resultant polyion [20]. The system of polyions, counterions, co-ions, and solvent molecules is stabilized by the association of counterions in the vicinity of the polyion skeleton. The source of this association is electrostatic and does not always indicate any direct binding of the fixed groups to the counterions by complexation. In order to develop useful insights with respect to the "polyelectrolyte effect," it is of interest to examine the pattern of transition from simple electrolyte to linear polyion, i.e., the contribution of size to the "polyelectrolyte effect." The pattern of transition from neutral polymer to linear polyion is also of interest. With such information the dependence of polyelectrolyte effect on linear charge density, or the linear charge separation on the polymer backbone, b, expressed in Å becomes accessible. Dextran samples containing carboxylate (Carboxymethyldextran, CmDx) or sulfate groups (Dextransulfate, DxS), the degree of substitution defining "b," were used to determine the dependence of the polyelectrolyte effect on their parameter. Their structures are shown in Fig. 1. A family of inorganic polyphosphates of linear and cyclic structures were selected on the basis of their stability in a neutral aqueous media to examine the nature of the transition from simple electrolyte to polyelectrolyte. Sodium ion activity measurements [21–23] combined with a ^{23}Na NMR [24-26] studies were employed to investigate the interaction between Na^+ ion and the polyanions produced by the dissociation of the sodium salts of both the ionic dextrans and the inorganic polyphosphates.

One of the simplest ways to observe the electrostatic effect inherent in polyion systems is to measure the *activity* of the counterion of the polyelectrolyte solution in the absence of simple salt. Single ion activities of these sodium salt solutions, a_{Na}, measured electrochemically by use of a Na^+ ion selective glass electrode can be expressed by the following equation;

$$a_{Na} = \gamma_{Na} C_{Na} \tag{1}$$

where γ_{Na} and C_{Na} correspond to the apparent activity coefficient and the total concentration of Na^+ ion present in the sample solution. Even though this electrochemical determination of the single ion activity coefficient includes the assumption that the liquid junction potential of the electrochemical cell is invariant during the measurement [22], this method is considered to be one of the most straightforward and reliable for the evaluation of the effective concentration of counterions in electro-lyte solutions [21–23]. The γ_{Na} values determined in this manner for in-organic polyphosphate samples of various degrees of polymerization are plotted in Fig. 2 against the phosphate concentration expressed on a monomer basis, Cp [27]. Since the samples used are, singly charged hypophosphite ion ($PH_2O_2^-$), P^I, a monomer analog of a long-chain polyphosphate ion, PP, oligo cyclic phosphate ion ($P_nO_{3n}^{n-}$), cP_n, and lin-ear polyphosphate ion with different chain lengths [28], P_n, the Cp val-ues are equated to C_{Na}. Sodium salt of hypophosphite, $NaPH_2O_2$, is an ex-ample of a simple electrolyte and the decrease in γ_{Na} of this salt with increasing C_{Na} is considered simply due to Debye-Hückel type interaction. It can be seen, as well that γ_{Na} approaches unity as C_{Na} decreases when the value of n is rather small. With an increase in the polymerization number, n, or the average degree of polymerization, ñ, i.e., the multipli-

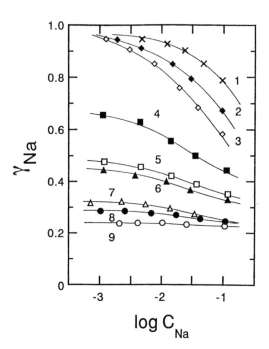

Figure 2 Apparent activity coefficients of Na^+ ions in aqueous solutions of Na^+ salts of hypophosphite, cyclic polyphosphates, and linear polyphosphates. 1:P^I; 2:cP_3; 3:cP_4; 4:$P_ñ$ (ñ = 15); 5:$P_ñ$ (ñ = 35); 6:$P_ñ$ (ñ = 50); 7:$P_ñ$ (ñ = 85); 8:$P_ñ$ (ñ = 101); 9:$P_ñ$ (ñ = 127, 150). The pH values of the sample solution are between 6 and 7.

cation of ionic groups by n P-O-P linkages, γ_{Na} decreases drastically to approach a constant value, ca. 0.2. For the linear polyphosphate ion whose n value is more than ca. 100, the γ_{Na} value remains constant of approximately, 0.2, irrespective of C_{Na}. This is one of the most important "polyelectrolyte effects" and the experimental evidence indicates that the electrostatic potential at the surface of the polymer is kept constant irrespective of the variation in the polymer concentration. With a sufficient chain length, the electrostatic effect is only dependent on the "b" value of the linear polyion. The γ_{Na} values measured for dextran sulfate (DxS) and carboxymethyldextran (CmDx) samples of various degrees of substitution are plotted against the linear charge density, $1/b$ (Å^{-1}) [29] in Fig. 3 to show this. In spite of the difference in the nature of the ionic groups of DxS and CmDx, the γ_{Na} values determined are only dependent on the $1/b$ value. The γ_{Na} values are also unaffected by the polymer concentration just as they are in the long-chain polyphosphate system and constancy of γ_{Na} has been frequently encountered with various polyelectrolyte systems [21–23]; the constant γ_{Na} value has been attributed to immobilization of a constant fraction of counterion by electrostatic interaction with the charged groups of the polyion while the fractional release of counterion to the solution from the polyion is independent of the mag-

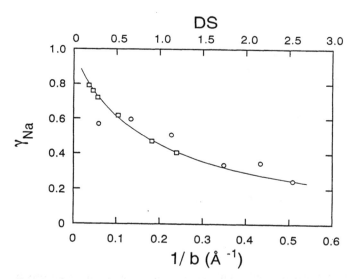

Figure 3 Apparent activity coefficients of Na^+ ions in aqueous solutions of Na^+ salts of DxS and CmDx samples with various degrees of substitution (Cp = 0.005 monomol dm^{-3}). (O) DxS; (□) CmDx. The pH values of the sample solution are between 6 and 7.

nitude of Cp [30]. According to Katchalsky and his co-workers [31], γ_{Na} determined electrochemically, is equal to the practical osmotic coefficient of the sodium salt of the polyion, $\phi_{p,Na}$, and is also uniquely related to the "b" value.

$$\gamma_{Na} = \phi_{p,Na} \tag{2}$$

It should be stressed that such "condensation" of counterion by polyion is determined just by the structural parameter that defines charge density along the length of the macromolecule. It is not influenced by external condition, such as Cp or the addition of salt. The fact that the colligative properties of salt and polyelectrolyte are found to be additive [32,33] when salt is added to the polyelectrolyte provides insight with respect to the uniqueness of $\phi_{p,Na}$ and γ_{Na}. Such behavior is attributable to the inaccessibility of the polyion, the "condensed" Na^+ ions and the solvent associated with the polyion domain, to the measurements being carried out. Their presence as a separate phase, however, is not detectable by the counterion activity measurement in the absence of simple salt.

The hydration state of the electrostatically "condensed" Na^+ ions is considered to be similar to that of the Na^+ ions in the bulk solution [34]. This is easily verified by measuring the ^{23}Na NMR spectra of Na^+-polyion mixture. All the spectra obtained for the sodium salts of inorganic polyphosphates and dextrans will give the chemical shift values close to 0 ppm with 0.1 mol dm^{-3} NaCl solution used as the reference. The peaks obtained for the oligoelectrolytes and the polyelectrolytes, however, are much broader than those of simple Na^+ salts such as NaCl. Due to a quadrupolar effect of the ^{23}Na nucleus (I = 3/2), the Na^+ ion relaxation mechanism is greatly influenced by the position of the Na^+ ion in an electric field gradient [35–38]. The peak width at half peak height ($W_{1/2}$) reflects the degree of territorial binding of the Na^+ ions to the polyion surface [39]. The plots of $W_{1/2}$ against the logarithm of polyphosphate chain length, shown in Fig. 4, are indeed consistent with the results obtained by the thermodynamic study just mentioned above (Fig. 5), i.e., the $W_{1/2}$ value for long-chain polyphosphate is insensitive to the Cp value, as expected by the γ_{Na} measurement, whereas the Cp effect is pronounced in the oligoelectrolyte system. The $W_{1/2}$ values determined for the samples of NaDxS and NaCmDx are plotted against the degree of substitution (DS) in Fig. 6. Both plots show discontinuities at a DS value of about 0.5; from this point a rapid increase in the $W_{1/2}$ value is observed as DS increases. It is of interest to recall that such a change at a particular linear charge density is predicted by *Manning's ion-condensation theory* (b = 7.14 Å at 25°C in an aqueous solution) [18].

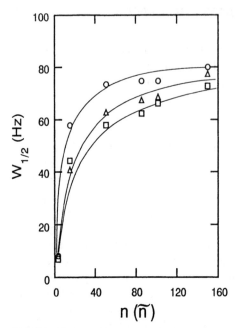

Figure 4 Correlation between the $W_{1/2}$ value and the degree of polymerization of sodium salts of inorganic polyphosphates. (□) $C_p = 0.005$ monomol dm^{-3}; (△) $C_p = 0.01$ monomol dm^{-3}; (○) $C_p = 0.05$ monomol dm^{-3}.

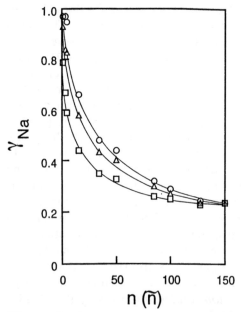

Figure 5 Chain length dependence of the apparent activity coefficients of Na^+ ions in aqueous solutions of Na^+ salts of polyphosphate. (○) $C_p = 0.001$ monomol dm^{-3}; (△) $C_p = 0.010$ monomol dm^{-3}; (□) $C_p = 0.100$ monomol dm^{-3}.

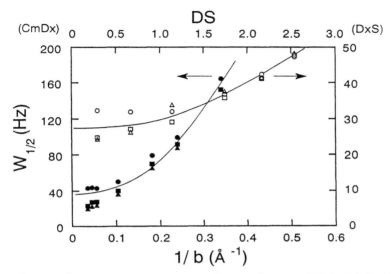

Figure 6 $W_{1/2}$ change with linear charge density of DxS and CmDx. ($\square$, $\blacksquare$) $C_p = 0.005$ monomol dm^{-3}; ($\triangle$, $\blacktriangle$) $C_p = 0.01$ monomol dm^{-3}; ($\bigcirc$, $\bullet$) $C_p = 0.05$ monomol dm^{-3}. Open symbols; DxS: Closed symbols; CmDx.

In the presence of excess 1:1 salt (B^+X^-), the binding equilibria of a polyion, P^-, and a small counterion, A^{z+}, are accessible to quantitative assessment, the *concentrations* of each chemical species being determined precisely by several established experimental procedures. The degree of the binding to polyions can generally be expressed by the average number of bound A^{z+} ions per fixed polyion group; monomer basis, θ_A, where θ_A is defined as the amount of bound A^{z+} ion, $(n_A)_b$, divided by the total amount of ionic groups fixed to the polyion, n_p.

$$\theta_A = \frac{(n_A)_b}{n_p} \tag{3}$$

For a water-soluble linear polyion system the value of θ_A can be determined by use of the total and the free concentrations of A^{z+} ions, expressed as C_A and $[A]$, respectively, and the total concentration of the polyion expressed on a monomer basis, C_p,

$$\theta_A = \frac{C_A - [A]}{C_p} \tag{4}$$

The binding equilibrium is expressed in the form of a binding isotherm; in many cases, θ_A is plotted against log $[A]$.

Several experimental methods based on different principles have been proposed for the study of polyion-small ion-binding equilibria; they include an electrochemical method, mainly through application of ion-specific electrodes [40–44], an equilibrium dialysis method [45], a dye method [46–51], and an ion-exchange distribution method [52–56]. In any case, the free A^{z+} ion concentration in the system under investigation is determined. They permits the evaluation of θ_A with Eq. (4). The binding quotient, K_A, used to express the overall (apparent) equilibrium is then given by Eq. (5).

$$K_A = \frac{\theta_A}{[A]} \tag{5}$$

K_A is a function of θ_A, e.g., K_A decreases with an increase in θ_A in the case of the interaction between negatively charged polymers and positively charged small ions, such as metal ions. The easiest way to avoid this complexity is through the determination of the intrinsic binding constant defined [19] as

$$K_A^0 = \lim_{\theta_A \to 0} \frac{\theta_A}{[A]} \tag{6}$$

The plots of $\theta_A/[A]$ ($=K_A$) versus θ_A are the well-known Scatchard plots, and the K_A^0 value is obtainable by extrapolating to $\theta_A = 0$.

Due to the strong electric field formed at the surface of the polyions, polyion-counterion interaction is enhanced. By taking into consideration the electrostatic free energy, the apparent binding constant measured experimentally, can generally be related to an intrinsic binding constant by the Boltzmann factor of $\exp(-Ze\psi/kT)$, where ψ and e represent the surface potential and the unit electric charge, respectively, and kT has its usual meaning. There are mainly two different approaches to estimate of the ψ value; the first uses a surface potential along the polymer plane (two-dimensional), and the second, an averaged potential identifiable with Donnan potential exercised by the solvent sheath of its polymer domain (three-dimensional). With the second operational picture linear polyions, their cross-linked gel analogs and ionic micelles are treated with a two-phase model, the volume of the second phase, being defined by the thickness of the solvent sheath of the polyion domain. The prerequisite for the two-phase model is, as has already been discussed previously by Marinsky [3,4], the permeability of the polyion domain by solvent and simple electrolyte.

Straightforward evidence for the validity of the two-phase model is provided by the pronounced salt concentration ($[B^+]$) dependence of the binding equilibrium of A^{Z+} ion to P^- ion. Since the binding is greatly suppressed by an increase in $[B^+]$, the binding isotherm, i.e., the plots of θ_A versus $\log[A]$, shift to the right when the neutral salt concentration is increased. The vertical displacement at each $\log[A]$ value for a particular change in the concentration level of B^+ remains essentially unchanged. Representative plots of this phenomenon are presented in Figs. 23, 40, and 56 to show this. By plotting the $\log[A]$ values at a specified θ_A value against the $\log[B]$ values, the straight line resolved is expected to have a slope equal to Z as long as the value of the polyion domain volume remain unaffected by the salt (B^+) concentration level. The following simple relationship is derived with the Gibbs-Donnan model to explain and justify this expectation. At equilibrium the distribution ratio of A^{Z+} and B^+ ion in the solution and solvent-sheathed polyion is provided by Eq. (7).

$$\frac{(a_A)_D}{a_A} = \left\{\frac{(a_B)_D}{a_B}\right\}^Z \tag{7}$$

where a corresponds to activity and D represents the polyion domain, i.e., the Donnan phase. For simplicity, activity is replaced by concentration, and Eq. (7) is transformed into the following equation;

$$\log[A] = Z\log[B] + \log\left\{\frac{[A]_D}{[B]_D^Z}\right\} \tag{8}$$

At a constant θ_A value, the $\{[A]_D/[B]_D\}^Z$ term, if it remains constant, must lead to the resolution of binding curves, if determined at different $[B]$ values, that parallel to each other with the distance between them expressed by $Z\log[B]$. Also, by use of the polyion domain volume, V_D, $[A]_D$ can be expressed as

$$[A]_D = \frac{k_f\theta_A n_p}{V_D} \tag{9}$$

where k_f corresponds to the fraction of the A^{Z+} ion that are "electrostatically bound" as distinguished from those chemically bound to the charged functionality repeated in the polyion domain and may be regarded constant irrespective of $[B]$ at $\theta_A \to 0$. It can have a value from 0 to 1, depending on the degree of chemical binding of the A^{Z+} ion to the ionic functionality repeated in the polymer molecule. In this equation n_p represents, as before, the total amount of the ionic groups repeated on the polyion expressed as monomoles. By substituting Eq. (9) into Eq. (8), and knowing that $\theta_A/[A] = K_A^0$ and that $[B]_D = n_p/V_D$ at $\theta_A \to 0$, we have

$$\log K_A^0 = -Z \log[B] + \log\left\{ (k_f)^{-1}\left(\frac{n_p}{V_D}\right)^{Z-1}\right\} \tag{10}$$

By differentiating $\log K_A^0$ with respect to $\log [B]$, Eq. (11) is obtained.

$$\frac{d(\log K_A^0)}{d(\log[B])} = -Z \tag{11}$$

The properties of such graphical representations of data compiled for the various combinations of $A^{Z+}/B^+/P^-$ systems have justified this interpretation.

In the two-phase model, two different modes of interaction with polyions are assigned to counterions [19,34]. One mode is attributed to purely electrostatic capture of the counterions by the smeared electrostatic potential radiating from the charged polymer backbone; this mode of counterion entrapment is called *ion-condensation* or *territorial binding*. As opposed to this, a direct binding of counterions to the oxygen atom(s) of the fixed negative charges of polyions can be classified as *site binding*. Only through the use of a microscopic measurement can the two kinds of ions be distinguished. For example, spectroscopic measurement can classify counterions as "free" or "bound." Since the hydration structures of the territorially bound counterions are essentially the same as in the bulk solution phase and they move freely within the volume of the polyion domain, their free state is identifiable, whereas a considerable change in the hydration structures of the "site-bound" counterions can be detected spectroscopically. For simplicity, the complexation equilibria of the polyion systems can be divided into two processes. First, the free A^{Z+} ions in the bulk solution are transferred into the polyion domain, and then the free A^{Z+} ion in the polyion domain reacts the ligand group on the polyion to form contact ion pairs or in some instances chelated complexes. The standard macroscopic measurements of free metal ion in the polyion/simple salt systems can be employed only indirectly to assign parameters to distinguish between the two kinds of metal ions. It is possible then, with the Gibbs-Donnan model to estimate the volume of the effective polyion domain, together with the amount of territorially bound A^{Z+} ions (metal ions in this review). This enables the analysis of the intrinsic metal complexation equilibria in the Donnan phase.

III. CARBOXYLATE POLYIONS AND THEIR GEL ANALOGS

A. Acid Dissociation Equilibria of Carboxymethyldextrans and Their Gel Analogs

In order to establish unified analytical treatment of the ion binding of linear polyions and their gel analogs, a study using the polyions and gel samples containing the same backbone was initiated. The weak-acid dextrans, CmDx and its CmDx gel analogs, i.e., CM-Sephadex C-25 and 50 (Pharmacia, Sweden) were chosen for this purpose. In order to examine the acid-dissociation properties of these polyacid systems in detail in the presence of excess of simple salt, e.g., NaCl, precise determination of the free H^+ concentration at equilibrium in the course of their neutralization with standard base was measured. Most of the p[H] determination were made using a glass electrode. In this research project, the calibration of the electrochemical cell was accomplished using Gran's plots [57] rather than standard buffers. To a portion (V_0 cm^3) of the H^+-ion form of the polyion solution, V_{NaOH} cm^3 standard NaOH solution (C_{NaOH} mol dm^{-3}) were added and the p[H] value together with the degree of dissociation of the polyion, α, were determined. The value of α defined as the ratio of dissociated carboxylate groups to the total (dissociated and associated) carboxylate groups, n_p, is calculated with the following equation:

$$\alpha = \frac{C_{NaOH}V_{NaOH} - [H](V_0 + V_{NaOH})}{n_p} \tag{12}$$

By use of the α value, an apparent acid dissociation constant, pK_{app}, is defined as

$$pK_{app} = p[H] - \log\left(\frac{\alpha}{1-\alpha}\right) \tag{13}$$

Representative plots of pK_{app} against α obtained for CmDx are shown in Figs. 7 and 8. The pK_{app} increases with an increase in α, this in turn, corresponds to an increase in the linear charge density of the polyion. Since CmDx molecules contain a carboxymethyl group as the charged functionality repeated in the polyion as shown in Fig. 1, the change in pK_{app} as a function of α cannot be due to a heterogeneity effect in the carboxylate groups, their pK_a values being structurally and statistically equivalent. The dependence of the pK_{app} value calculated with Eq. (13) on α reflects the contribution of a Donnan potential term, e.g., "polyelectrolyte effect," to the pK_{app} values compiled. Such nonideal acid-dissociation behavior of the polyacid arises from the use of the free H^+-ion concentration in the bulk

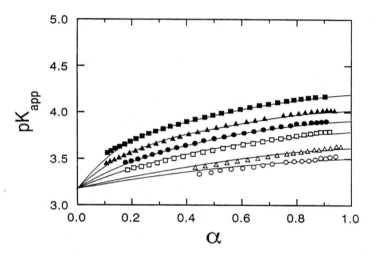

Figure 7 Dependence of pK_{app} on the degree of dissociation of CmDx at Cs = 0.1 mol dm^{-3}. (O) DS = 0.20; (△) DS = 0.50; (□) DS = 0.80; (●) DS = 0.96; (▲) DS = 1.26; (■) DS = 1.70.

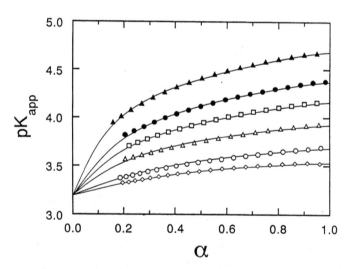

Figure 8 Effect of excess salt concentration on the pK_{app} vs. α plots for CmDx (DS = 1.70). (▲) Cs = 0.02 mol dm^{-3}; (●) Cs = 0.05 mol dm^{-3}; (□) Cs = 0.10 mol dm^{-3}; (△) Cs = 0.20 mol dm^{-3}; (O) Cs = 0.50 mol dm^{-3}; (◇) Cs = 1.00 mol dm^{-3}.

solution phase in place of the free H^+-ion concentration at the reaction site. According to the Gibbs-Donnan model, the free H^+-ion concentration at the reaction site, once corrected for nonideality, is regarded as equal to the free H^+-ion concentration in the polymer domain, i.e., in the Donnan phase, $[H]_D$. By use of $[H]_D$ instead of $[H]$, the intrinsic acid dissociation constant, pK_0, no longer influenced by the polyelectrolyte effect, is defined by Eq. (14)

$$pK_0 = p[H]_D - \log\left(\frac{\alpha}{1-\alpha}\right) \tag{14}$$

The pK_0 value, hypothetical and constant by its definition, is determinable by extrapolating the experimental pK_{app} values to intercept they coordinate at $\alpha = 0$, where the polyelectrolyte effect has diminished to zero at $\alpha = 0$. In the case of CmDx, the pK_0 value is estimated to be 3.20 as shown in Figs. 7 and 8, respectively. This value is quite close to the pK_a value of methoxy acetic acid ($pK_a = 3.31$, $I = 0.1(NaClO_4)$) [58], the monomer analog of CmDx. Combining Eqs. (13) and (14), we obtain Eq. (15).

$$p\left(\frac{[H]_D}{[H]}\right) = pK_{app} - pK_0 = \Delta pK \tag{15}$$

This equation indicates that the ΔpK term defined at a specified α value reflects the free H^+-ion concentration ratio between the Donnan phase and the bulk solution phase.

It is noteworthy that pK_{app} is greatly dependent on the added salt concentration, Cs, as shown in Fig. 8. Approximately a linear relationship is obtained between the ΔpK term and the log Cs value at a specified α. The negative slope of the straight-line portion increases with an increase in α. In sufficiently high α regions, the slope, $d(\Delta pK)/d(\log Cs)$, approaches the value of "-1" as shown in Fig. 9; this relationship is identical with those observed for the gel (CM-Sephadex C-25 and C-50) systems. This similarity in the salt concentration dependence of the H^+ ion-binding properties observed with both the polyion gel (ion exchanger) and the linear polyion systems does indeed indicate that the H^+ ion binding to CmDx samples of sufficiently high linear charge density may be ascribed just it is for the gel analog by substituting the polyion domain for the gel phase. Contrary to the large dependence of ΔpK in the low-Cs regions, it is found to be small at all the α value when the Cs value is higher than ca. 0.5 mol dm^{-3} (Fig. 8). In this concentration region, invasion of the

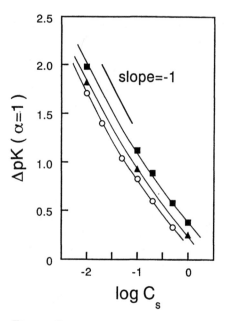

Figure 9   Relationship between ΔpK and log Cs for CmDx and the gel analogs. (○) CmDx (DS = 1.26); (■) CM-Sephadex C-25; (▲) CM-Sephadex C-50. The ΔpK values are determined by extrapolation to $\alpha = 1$ for respective pK_{app} vs. α plots.

Donnan phase by the supporting electrolyte (NaCl) is pronounced, and the Donnan potential almost disappears. The α versus p[H] plots, determined at various Cs values are shown in Fig. 10; approximately a parallel relationship is observed. By increasing Cs, the curves are shifted to the left, however, the shift is sizably less pronounced when Cs > 0.5 mol dm^{-3}. At such high concentrations of salt, the H^{+} ion binding to carboxylate groups of the CmDx molecules is directly measurable because ΔpK approaches zero. The limiting curve resolved is attributable to the invasion of the polyion domain by enough salt to reduce the magnitude of the Donnan potential (ΔpK) term almost to be zero. This phenomenon is described as a *wall of site binding*, and a reasonable projection of the site-binding equilibrium of CmDx.

   The effect of the degree of substitution (DS) of CmDx samples on the pK_{app} versus α plots are shown in Fig. 7. The pK_{app} values at a specified α value increase with the DS value of the CmDx sample, indicating that the linear charge density of the polyion is the predominant factor influencing the shape of the curve. The linear charge separation of the

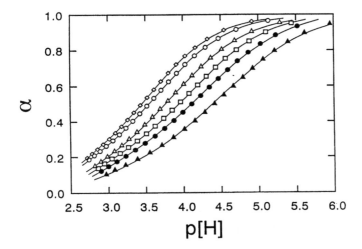

Figure 10 α vs. p[H] plots determined as a function of Cs for CmDx (DS = 1.70). (▲) Cs = 0.02 mol dm^{-3}; (●) Cs = 0.05 mol dm^{-3}; (□) Cs = 0.10 mol dm^{-3}; (△) Cs = 0.20 mol dm^{-3}; (○) Cs = 0.50 mol dm^{-3}; (◇) Cs = 1.00 mol dm^{-3}

CmDx molecule, b, expressed in Å [17], is determined by both the DS and α values. By assuming the structurally based unit length of the dextran chain to be 5 Å (see Fig. 1), the b value, is expressed as

$$b = \frac{5}{DS \times \alpha} \tag{16}$$

When the pK$_{app}$ values determined for samples of various DS values (Fig. 7) are plotted against α/b, they lie approximately on the same curve as shown in Fig. 11. This indicates ΔpK can be expressed simply by the linear charge density [59].

In contrast to the H$^+$ ion binding, the binding of alkali metal ions, such as Na$^+$ ions to CmDx is considered to be an example of purely electrostatic, or territorial, binding. Since Na$^+$ ion is monovalent, the Na$^+$ ion concentration in the polyion domain, [Na]$_D$, can be expressed approximately by use of Eqs. (7) and (15) or by using Eq. (17):

$$[Na]_D = 10^{\Delta pK}[Na] \tag{17}$$

where [Na] is the Na$^+$ ion concentration in the bulk solution phase. The value of [Na]$_D$ can also be expressed in the following way by use of the Donnan phase volume term, V$_D$ (dm^3) as

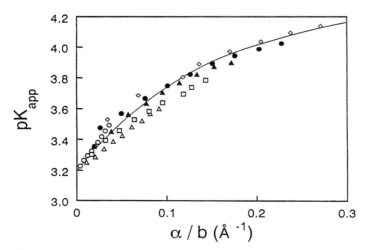

Figure 11 Convergence of the pK_{app} vs. α plots (Cs = 0.10 mol dm^{-3}). (O) DS = 0.20; ($\triangle$) DS = 0.50; ($\square$) DS = 0.80; ($\blacktriangle$) DS = 0.96; ($\bullet$) DS = 1.26; ($\triangle$) DS = 1.70.

$$[Na]_D = \frac{\alpha(1-\phi_{p,Na})n_p}{V_D} + (C_{NaCl})_D \tag{18}$$

where $\phi_{p,Na}$ is the practical osmotic coefficient of the Na$^+$ salt of the polyion, and indicates the fraction of Na$^+$ ion that escapes from the polyion domain. The quantity, $(C_{NaCl})_D$, represents the concentration of salt, NaCl, that invades the Donnan phase. Since $(C_{NaCl})_D$ can be expressed approximately with the equation

$$(C_{NaCl})_D = 10^{-\Delta pK}[Na] \tag{19}$$

the value of V_D can be approximated [6,15,42,43] by combining Eqs. (17)–(19) as shown:

$$V_D/(\alpha n_p) = (1-\phi_{p,Na})(10^{\Delta pK} - 10^{-\Delta pK})^{-1}[Na]^{-1} \tag{20}$$

The $(1-\phi_{p,Na})n_p$ term is used to account for the difference in the equilibrium of simple salt with a linear polyelectrolyte and its cross-linked gel analog [60]. The release of territorially bound counterion from the polyion domain to the solution leads to a gain of negative and positive charges, respectively, by the solvent sheath of the polyelectrolyte and the salt solution. Whereas the two separate phases in the gel/salt system remain electroneutral, this is not the case with the linear polyelectrolyte/salt system. This difference is resolved, however, in the development of the

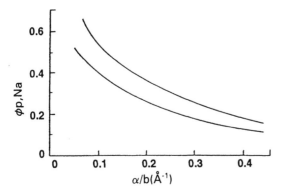

Figure 12 Practical osmotic coefficient, $\phi_{p,Na}$, of the salt-free polyelectrolyte solutions as a function of linear charge density, α/b. The upper and the lower curves correspond to the $\phi_{p,Na}$ values of polyions of polysaccharide and polyvinyl polymer backbones, respectively.

expression for the Donnan potential term in both systems. By considering single ions in this development the condition of electroneutrality is no longer applicable in either system. The equality of their electrochemical potentials in the two phases is used as the criterion for equilibrium and the analytical expressions for the Donnan potential term in both of the two phase systems are equally valid.

Use of the $\phi_{p, Na}$ values available for the salt-free polyelectrolyte to determine the fractional release of Na^+ to solution phase is justified by the experimentally demonstrated additivity of the colligative properties of simple salt and linear polyelectrolyte when they are mixed together. For the linear polyion systems, it has already been shown that $\phi_{p, Na}$ is a fairly selective function of the linear charge separation of the polymer backbone, b, expressed in Å [6,15,42,43]. Such data, compiled for polysaccharides and linear polyelectrolytes with polyvinyl backbone by Katchalsky and co-workers are presented in Fig. 12 by plotting $\phi_{p, Na}$ versus α/b. By use of the ΔpK terms (Fig. 13) and the $\phi_{p, Na}$ values, interpolated from the upper curve in Fig. 12 for polysaccharides, the specific Donnan phase volume, $V_D/(\alpha n_p)$, for CmDx has been calculated with Eq. (20). These $V_D/(\alpha n_p)$ values are plotted against α/b in Fig. 14. The $V_D/(\alpha n_p)$ values obtained for polyions of sufficiently high α/b value, i.e., high linear charge density, showed a little or no salt concentration dependence. The $V_D/(\alpha n_p)$ values for the polyions of high charge density are approximately 1 dm^3 monomol^{-1}. This result indicates that the Donnan phase (polyion domain) is comparable to a concentrated electrolyte solution, whose ionic strength is ca. 1 mol dm^{-3}. With a decrease in α/b, the specific Donnan

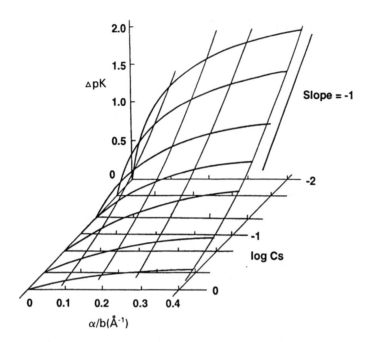

Figure 13 Universal curve for ΔpK as a function of α/b (Å^{-1}). The smooth curves are obtained by using data compiled for this review and for earlier work [59].

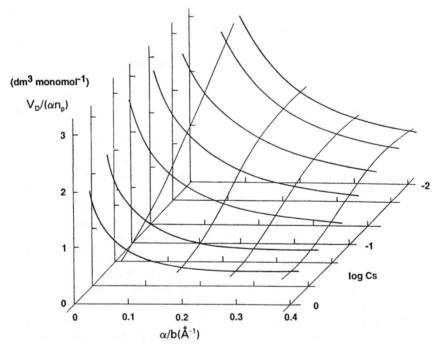

Figure 14 Plots of calculated $V_D/(\alpha n_p)$ values against α/b (Å^{-1}).

phase volume increases markedly. Also, it is of interest to note that the computed Donnan phase volume is greatly dependent on the salt concentration at these lower α/b values as well. This may indicate that the cohesiveness of the polyion, at low charge density in these CmDx samples, is quite sensitive to the added salt concentration [61].

The CmDx gels, produced by cross-linking carboxymethyldextran with a glycerol-1,3 ether linkage, are commercially available from Pharmacia, Sweden, as CM-Sephadex. Gels of two different degrees of cross-linking, i.e., C-25 and C-50, are available. The C-25 gel has a much higher degree of cross-linking than the C-50, and is "more rigid." The H^+ exchange capacity per gram of an dry gel in both instances is quite similar to each other, indicating an identical DS value of about unity [6].

The difference between the pK_{app} versus α plots determined for the C-25 and C-50 gels (Figs. 15 and 16) in spite of the equivalent DS value is attributable to the difference in uptake of solvent by the two gels. The following procedure was used to facilitate measurement of this parameter for examination of the validity of this projection. The CM-Sephadex gels have been produced as a packing material for liquid chromatography and one needs to subtract the void volume from the apparent gel phase volume [62] in order to determine the net gel phase volume because of the well-defined sphere gel. The dead volume due to the dextran backbone is considered then to resolve *the net volume of solvent in the gel phase* (V_g). For the purpose of the void volume calculation, the concentration

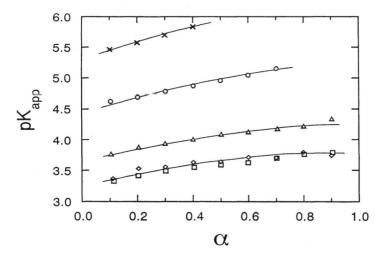

Figure 15 Depedence of pK_{app} on the degree of dissociation of CM-Sephadex C-25 gel. ($\times$) $Cs = 0.001$ mol dm^{-3}; ($\bigcirc$) $Cs = 0.01$ mol dm^{-3}; ($\triangle$) $Cs = 0.10$ mol dm^{-3}; ($\square$) $Cs = 1.00$ mol dm^{-3}; ($\diamond$) $Cs = 2.00$ mol dm^{-3}.

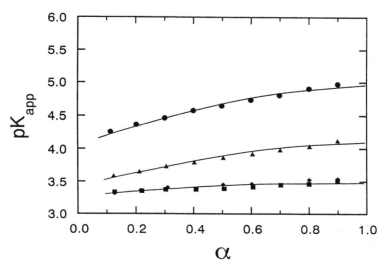

Figure 16 Dependence of pK_{app} on the degree of dissociation of CM-Sephadex C-50 gel. (●) Cs = 0.01 mol dm^{-3}; (▲) Cs = 0.10 mol dm^{-3}; (■) Cs = 1.00 mol dm^{-3}; (◆) Cs = 2.00 mol dm^{-3}

affected in a long-chain polyphosphate has been utilized [63] and the dextran skeleton volume has been calculated by the water taken up by accurately weighed quantity of the bone dry gel using the specific gravity of dextran, i.e., 2.0 g cm^{-3} and 1.5 g cm^{-3} for C-25 and C-50 gels, respectively. The $Vg/(\alpha n_p)$ values determined in this way at different salt concentrations are plotted against α for both systems (Figs. 17 and 18). The similarity in shape of the specific volume versus linear charge density curves resolved for linear polyion system (Fig. 14) with the shape of the curve ($V_D/\alpha n_p$) for the corresponding polyion gel systems (Figs. 17 and 18) is quite striking. This duplication provides strong supports for the validity of the computational procedure used to characterize the linear carboxylate polyions within the context of the present model. In contrast to the insensitivity of the C-25 gel volume to change in the salt concentration, the lowly cross-linked C-50 gel responds just like its linear analog. Measurements of Vg for the flexible C-50 gel, when compared to the estimate of V_D for the linear CmDx polyions at the same experimental conditions, are sizably larger. This result was unexpected and an alternate approach to the estimate of solvent uptake by the C-50 gel and its linear analog was employed next to clarify this situation.

Further documentation of the above comparison of the linear polyion systems with their gel analogs has been reached with Donnan

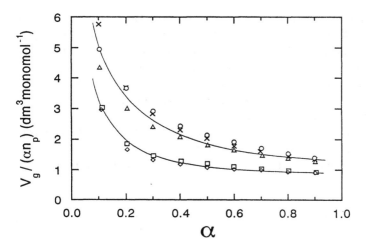

Figure 17   Plots of calculated $V_g/(\alpha\ n_p)$ values against α for CM-Sephadex C-25 gel. ($\times$) $C_s = 0.001$ mol dm^{-3}; (○) $C_s = 0.01$ mol dm^{-3}; (△) $C_s = 0.10$ mol dm^{-3}; (□) $C_s = 1.00$ mol dm^{-3}; (◇) $C_s = 2.00$ mol dm^{-3}.

phase volume estimates based upon apparent acid-dissociation constant estimates as a function of salt concentration level. In the case of the acid dissociation of CmDx, $\phi_{p,Na}$ approaches unity, as the polyion neutralization proceeds ($\alpha \to 0$), and the apparent acid dissociation constant approaches the intrinsic value (Figs. 17 and 18). Even though the $\phi_{p,Na}$ values

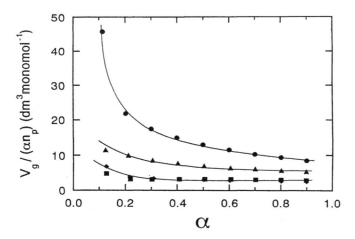

Figure 18   Plots of calculated $V_g/(\alpha n_p)$ values against α for CM-Sephadex C-50 gel. (●) $C_s = 0.01$ mol dm^{-3}; (▲) $C_s = 0.10$ mol dm^{-3}; (■) $C_s = 1.00$ mol dm^{-3}; (◆) $C_s = 2.00$ mol dm^{-3}.

of the gel system remain 0 at any α value and Donnan phase of the polyion gels still exists even at $\alpha = 0$, the apparent acid dissociation constant still approaches its intrinsic value as α approaches zero. The direct determination of the pK_0 value by the extrapolation of the pK_{app} at $\alpha = 0$ is, therefore, feasible in the CmDx gel systems as well. Because the shape of the curves, as $\alpha \to 0$, is less easily resolvable, its assessment is somewhat more susceptible to error, however.

The gel phase volume determined for the cross-linked CmDx gels in contact with NaCl at various concentration levels can be employed in the following equation to evaluate $[Na]_g$, the Na^+ ion concentration in the gel phase.

$$[Na]_g = \frac{\alpha n_p}{V_g} + (C_{NaCl})_g \tag{21}$$

where $(C_{NaCl})_g$ corresponds to the concentration of NaCl imbibed by the gel phase and calculable with the simple Donnan relation. According to Eq. (7), the activity ratio of H^+ ion and Na^+ ion between the two phases can be expressed as

$$\frac{(a_H)_g}{a_H} = \frac{(a_{Na})_g}{a_{Na}} \tag{22}$$

Assuming the activity coefficient quotient $\{(\gamma_{Na})_g/(\gamma_H)_g\}/\{\gamma_{Na}/\gamma_H\}$ is unity, Eq. (22) can be replaced by the equation

$$\frac{[H]_g}{[H]} = \frac{[Na]_g}{[Na]} \tag{23}$$

The pK^*_{app} value resolved by the use of the H^+ ion concentration in the gel phase, $[H]_g$, now applies directly to the dissociation of the carboxylic acid moiety (methoxy acetic acid) repeated throughout the matrix of the CmDx gel.

$$pK^*_{app} = p[H]_g - \log\left(\frac{\alpha}{1-\alpha}\right) \tag{24}$$

By combining Eqs. (21) and (23), the $p[H]_g$ term becomes accessible in terms that are measurable.

$$p[H]_g = p[H] - p[Na] + \log\left(\frac{\alpha n_p}{V_g} + (C_{NaCl})_g\right) \tag{25}$$

and pK_{app}^* can be expressed by use of pK_{app} as shown

$$pK_{app}^* = pK_{app} - p[Na] + \log\left(\frac{\alpha n_p}{V_g} + (C_{NaCl})_g\right) \qquad (26)$$

With V_g experimentally accessible one obtains a direct estimate of the Donnan potential term, $[Na]_g/[Na]$, for comparison with values of the term based on extrapolation of pK_{app} versus α curves compiled for Sephadex gels. The pK_{int} value extrapolated from the curves was combined with the pK_{app} values in Eq. (15) to obtain the Donnan potential terms for this comparison. The plots of $\log([Na]_g/[Na])$ against α that are presented in Fig. 19 for the C-25 and C-50 systems are based on the direct measurement of V_g. It is obvious that the magnitude of the $([Na]_g/[Na])$ term is sizable for the more rigid gel. The inflexibility of the C-25 gel matrix apparently prevent solvent entry that would otherwise occur at the α/b value assignable to the polyion with its degree of substitution corresponding to 1.05. The Donnan potential term, $([Na]_g/[Na])$, for the loosely cross-linked C-50 gel system is much smaller with solvent uptake approaching the values expected at α/b. As shown in Fig. 20, the pK_{app}^* values obtained with Eq. 25 remain constant, ca. 3.3–3.5 over the a range of study for the C-25 system. This result shows the Donnan potential term for the C-25/simple salt system is adequately expressed by Eq. (22).

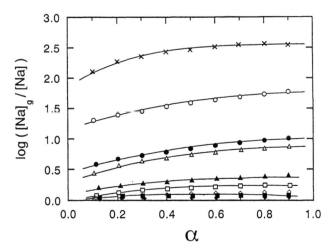

Figure 19 Log($[Na]_g/[Na]$) term as a function of α. ($\times$) Cs = 0.001 mol dm^{-3}; (O, ●) Cs = 0.01 mol dm^{-3}; ($\triangle$, $\blacktriangle$) Cs = 0.10 mol dm^{-3}; ($\square$, ■) Cs = 1.00 mol dm^{-3}; ($\lozenge$, ◆) Cs = 2.00 mol dm^{-3}. Open symbols; C-25: closed symbols; C-50.

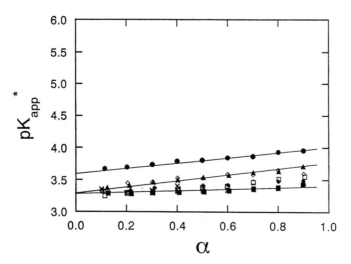

Figure 20 Constancy of the pK^*_{app} values of CM-Sephadex gels. (×) Cs = 0.001 mol dm^{-3}; (○, ●) Cs = 0.01 mol dm^{-3}; (△, ▲) Cs = 0.10 mol dm^{-3}; (□, ■) Cs = 1.00 mol dm^{-3}; (◊, ◆) Cs = 2.00 mol dm^{-3}. Open symbols C-25: Closed symbols; C-50.

The values of 3.3 to 3.6 obtained for the pK^*_{app} value to be assigned to the C-25 gel is close to the pK value of 3.31 (I = 0.1(NaClO$_4$)) published for methoxyacetic acid [58], the weak carboxylic acid that most closely resembles the weak-acid functionality repeated in the cross-linked carboxymethyldextran. It can be concluded, on the basis of this observation, that Eq. (25) provides an adequate resolution of pK_{int}. Indeed, the resemblance between pK^*_{app} and pK_{int} has led to the use of statistical arguments for the rationalization of the absence of nonideality in the $\alpha/(1 - \alpha)$ term. The pK^*_{app} values for the highly swollen gel of C-50 at Cs = 0.01 mol dm^{-3}, on the other hand, increases with increasing α. The extrapolated pK^*_{app} values of 3.3–3.6 at α = 0 are consistent with the values of 3.3–3.6 obtained for the C-25 gel over the α range studied.

The rise in pK^*_{app} with α is unexpected because there is no reason to expect the nonideality characteristics of the $\alpha/(1 - \alpha)$ and [H]$_g$ terms in Eq. (25) to change as drastically as this result would appear to imply. As a consequence, the unexpected rise in pK^*_{app} has to be attributed to overestimate of V_g in the course of its measurement. Such a possibility has been attributed to the sizable macroporosity of the C-50 gel. Apparently, long-chain polyphosphate ions, the macromolecule used to monitor solvent uptake by the gel [43] through its concentration change in salt solution used to equilibrate with accurately weighed bone dry C-50 samples

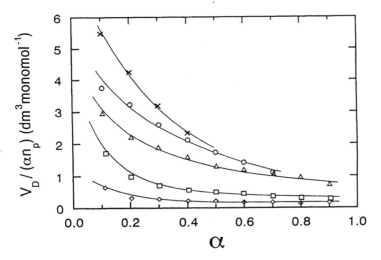

Figure 21 Plots of calculated $V_D/(\alpha n_p)$ values against α for CM-Sephadex C-25 gel at different NaCl concentration levels. ($\times$) Cs = 0.001 mol dm^{-3}; ($\bigcirc$) Cs = 0.01 mol dm^{-3}; ($\triangle$) Cs = 0.10 mol dm^{-3}; ($\square$) Cs = 1.00 mol dm^{-3}; ($\Diamond$) Cs = 2.00 mol dm^{-3}.

were essentially excluded, while the bulk electrolyte was not from the largest pores of the C-50 gel which were considered to function as an extention of the solution phase. The failure of the larger pore to exhibit separate phase properties while still excluding the transfer of macromolecules would lead to overestimates of solvent transfer, V_g, as a result, the observed increase in pK_{app}^* with α is understandable on this basis.

The gel volumes of the CM-Sephadex C-25 and C-50 polyion gel analogs of CmDx have also been calculated with Eq. (20) using the pK_0 value of 3.20, which has been assigned to the CmDx systems. The $V_D/(\alpha n_p)$ values thus calculated by use of the pK_{app} values (Figs. 15 and 16) are plotted against α in Figs. 21 and 22. By comparing Fig. 21 with Fig. 17, it is apparent that the V_D values are quite consistent with the V_g values for the C-25 system. As can be seen by comparing Fig. 22 and Fig. 18, however, the V_D values calculated for the C-50 system is far smaller than the corresponding V_g values to support the above rejection of the V_g values.

B. Metal Complexation Equilibria of Carboxymethyldextrans and Their Gel Analogs

The weak carboxylic acid that is repeated in linear polyelectrolyte such as the carboxymethyldextrans form complexes with metal ions. A high degree of polymerization assures statistical equivalence of the weak-acid

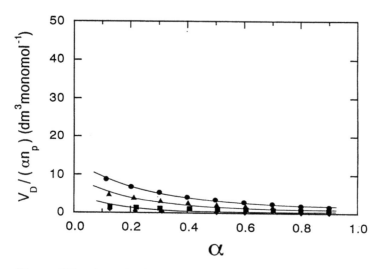

Figure 22 Plots of calcualted $V_D/(\alpha\, n_p)$ values against α for CM-Sephadex C-50 gel at different NaCl concentration levels. ($\bullet$) Cs = 0.01 mol dm^{-3}; ($\blacktriangle$) Cs = 0.10 mol dm^{-3}; ($\blacksquare$) Cs = 1.00 mol dm^{-3}; ($\blacklozenge$) Cs = 2.00 mol dm^{-3}.

molecule repeated in the polyion. On this basis one can expect the intrinsic dissociation properties of the repeating weak-acid functionality and its metal complexes to duplicate those of the weak-acid molecule most closely resembling the one repeated in the polyelectrolytes. This aspect is examined next. As has been discussed in Sec. II, the chemical form and the binding constant of a particular metal complex formed at the surface of a polyion is accessible with Eq. (6). Figure 23 provides an example of a binding isotherm resolved for the CmDx(α = 1)/Ca^{2+} system examined in the presence of excess NaCl. The log K_{Ca} values calculated with Eq. (5) are plotted against θ_{Ca} (Fig. 24). The log K_{Ca}^0 values, obtained by extrapolations of θ_{Ca} = 0, for CmDx samples of various linear charge densities, are plotted against log Cs (Fig. 25). It is apparent that the log K_{Ca}^0 values are quite sensitive to the Cs values and that the slopes of the log-log plots are approximately -2. This result is consistent with expectation since the K_{Ca}^0 term includes the Donnan potential term to the power of 2, the valence of Ca^{2+} ion. Also, it can be seen that the K_{Ca}^0 values are dependent on the linear charge density (1/b) of the CmDx sample; this is a reflection of the dependence of the Donnan term, V_D, on the linear charge density (1/b). Similar trends can also be seen in the binding of the CmDx gel analogs. The Ca^{2+} ion binding isotherms of CM-Sephadex gels of different degrees of cross-linking are presented in Figs. 26 and 27. It can be

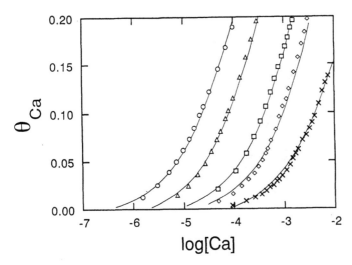

Figure 23 Binding isotherms for Ca^{2+}-CmDx (DS = 1.26, α = 1) at different NaCl concentration levels. (○) Cs = 0.02 mol dm⁻³; (△) Cs = 0.05 mol dm⁻³; (□) Cs = 0.10 mol dm⁻³; (◇) Cs = 0.20 mol dm⁻³; (×) Cs = 0.50 mol dm⁻³.

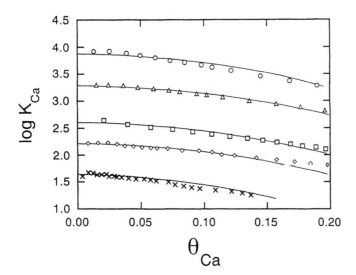

Figure 24 Determination of the log K_{Ca}^0 values for Ca^{2+}-CmDx (DS = 1.26, α = 1) at different NaCl concentration levels. (○) Cs = 0.02 mol dm⁻³; (△) Cs = 0.05 mol dm⁻³; (□) Cs = 0.10 mol dm⁻³; (◇) Cs = 0.20 mol dm⁻³; (×) Cs = 0.50 mol dm⁻³.

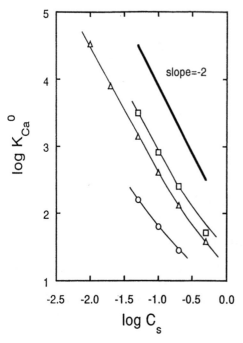

Figure 25 Relationship between log K_{Ca}^0 and log Cs for CmDx ($\alpha = 1$). (○) DS = 0.50; (△) DS = 1.26; (□) DS = 1.70.

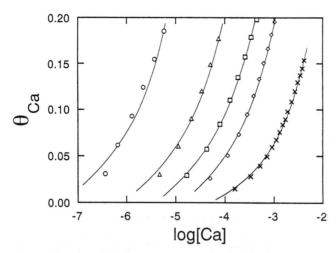

Figure 26 Binding isotherms for Ca^{2+}-CM-Sephadex C-25 ($\alpha = 1$) at different NaCl concentration levels. (○) Cs = 0.01 mol dm^{-3}; (△) Cs = 0.05 mol dm^{-3}; (□) Cs = 0.10 mol dm^{-3}; (◇) Cs = 0.20 mol dm^{-3}; (×) Cs = 0.50 mol dm^{-3}.

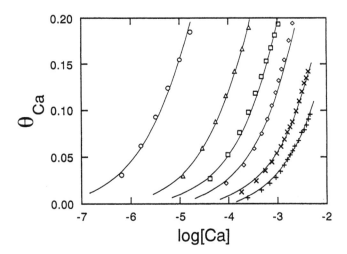

Figure 27 Binding isotherms for Ca^{2+}-CM-Sephadex C-50 ($\alpha = 1$) at different NaCl concentration levels. (O) Cs = 0.01 mol dm^{-3}; (△) Cs = 0.05 mol dm^{-3}; (□) Cs = 0.10 mol dm^{-3}; (×) Cs = 0.20 mol dm^{-3}; (+) Cs = 0.50 mol dm^{-3}.

seen that the shapes of the isotherms (Figs. 23, 26, and 27) are quite similar to each other. The log K_{Ca}^0 values resolved using these binding data are plotted against the log Cs values (Fig. 28). Since the DS value for the CM-Sephadex gels have been determined to be ca. unity, the log K_{Ca}^0 values obtained with the linear CmDx sample (DS = 1.26) are also plotted in the same figure for comparison. The remarkable similarities noted in the previous section dealing with the acid dissociation properties of the carboxymethyldextrans (Fig. 9), repeated in these studies as well. All the three plots are linear in the low-Cs region and the slope of the three lines is approximately −2, indicating that the Ca^{2+} complexation equilibria in the linear polyion (CmDx) systems and their gel analogs are controlled by the same factor, the volume of solvent that defines the concentration of their respective separate phase properties.

The linear CmDx, just like to cross-linked gel analogs, CM-Sephadex C-25 and C-50 defines a separate phase when mixed with simple salt solutions. The volume, V_g, that characterizes the separate phase they define is determined by the charge density, i.e., the distance in Å units that separate the neighboring charges along the polyelectrolyte length, α/b. In the gel the crosslinking exerts restraint on solvent uptake. This leads to smaller gel volumes than would otherwise arise at a particular α/b value. The C-25 being a rigid gel is smaller in solvent uptake at a particular α/b than the more flexible C-50 whose solvent uptake is smaller than for the

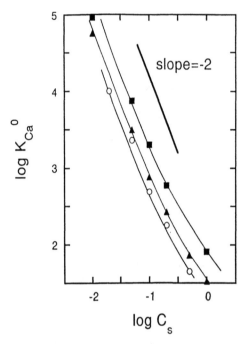

Figure 28 Relationship between log K_{Ca}^0 and log Cs for CmDx and its gel analogs ($\alpha = 1$). ($\bigcirc$) CmDx (DS = 1.26); ($\blacksquare$) CM-Sephadex C-25; ($\blacktriangle$) CM-Sephadex C-50.

linear CmDx with a somewhat higher charge density. A comparison of ΔpK and K_{Ca}^0 versus log plots in Figs. 9 and 28 that were obtained with CM-Sephadex C-25 (DS = 1.05), C-50 (DS = 1.05), and CmDx (DS = 1.26) shows that the pattern of behavior is consistent with the above.

For the precise examination of the complexation equilibria in linear carboxylate polyion systems, the K_M^0 values of the Ag^+-CmDx and the Ca^{2+}-CmDx binding equilibria have been examined at various α values [42,43]. In this case, the K_M values resolved in the presence of trace-level concentrations of metal ion have been substituted for the intrinsic constant, K_M^0 values. Concurrent measurements of p[H] and p[M], at equilibrium, of the $M^{z+}/(Na^+, H^+)CmDx/Na^+$ (excess) system enabled the simultaneous analyses of the acid-dissociation and the metal complexation equilibria. The log K_{Ca}^0 and the log K_{Ag}^0 values determined on the metal association studies are plotted versus α in Fig. 29. The increase in the log K_M^0 value with α is more pronounced with the higher valent metal ion. It should be pointed out as well that log K_{Ca}^0 is greatly influenced by the added salt concentration in the Ca^{2+} ion-binding system.

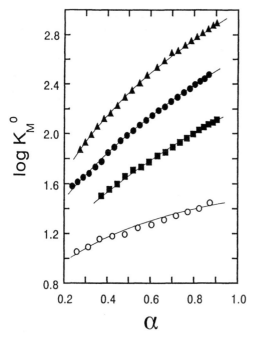

Figure 29 Log K_M^0 as a function of α for Ag$^+$-CmDx-NaClO$_4$ and Ca^{2+}-CmDx-NaCl systems (DS = 1.26): M = Ag$^+$; (○) Cs = 0.10 mol dm^{-3}:M = Ca^{2+}; (▲) Cs = 0.05 mol dm^{-3}; (●) Cs = 0.10 mol dm^{-3}; (■) Cs = 0.20 mol dm^{-3}.

The following relationship is derived for the distributions of ions M^{Z+} and Na$^+$ between the polyion domain (Donnan phase) and the bulk solution phase with the Gibbs-Donnan model.

$$\frac{(a_M)_D}{a_M} = \left\{ \frac{(a_{Na})_D}{a_{Na}} \right\}^Z \tag{27}$$

The concentration ratio of M^{Z+} ion between the two phases is then expressed as

$$\frac{[M]_D}{[M]} = \left\{ \frac{\gamma_M}{(\gamma_M)_D} \right\} \left\{ \frac{(\gamma_{Na})_D}{\gamma_{Na}} \right\}^Z \left\{ \frac{[Na]_D}{[Na]} \right\}^Z$$

$$= G \left\{ \frac{[Na]_D}{[Na]} \right\}^Z \tag{28}$$

where G represents the activity coefficient quotient of the pairs of ions in the two phases. The activity coefficients in the Donnan phase, $(\gamma_i)_D$, have been estimated by using the ionic strength in the Donnan phase, I_D, the mean activity coefficients [64] reported in the literature at that ionic strength and on the assumption that $\gamma_{K^+} = \gamma_{Cl^-}$. The I_D is calculable with the following equation [6,7,15,16,42,43], while $(\gamma_{M^{z+}})_D$ and $(\gamma_{Na^+})_D$ at I_D are obtained by dividing $(\gamma_{\pm MCl_z})^{Z+1}$ by $(\gamma_{\pm KCl})^Z$ and $(\gamma_{\pm NaCl})^2$ by $(\gamma_{\pm KCl})$ at I_D.

$$I_D = \left(1 - \frac{\phi_{p,Na}}{2}\right)\frac{\alpha n_p}{V_D} \tag{29}$$

Since the term $[Na]_D/[Na]$ can be expressed as

$$\frac{[Na]_D}{[Na]} = 10^{\Delta pK} \tag{30}$$

the free metal ion concentration in the Donnan phase, $[M]_D$, is relatable to the bulk electrolyte solution, $[M]$, by the experimental evaluation of ΔpK.

$$[M]_D = G10^{Z\Delta pK}[M] \tag{31}$$

The intrinsic binding constant, K_M^0 has been defined as

$$K_M^0 = \frac{\theta_M}{[M]} \qquad (\theta_M \rightarrow 0) \tag{32}$$

The value of θ_M can be expressed as the sum of free, $(n_M)_D$, and complexed ion, $\Sigma(n_{MAi})_D$, in the polyion domain,

$$\theta_M = \frac{(n_M)_D + \Sigma(n_{MAi})_D}{\alpha n_p} \tag{33}$$

By combining Eqs. (32) and (33), and by relating the quantity of the complexed species in the Donnan phase to their mass-action-based expression, the following equation is obtained:

$$K_M^0 = \frac{([M]_D + \Sigma(\beta_i)_D[M]_D[A]_D^i)V_D}{[M]\alpha n_p} \tag{34}$$

where $(\beta_i)_D$ corresponds to the overall stability constant of the successive complexes of M^{z+} ion with the functionality (ligand) repeated in the polyion molecule, A^-, as shown:

$$(\beta_i)_D = \frac{[MA_i]_D}{[M]_D[A]_D^i} \tag{35}$$

where $[A]_D$ represents the free ligand concentration defined as

$$[A]_D = \frac{\alpha n_p}{V_D} \tag{36}$$

By rearranging Eq. (34) we obtain [42,43]

$$\begin{aligned}
F_1 &= K_M^0 \left(\frac{[M]_D}{[M]}\right)^{-1} - \frac{V_D}{\alpha n_p} \\
&= K_M^0 G^{-1} 10^{-Z\Delta pK} - \frac{V_D}{\alpha n_p} \\
&= \Sigma(\beta_i)_D[A]_D^{i-1}
\end{aligned} \tag{37}$$

Since the ΔpK and the V_D terms are determinable with the p[H] measurements, F_1 can be expressed as a function of $[A]_D$. The F_1 values are plotted against $[A]_D$ for both the Ag^+/CmDx and the Ca^{2+}/CmDx systems in Fig. 30. In spite of the substantial variation in $[A]_D$, constant F_1 values have

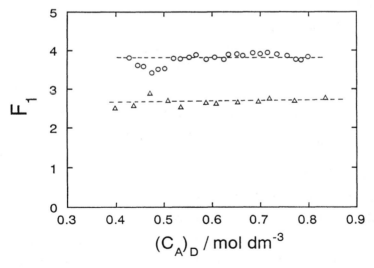

Figure 30   Constancy of F_1 values in Ag^+-CmDx-$NaClO_4$ and Ca^{2+}-CmDx-NaCl systems (DS = 1.26). Cs = 0.10 mol dm^{-3}; ($\triangle$) Ag^+ system; ($\bigcirc$) Ca^{2+} system.

been obtained in both systems, indicating that the formation of one-to-one complexes (AgA^0 and CaA^+) is predominant in both systems. The respective stability constants, $(\beta_1)_D$, are 2.7 and 3.8 $mol^{-1} dm^3$. These constants are in good agreement with those of the corresponding one-to-one complexes with acetate ligand, i.e., $\beta_1(AgAc^0) = 2.3$ ($I = 3$, 25°C) [65], and $\beta_1(CaAc^+) = 3.4$ ($I = 0.15$, 25°C) [66].

The fact that the stability constants of monodentate ligand complexes of polyion systems are comparable in magnitude with those of the simple molecule resembled, once corrected for the electrostatic effect, is consistent with expectation based on the statistical arguments discussed earlier. The absence of multidentate complex formation is also predictable. Whereas the formation of MA^+ is not affected by the low accessibility of the polyion species, MA^+ and A^-, their nonideality, f_{MA^+} and f_{A^-}, canceling in the mass action expression for the formation of MA^+ ($f_{MA^+}/f_{A^-} = 1$), this is not the case for the bidentate species, because $f_{MA_2}/(f_{A^-})^2 = 1/f_{A^-}$, the nonideality term for $1/A^-$ remaining uncanceled. The tendency for bidentate complex formation is, on this basis alone, a factor of f_{A^-} less likely.

The quantity of "territorially bound" metal ions at each experimental condition is accessible with the assessment of V_D.

$$(n_M)_D = [M]_D V_D \tag{38}$$

This allows estimates of the quantity of site-bound metal ions. For the monodentate complex $(MA)^{Z-1}$ the quantity of complex in the Donnan phase, $(n_{MA})_D$, is

$$(n_{MA})_D = \theta_M \alpha n_p - (n_M)_D \tag{39}$$

The apparent stability constant of the monodentate ligand complex, $(\beta_1)_{app}$, is provided by use of $[M]$ as shown:

$$(\beta_1)_{app} = [MA]_D [M]^{-1} [A]_D^{-1} \tag{40}$$

Since the concentration ratio, $[MA]_D/[A]_D$, can be substituted for by $(n_{MA})_D/(\alpha n_p)$ in the presence of trace-level concentrations of metal ions, $(\beta_1)_{app}$ is calculable directly as

$$(\beta_1)_{app} = (n_{MA})_D [M]^{-1} (\alpha n_p)^{-1} \tag{41}$$

The $(\beta_1)_{app}$ value is relatable to $(\beta_1)_D$ by

$$\log (\beta_1)_{app} - \log G = \log (\beta_1)_D + Z(pK_{app} - pK_0) \tag{42}$$

By plotting $\log (\beta_1)_{app} - \log G$ against pK_{app}, one should expect to resolve a straight line with a slope of Z, the charge of the metal ion. Also, by extrapolating the left hand side of Eq. (42) to $pK_{app} = pK_0$ (=3.20), the

$\log (\beta_1)_{app}$ value should correspond to $\log (\beta_1)_D$ value, the G term in Eq. (42) equaling unity when $\alpha = 0$. This prediction is shown to be valid by plotting the binding data obtained for the Ag^+-CmDx and the Ca^{2+}-CmDx systems (Fig. 31) in this manner. The slopes of the respective lines are 1 and 2, the valence of the Ag^+ and Ca^{2+} ions, respectively. Furthermore, when the binding data for the Ca^{2+}-CmDx systems determined at different concentration level of excess salt are examined in the same way, as shown in Fig. 32, the $\log (\beta_1)_{app} - \log G$ plots lie exactly on the same line, showing that the added salt concentration effect on the binding equilibria is fully corrected by the Gibbs-Donnan model.

The $\log K_M^0$ values obtained for the binding of Zn^{2+} ion to C-25 and C-50 gels together with the pK_{app} values determined concurrently are plotted against α (Fig. 33). As noted earlier the much higher degree of binding of H^+ and Zn^{2+} ion binding to the C-25 gel than the C-50 gel is apparently due to the higher degree of cross-linking in the C-25 gel. The extra resistance to solvent transfer, apparently a consequence of the higher rigidity of the C-25 gel because of its higher degree of cross-linking is undoubtedly responsible for the lower V_g values which lead to the larger Donnan terms responsible for the apparent increase in binding. The $\log K_{Zn}^0$ values have been used together with the pK_{app} values to calculate the $\log (\beta_1)_{app}$ values for the binding of Zn^{2+} ion to the gels with Eq. (42), in a manner similar to this exercise with the CmDx systems. In

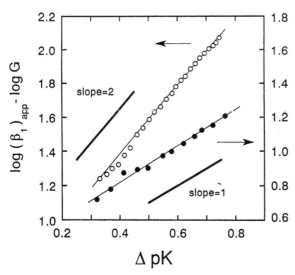

Figure 31 Relationship between $(\log (\beta_1)_{app} - \log G)$ and pK_{app} for CmDx (DS = 1.26). $Cs = 0.10$ mol dm^{-3}; (●) Ag^+ system; (○) Ca^{2+} system.

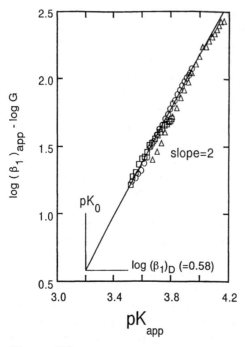

Figure 32 Relationship between $(\log\,(\beta_1)_{app} - \log\,G)$ and pK_{app} for CmDx (DS = 1.26). Ca^{2+} system; ($\triangle$) Cs = 0.05 mol dm^{-3}; ($\bigcirc$) Cs = 0.10 mol dm^{-3}; ($\square$) Cs = 0.20 mol dm^{-3}.

the gel system, V_D and I_D are calculable with Eqs. (20) and (29). The values of $\log\,(\beta_1)_{app} - \log\,G$ are plotted against pK_{app} in Fig. 34. It is clear that the plots obtained in both gel systems are located on the same straight line despite the difference in the degree of cross-linking, and that the slope of the line is 2, the charge of the Zn^{2+} ion, as expected. This suggests that in both gel systems the formation of the monodentate ZnA^+-type complex is predominant. The value of $\log\,(\beta_1)_{app}$ at $pK_{app} = 3.20$ is identifiable with $\log(\beta_1)_D$ (=0.65), the stability constant of the monodentate ligand complex. This value is quite close to the stability constant of the one-to-one complex of Zn^{2+} ion with acetate ions, i.e., $\log\,\beta_1 = 0.63$ (I = 1, 25°C) [65]. One can conclude that despite the inaccessibility to direct measurement of the interaction of counterions with repeating functionality of cross-linked polyions it is accounted for quantitatively by correcting for the Donnan potential term applicable to these two phase systems.

In order to compare the $(\beta_1)_D$ values determined for the linear polyion systems with their cross-linked gel analogs, results of the Ca^{2+} ion binding (Figs. 25 and 28) data obtained for both in their fully dissociated

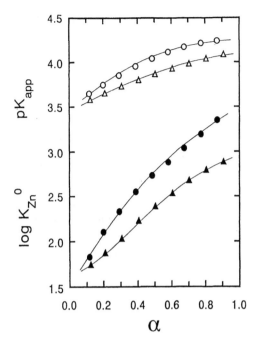

Figure 33 pK_{app} and log K_{Zn}^0 as a function of α for CM-Sephadex gel. Cs = 0.10 mol dm^{-3}; pK_{app}; (O) C-25; ($\triangle$) C-50: log K_{Zn}^0; ($\bullet$) C-25); ($\blacktriangle$) C-50.

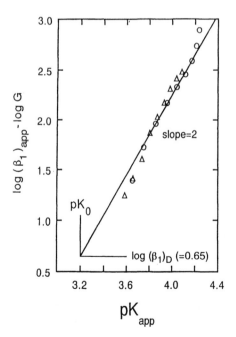

Figure 34 Relationship between (log $(\beta_1)_{app}$ – log G) and pK_{app} for Zn^{2+}/CM-Sephadex gel systems. Cs = 0.10 mol dm^{-3}; (O) C-25; ($\triangle$) C-50.

form are analyzed by the same procedure. The plots of $\log(\beta_1)_{app} - \log G$ against ΔpK, obtained for the CmDx and the gel analogs are shown in Figs. 35 and 36, respectively. All the plots determined with the fully dissociated samples lie on the same line in both figures, showing that the model is equally applicable to linear polyion and its gel analogs.

The calculation procedures ascribed above have practical importance. They can be employed to predict the distribution ratio of a metal ion, D_M^0, between the bulk solution phase and the ion-exchanger phase. The distribution profiles at various pH's and salt concentrations can be estimated just by use of the unique microscopic stability constant, $(\beta_1)_D$, if the monodentate complex formation reaction is predominant. With D_M^0 defined as the ratio of the amount of the bound metal ions to the total capacity of the carboxylate ion exchanger is related to the intrinsic binding constant, K_M^0, as shown

$$D_M^0 = \alpha K_M^0 \tag{43}$$

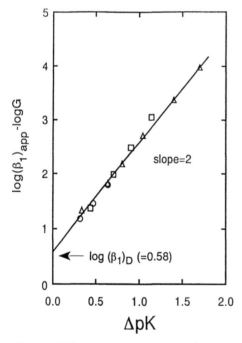

Figure 35 Relationship between $(\log (\beta_1)_{app} - \log G)$ and ΔpK of Ca^{2+}-CmDx ($\alpha = 1$) systems. (O) DS = 0.50; (Δ) DS = 1.26; ($\square$) DS = 1.70. Original data are shown in Fig. 25.

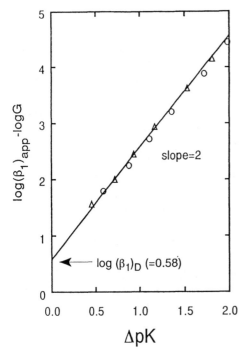

Figure 36 Relationship between $(\log (\beta_1)_{app} - \log G)$ and ΔpK of Ca^{2+}-CM-Sephadex gel $(\alpha = 1)$ systems. (O); C-25: (Δ); C-50. Original data are shown in Fig. 28.

it can be expressed as a function of $(\beta_1)_D$.

$$D_M^0 = \frac{(n_M)_D + (n_{MA})_D}{n_p[M]}$$

$$= \frac{[M]_D}{[M]} \left(\frac{V_D}{\alpha n_p} + \alpha (\beta_1)_D \right)$$

$$= G10^{Z\Delta pK} \left(\frac{V_D}{\alpha n_p} + \alpha (\beta_1)_D \right) \qquad (44)$$

Once the ΔpK and V_D terms for the ion exchanger gels are defined at specific Cs and α values, i.e., at specified p[H] values, then the respective D_M^0 values for a metal ion, M^{Z+}, are calculable by use of the two terms together with the unique $(\beta_1)_D$ value. To demonstrate this capability the

pK_{app} versus α profiles shown in Fig. 15 have been employed in Eq. (44) to calculate D_{Zn}^0 values for the CM-Sephadex C-25 gel. The $(\beta_1)_D$ value assigned to the Zn^{2+} complex with the carboxylate group of the C-25 gel was assumed to be 5 mol^{-1} dm^3. It can be seen that all the plots of log D_{Zn}^0 against p[H] are consistent with the calculated values (the solid lines in Fig. 37) over the ionic strength range studied in this experiment, i.e., $I = 0.1$–1.0 mol dm^{-3}. The same $(\beta_1)_D$ value ($=5$ mol^{-1} dm^3) is also used for the prediction of the distribution behavior of Zn^{2+} ion to the loosely cross-linked CM-Sephadex C-50 gel (Fig. 38). In addition to the distribution profile of Zn^{2+} ion, the log D_M^0 values calculated for other divalent metal ions are also plotted against p[H] in the same figure. It is apparent that the p[H] dependence of the log D_M^0 value is more pronounced the higher the $(\beta_1)_D$ value and it can be concluded that the log D_M^0 variation with p[H] can be satisfactorily predicted with the present model. It is noteworthy that the binding of Mg^{2+} ion to the carboxylate group of the C-50 gel is apparently territorial, i.e., $(\beta_1)_D = 0$.

With the unique formation of monodentate complexes with metal ion demonstrated the $(\beta_1)_D$ values of their complexes with the CM-Sephadex gels can be determined with Eq. (45) by use of respective K_M^0 values, together with the predetermined ΔpK and $[A]_D^{-1}$ terms.

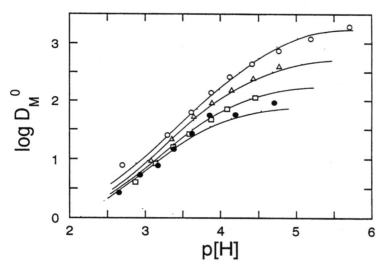

Figure 37 p[H] dependence of log D_{Zn}^0 values (CM-Dephadex C-25/NaClO$_4$). (O) Cs = 0.10 mol dm^{-3}; ($\triangle$) Cs = 0.20 mol dm^{-3}; ($\square$) Cs = 0.50 mol dm^{-3}; (●) Cs = 1.00 mol dm^{-3}. The solid lines refer to the calculated curves by use of $(\beta_1)_D$ of 5 mol^{-1} dm^3.

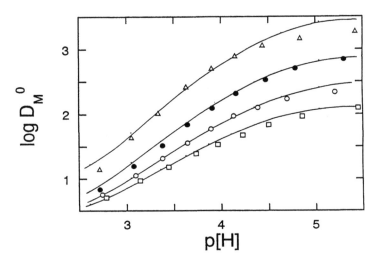

Figure 38 p[H] dependence of log D_M^0 values (CM-Sephadex C-50/NaClO$_4$ (Cs = 0.10 mol dm^{-3})). ($\square$) M = Mg^{2+}(0); ($\bigcirc$) M = Co^{2+} (1); ($\bullet$) M = Zn^{2+}(5); ($\triangle$) M = Cd^{2+} (21). The solid lines refer to the calculated curves. The $(\beta_1)_D$ values used for the calculation are shown in parentheses.

$$(\beta_1)_D = K_M^0 G^{-1} 10^{-Z\Delta pK} - \left(\frac{V_D}{\alpha n_p}\right) \tag{45}$$

The $(\beta_1)_D$ values as estimated for the complexes of various metal ions with CM-Sephadex C-25 and C-50 gels at $\alpha = 1$ are listed in Tables 1 and 2. It is apparent that the $(\beta_1)_D$ values of respective metal complexes determined for both gel systems are, within experimental error, in agreement with each other. This documents the validity of the Gibbs-Donnan model for interpretation of the electrostatic effect encountered in the cross-linked polyion systems.

The log$(\beta_1)_D$ values determined for the CM-Sephadex C-25 gel system are plotted against the log β_1^* values reported in the literature for the corresponding complexes with acetate, methoxyacetate, and ethoxyacetate ions (Fig. 39) for comparison. The scatter in the reported log β_1^* values is due to the fact that these values were determined by different research groups with various methods and at different ionic strengths. By taking their aspect into consideration, the degree of agreement between the $(\beta_1)_D$ and β_1^* values that the straight line in this figure projects indicates the complete accord with statistically based projection. On this basis we can conclude that by use of the β_1^* value reported for the monomer most closely resembled; then the log D_M^0 change with p[H] and Cs should

Table 1   Assignment of $(\beta_1)_D$ Values to Complexes of Fully Dissociated CM-Sephadex C-25 Gels

Metal	$\log K_M^0$	$(\beta_1)_D$ $(mol^{-1}\,dm^3)$
M^{Z+}/C-25/$NaClO_4$ ($Cs = 0.10$ mol dm^{-3}), $\Delta pK = 1.11$, $[A]_D^{-1} = 0.774$ mol^{-1} dm^3		
Mg^{2+}	2.45 ± 0.04	0
Ca^{2+}	3.28 ± 0.04	3.2 ± 0.4
Sr^{2+}	3.02 ± 0.04	1.1 ± 0.2
Co^{2+}	2.96 ± 0.04	1.1 ± 0.2
Ni^{2+}	3.04 ± 0.04	1.5 ± 0.2
Zn^{2+}	3.50 ± 0.04	5.8 ± 0.6
Cd^{2+}	3.98 ± 0.04	18 ± 2
Cu^{2+}	4.70 ± 0.04	99 ± 10
Ag^{2+}	2.00 ± 0.04	5.1 ± 0.6
M^{Z+}/C-25 gel/$NaClO_4$ ($Cs = 0.01$ mol dm^{-3}), $\Delta pK = 1.98$, $[A]_D^{-1} = 1.05$ mol^{-1} dm^3		
Mg^{2+}	4.47 ± 0.04	0.8 ± 0.2
Ca^{2+}	5.09 ± 0.04	3.1 ± 0.4
Ag^{2+}	2.96 ± 0.04	5.9 ± 0.7

be quantitatively predictable with the present model. The most important conclusion derived from the agreement of the intrinsic binding constants determined with the present model with those of the corresponding metal complexes of the simple molecule most closely resembles the repeating functionality in the polyion domain is that the ionic atmosphere in the polyion domain can be approximated by a homogeneous concentrated solution composed of fixed ions, counterions, and coions.

Table 2   Assignment of $(\beta_1)_D$ Values to Complexes of Fully Dissociated DM-Sephadex C-50 Gels

Metal	$\log K_M^0$	$(\beta_1)_D$ $(mol^{-1}\,dm^3)$
M^{Z+}/C-50/$NaClO_4$ ($Cs = 0.10$ mol dm^{-3}), $\Delta pK = 0.94$, $[A]_D^{-1} = 1.18$ mol^{-1} dm^3		
Mg^{2+}	2.02 ± 0.04	0
Ca^{2+}	2.90 ± 0.04	3.7 ± 0.5
Sr^{2+}	2.52 ± 0.04	0.8 ± 0.2
Co^{2+}	2.68 ± 0.04	1.7 ± 0.3
Ni^{2+}	3.88 ± 0.04	3.4 ± 0.5
Zn^{2+}	3.15 ± 0.04	7.4 ± 0.9
Cd^{2+}	3.67 ± 0.04	26 ± 3
Cu^{2+}	4.43 ± 0.04	158 ± 15
Ag^{2+}	1.82 ± 0.04	3.7 ± 0.5

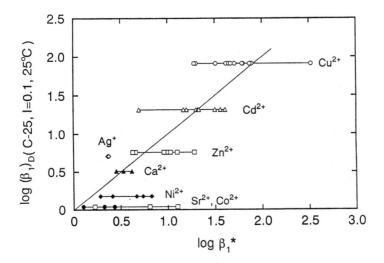

Figure 39 Relationship between log $(\beta_1)_D$ (CM-Sephadex C-25, Cs = 0.1 mol dm^{-3}) and log β_1^*. (●) M = Sr^{2+}; (□) M = Co^{2+}; (◆) M = Ni^{2+}; (▲)M = Ca^{2+}; (□) M = Zn^{2+}; (△) M = Cd^{2+}; (○) M = Cu^{2+}; (◇) M = Ag$^+$. *All these values are listed in *Critical Stability Constants*, vol. 3 (1977), compiled by A. E. Martell and R. M. Smith (Plenum Press).

IV. SULFATE AND SULFONATE POLYIONS AND THEIR GEL ANALOGS

A. Metal Ion-Binding Equilibria of Dextransulfates and Sulfopropyl Dextran Gels

The metal ion-binding to sulfate or sulfopropyl groups of strong-acid linear polyions and their cross-linked gel analogs, unlike the carboxylate polyions, is considered to be purely electrostatic. In the previous section, it has been shown that the V_D values of weak acid polyions can be estimated by using of ΔpK as a probe and the V_D values have been expressed as a function of the degree of acid dissociation and the concentration of the added salt. In the case of strong-acid polyions, H$^+$ ion binding to fixed ionic groups is in the territorial binding mode, and the V_D value is not determinable by the computational procedure proposed for the weak-acid polyion systems. An alternative approach to determination of the hypothetical Donnan phase volume had to be introduced in order to apply the Gibbs-Donnan concept to the counterion-binding equilibria of the strong-acid linear polyions and their gel analogs. The computational approach for estimate of V_D values for the linear polyelectrolyte dextran sulfate (DxS) and the cross-linked sulfopropyl dextran gels (SP-Sephadex

(Pharmacia)) was affected by examining multivalent metal ion, Ca^{2+} - and La^{3+} - [46] binding equilibria in the presence of different sodium salt concentration levels. Using the K_{Ca}^0 and K_{La}^0 values resolved with these data, an iterative approach, described in the text that follows, was used to obtain the V_D/n_p value applicable to the particular system under investigation.

The V_D values of the strong-acid polyions, analyzed as described above, for the *asymmetrical ion-exchange* equilibria of M^{Z+} ions ($Z = 2$, 3) and Na^+ ions between the two phases assume that the binding of all ions is purely electrostatic in nature. The representative binding isotherms for Ca^{2+} ion and the DxS and SP-Sephadex C-25 and C-50 gels are presented in Figs. 40, 41, and 42, respectively. In spite of their substantial difference in physical appearance, i.e., DxS is completely water soluble, whereas the SP-Sephadex gels are water insoluble, the shapes of these binding isotherms are quite similar to each other; in particular, the resemblance in the salt concentration dependence of the binding isotherms is noticeable. Plots of log K_{Ca} against θ_{Ca} are presented as an alternate expression of the binding isotherms in Figs. 43 and 44. As was observed for the Ca^{2+} ion binding to CmDx in the presence of simple salt (Fig. 24), the log K_{Ca} values decrease gradually with an increase in θ_{Ca}, due to

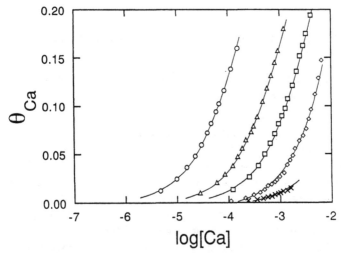

Figure 40 Binding isotherms for Ca^{2+}-DxS (Pharmacia) at different NaCl concentration levels. DS value is estimated to be ca. 2. (○) $Cs = 0.02$ mol dm^{-3}; (△) $Cs = 0.05$ mol dm^{-3}; (□) $Cs = 0.10$ mol dm^{-3}; (◇) $Cs = 0.20$ mol dm^{-3}; (×) $Cs = 0.50$ mol dm^{-3}.

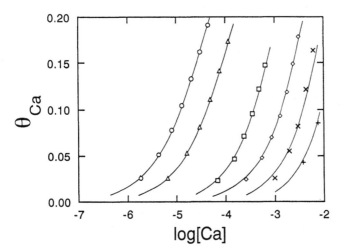

Figure 41   Binding isotherms for Ca^{2+}-SP-Sephadex C-25 at different NaCl concentration levels. (O) Cs = 0.01 mol dm^{-3}; ($\triangle$) Cs = 0.02 mol dm^{-3}; ($\square$) Cs = 0.05 mol dm^{-3}; ($\lozenge$) Cs = 0.10 mol dm^{-3}; ($\times$) Cs = 0.20 mol dm^{-3}; (+) Cs = 0.50 mol dm^{-3}.

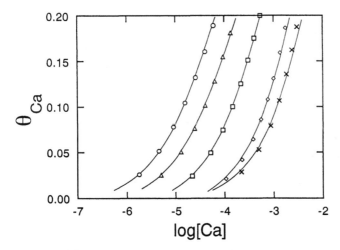

Figure 42   Binding isotherms for Ca^{2+}-SP-Sephadex C-50 at different NaCl concentration levels. (O) Cs = 0.005 mol dm^{-3}; ($\triangle$) Cs = 0.01 mol dm^{-3}; ($\square$) Cs = 0.02 mol dm^{-3}; ($\lozenge$) Cs = 0.05 mol dm^{-3}; ($\times$) Cs = 0.10 mol dm^{-3}.

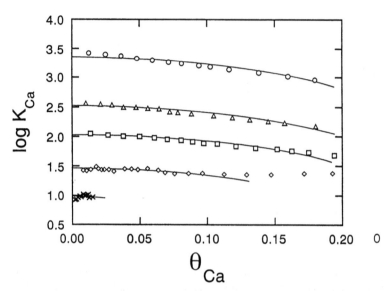

Figure 43   Determination of the log K_{Ca}^0 values for Ca^{2+}-DxS (Pharmacia) at different NaCl concentration levels. (○) Cs = 0.02 mol dm^{-3}; (△) Cs = 0.05 mol dm^{-3}; (□) Cs = 0.10 mol dm^{-3}; (△) Cs = 0.20 mol dm^{-3}; (×) Cs = 0.50 mol dm^{-3}.

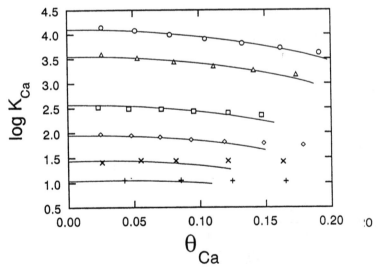

Figure 44   Determination of the log K_{Ca}^0 values for Ca^{2+}-SP-Sephadex C-25 at different NaCl concentration levels. (○) Cs = 0.01 mol dm^{-3}; (△) Cs = 0.02 mol dm^{-3}; (□) Cs = 0.05 mol dm^{-3}; (△) Cs = 0.10 mol dm^{-3}; (×) Cs = 0.20 mol dm^{-3}; (+) Cs = 0.50 mol dm^{-3}.

anticooperativity. This indicates that the apparent binding is due to entropic, i.e., the binding is influenced by the difference between the local concentration of Na^+ ion in the vicinity of the polyion skeleton and the Na^+ ion concentration in the bulk solution. The curves obtained at different ionic strengths are approximately parallel to each other shows that anticooperativity is most probably expressed as a function of θ_{Ca} [19]. The logarithmic value of the intrinsic binding constant, $\log K_{Ca}^0$, can be determined by extrapolating these curves to $\theta_{Ca} = 0$. The $\log K_M^0$ ($M = Ca^{2+}$ and La^{3+} [46], respectively) values so obtained, are plotted in Figs. 45 and 46 against the log Cs values for DxS (Pharmacia) and SP-Sephadex gel-binding systems, respectively. The slopes of the linear plots resolved for metal ion-binding systems are approximately $-Z$, Z corresponding, as before, to the charges of the respective metal ions investigated. These results indicate that the metal interactions are expressible as cation-exchange reactions. They indicate, as well that the counterion binding equilibria of the strong-acid linear polyelectrolytes are also expressible by the Donnan-based concept just like the polyion gel, i.e., ion-exchanger systems.

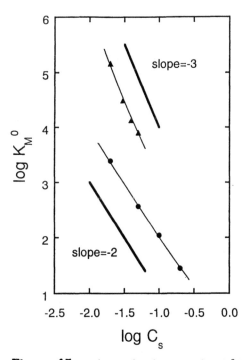

Figure 45   Relationship between $\log K_M^0$ and log Cs for DxS (Pharmacia). (○); $M = Ca^{2+}$: (●); $M = La^{3+}$.

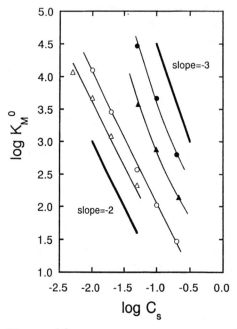

Figure 46 Relationship between log K_M^0 and log Cs for SP-Sephadex gels. (O, ●) C-25 gel; (Δ, ▲) C-50 gel. Open symbols; M = Ca^{2+}: closed symbols; M = La^{3+}.

As has already been shown in Eq. (7), the following equation can be derived for the M^{Z+}/Na^+ ion-exchange system

$$\frac{(a_M)_D}{a_M} = \left\{ \frac{(a_{Na})_D}{a_{Na}} \right\}^Z \tag{46}$$

where $(a)_D$ corresponds, as before, to activity in the Donnan phase. Since it is reasonable to assume that the binding of M^{Z+} and Na^+ ions to the strong acid polyions is purely electrostatic, $(a_M)_D$ and $(a_{Na})_D$ can be expressed by the following equations:

$$(a_M)_D = (\gamma_M)_D[M]_D = \frac{(\gamma_M)_D \theta_M n_p}{V_D} \tag{47}$$

$$(a_{Na})_D = (\gamma_{Na})_D[Na]_D = \frac{(\gamma_{Na})_D(1 - \phi_{p,Na} - Z\theta_M)n_p}{V_D} \tag{48}$$

where $(\gamma_M)_D$ and $(\gamma_{Na})_D$ stand for the activity coefficients of M^{Z+} and Na^+ ions in the Donnan phase, respectively. By substituting Eqs. (47) and (48) into Eq. (46), and by rearranging Eq. (46), we obtain Eq. (49):

$$(49)$$

$$\left(\frac{V_D}{n_p}\right)^{Z-1} = \frac{\gamma_M/(\gamma_M)_D}{\{\gamma_{Na}/(\gamma_{Na})_D\}^Z} ([M]/[Na]^Z)(1 - \phi_{p,Na} - Z\theta_M)^Z\theta_M^{-1}$$

As $\theta_M \rightarrow 0$, the value of V_D/n_p in the presence of negligible amount of M^{Z+} ion can be expressed by using the K_M^0 value as shown:

$$\left(\frac{V_D}{n_p}\right)^{Z-1} = \frac{\gamma_M/(\gamma_M)_D}{\{\gamma_{Na}/(\gamma_{Na})_D\}^Z} (K_M^0)^{-1}[Na]^{-Z}(1 - \phi_{p,Na})^Z \qquad (50)$$

In the asymmetrical ion-exchange equilibrium system, where $Z \geq 2$, the volume term in the left-hand side of Eq. (50) remains and its value can be determined by use of the values of K_M^0 and [Na]. In the case of polyion gels, electroneutrality persists and there is no ϕ_p term to consider, whereas in the case of the linear polyelectrolyte, electroneutrality is not reached in the Donnan phase and the $\phi_{p,Na}$ term is needed to describe the site vacancy of the polyions [16]. The $\phi_{p,Na}$ values used in this calculation were taken from the values determined for these salt-free polyions and reported by Katchalsky and his co-workers (Fig. 12) [31]. It is reasonable to justify the assumption of $\phi_{p,Na}$ uniqueness in the presence of added salt with the "additivity rule."

The single-ion activity coefficients of M^{Z+} and Na^+ ions in the Donnan phase in Eq. (50) have been estimated by use of the published mean activity coefficients of KCl, NaCl, $CaCl_2$, and $LaCl_3$, respectively [64] as described earlier. The ionic strength in the Donnan phase, I_D, needed for these computations of the $(\gamma_i)_D$ values, is calculated by assuming homogeneous distribution of ions in this phase and

$$I_D = \left(1 - \frac{\phi_{p,Na}}{2}\right)\frac{n_p}{V_D} \qquad (51)$$

As the first approximation, the first V_D/n_p value is calculated by assuming the activity coefficient quotient in Eq. (50) is unity. Then, by use of the first approximated V_D/n_p value, the first I_D value is calculated with Eq. (51) to permit computation of the activity coefficients of the respective species in the Donnan phase. The second approximated V_D/n_p value is then obtained with Eq. (50). By this iterative procedure, a self-consistent V_D/n_p value is finally determined.

The V_D/n_p value determined for the SP-Sephadex gel systems in this way are listed in Table 3. In spite of the substantial difference in the K_{Ca}^0 and the K_{La}^0 values, the V_D/n_p values determined using the exchange reactions of the cations of different charges as probes in this way are, within experimental error, self-consistent at a specified Cs (=[Na]). Also, note that these V_D values are obtained by using activity coefficients of Ca^{2+} and La^{3+} ions calculated at ionic strength values estimated with Eq. (51) by regarding the Donnan phase as a homogeneous and concentrated electrolyte solution [16].

Since the boundary between the two phases is well defined, the specific gel phase volume (V_g/n_p) is measurable by the procedure described in Sec. III.A. These V_g/n_p values for the SP-Sephadex gels are also listed in Table 3 for comparison with the Eq. (50)-based values. We can see that the V_g/n_p values are quite dependent on the degree of cross-linking; those for C-25 gel are relatively invariant with the change in the concentration of NaCl, with only a small volume decrease observed in the high salt concentration range. In the highly swollen C-50 gel system there is a pronounced volume dependence on the salt concentration. By comparing V_D with the corresponding V_g, it is apparent that the V_g/n_p values for the C-25 gels are in reasonable agreement with the corresponding V_D/n_p estimates even though the V_D/n_p value for the C-25 system is a little smaller than the corresponding V_g/n_p. In the C-50 system, however, the net gel phase volume by direct measurement is much larger than the Donnan phase volume computed by using the multivalent metal ion-binding equilibria as probes. This sizable discrepancy is consistent with the sizable discrepancy in the parallel set of volume assessments for the highly swollen weak-acid gel (CM-Sephadex C-50) system and the terms can be attributed to error in the direct measurement [67]. As has already been discussed in Sec. III.A, such overestimate of V_g/n_p has been presumed to be a consequence of the inaccessibility of the polyphosphate macromolecule to the macroporous gel so far used to monitor solvent uptake by bone dry gel from salt solution equilibrated with it.

The $\phi_{p,Na}$ values needed for the computation of the V_D/n_p values associated with the linear polyion, DxS, were obtained by interpolation of the $\phi_{p,Na}$ versus α/b plots accessible for ionic polysaccharides (Fig. 12) where the b values were calculated with Eq. (16). The log K_{Ca}^0 and the log K_{La}^0 values determined for the commercially available DxS sample (Pharmacia), with a DS value of ca. 2, were also used at under various NaCl concentration levels. The K_{Ca}^0 values were determined in complexation studies facilitated by potentiometric measurements, whereas the K_{La}^0 values due to Mattai and Kwak [46], were obtained by a dye method. In spite of the difference in valence of the counterions and the binding constants, the V_D/n_p values resolved for both systems are consistent with

Table 3 Comparison of V_D/n_p and V_g/n_p Values Calculated for SP-Sephadex Gels

| | V_D/n_p Determination | |
| | C-25 gel (Ca^{2+} ion-binding equilibria as probes) | |
Cs (mol dm^{-3})	$\log K_{Ca}^0$	V_D/n_p (dm^3 monomol^{-1})
0.01	4.10 ± 0.04	1.21 ± 0.08
0.02	3.46 ± 0.04	1.23 ± 0.08
0.05	2.57 ± 0.04	1.35 ± 0.08
0.10	2.02 ± 0.04	1.17 ± 0.08
0.20	1.47 ± 0.04	1.01 ± 0.07
	C-25 Gel (La^{3+} ion-binding equilibria as probes)	
Cs (mol dm^{-3})	$\log K_{La}^0$	V_D/n_p (dm^3 monomol^{-1})
0.05	4.47 ± 0.04	1.11 ± 0.03
0.10	3.67 ± 0.04	0.98 ± 0.03
0.20	2.80 ± 0.04	0.91 ± 0.03
	C-50 gel (Ca^{2+} ion-binding equilibria as probes)	
Cs (mol dm^{-3})	$\log K_{Ca}^0$	V_D/n_p (dm^3 monomol^{-1})
0.005	4.07 ± 0.04	4.12 ± 0.36
0.01	3.68 ± 0.04	2.58 ± 0.22
0.02	3.09 ± 0.04	2.40 ± 0.20
0.05	2.33 ± 0.04	2.09 ± 0.20
	C-50 gel (La^{3+} ion-binding equilibria as probes)	
Cs (mol dm^{-3})	$\log K_{La}^0$	V_D/n_p (dm^3 monomol^{-1})
0.052	3.58 ± 0.04	1.99 ± 0.06
0.099	2.88 ± 0.04	1.68 ± 0.05
0.216	2.14 ± 0.04	1.27 ± 0.03
	V_g/n_p Determination	
	C-25 gel	
Cs (mol dm^{-3})		V_g/n_p (dm^3 monomol^{-1})
0.01		1.96
0.02		1.77
0.05		1.51
0.10		1.31
0.20		1.46
0.50		1.44
	C-50 gel	
Cs (mol dm^{-3})		V_g/n_p (dm^3 monomol^{-1})
0.01		15.00
0.02		12.88
0.05		9.27
0.10		6.37
0.20		5.39
0.50		4.51

each other. They are listed in Table 4. It is of special interest to compare the V_D/n_p values [29] compiled for DxS samples of various degrees of substitution through use of the Ca^{2+} ion-binding equilibria with the corresponding $V_D/\alpha n_p$ values determined earlier by use of the acid-dissociation properties of CmDx. The log K_{Ca}^0 values are listed together with the V_D/n_p values determined at various salt concentrations in Table 5. Both computed V_D/n_p values are plotted against the linear charge density of the linear polyions, α/b (Fig. 47); the α value is unity in the case of the strong-acid polyion systems. By comparing Fig. 47 with Fig. 14, it can be seen that ionic (carboxylate and sulfate) dextrans of sufficiently high charge density (the α/b value is higher than ca. 0.2) have similar Donnan phase volume even though the fixed ionic groups are different. This indicates that the dominant factor affecting on the ionic dextran–metal ion-binding equilibria is the linear charge density of the polyion, i.e., the α/b value. However, in the lower α/b regions, i.e., α/b is lower than ca. 0.2, the V_D values for the DxS and CmDx sample decrease differently with decrease in α/b.

The drastic decrease in $V_D/\alpha n_p$ values with respect to a decrease in α/b from ca. 0.2 found in the Ca^{2+}-DxS binding system is consistent with a "polyelectrolyte phase volume" calculation based on a "two-variable theory" proposed by Manning [19] for the linear polyion/counterion/supporting ion (excess) system. In this theory, the total free energy of the system is considered to be the sum of the electrostatic repulsions among the negative charges fixed on the polymer, and the mixing entropy of the counterions and the solvent between the polyelectrolyte phase and the

Table 4 Computation of V_D/n_p for DxS (Pharmacia) ($\phi_{p,Na}$ is estimated to be 0.2)

Cs (mol dm^{-3})	(Ca^{2+} ion-binding equilibria as probes)	
	log K_{Ca}^0	V_D/n_p (dm^3 monomol^{-1})
0.02	3.39 ± 0.04	0.97 ± 0.07
0.05	2.56 ± 0.04	0.97 ± 0.06
0.10	2.04 ± 0.04	0.80 ± 0.05
0.20	1.45 ± 0.04	0.74 ± 0.05
Cs (mol dm^{-3})	(La^{3+} ion-binding equilibria as probes)	
	log K_{La}^0[46]	V_D/n_p (dm^3 monomol^{-1})
0.02	5.16 ± 0.05	1.30 ± 0.04
0.03	4.50 ± 0.05	1.39 ± 0.04
0.04	4.13 ± 0.05	1.35 ± 0.04
0.05	3.90 ± 0.05	1.27 ± 0.04

Table 5 Computation of V_D/n_p for DxS[a]

Cs (mol dm^{-3})	log K_{Ca}^0	V_D/n_p (dm^3 monomol^{-1})
DS = 0.29	(b = 17.2 Å, $\phi_{p,Na}$ = 0.670)	
0.01	3.38	0.73
0.02	2.75	0.74
0.05	2.05	0.56
0.10	1.67	0.34
0.20	−	−
DS = 0.67	(b = 7.50 Å, $\phi_{p,Na}$ = 0.459)	
0.01	3.49	1.43
0.02	2.80	1.65
0.05	2.13	1.18
0.10	1.73	0.72
0.20	−	−
DS = 1.14	(b = 4.39 Å, $\phi_{p,Na}$ = 0.299)	
0.01	3.54	2.08
0.02	2.91	2.11
0.05	2.19	1.67
0.10	1.79	1.02
0.20	−	−
DS = 1.74	(b = 2.87 Å, $\phi_{p,Na}$ = 0.194)	
0.01	3.78	1.63
0.02	3.12	1.76
0.05	2.36	1.52
0.10	1.93	0.99
0.20	1.44	0.73
DS = 2.16	(b = 2.31 Å, $\phi_{p,Na}$ = 0.149)	
0.01	3.89	1.45
0.02	3.23	1.53
0.05	2.45	1.39
0.10	2.00	0.94
0.20	1.49	0.73
DS = 2.54	(b = 1.97 Å, $\phi_{p,Na}$ = 0.115)	
0.01	3.95	1.38
0.02	3.30	1.43
0.05	2.53	1.25
0.10	2.07	0.87
0.20	1.61	0.61

[a]The DxS samples whose DS values are 0.29, 0.67, and 1.14 precipitated in a 0.20 mol dm^{-3} NaCl solution.

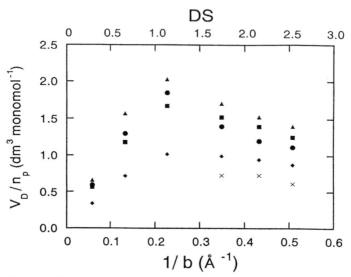

Figure 47 Plots of calculated V_D/n_p against $1/b$ (Å^{-1}). (●) Cs = 0.01 mol dm^{-3}; (▲) Cs = 0.02 mol dm^{-3}; (■) Cs = 0.05 mol dm^{-3}; (◆) Cs = 0.10 mol dm^{-3}; (×) Cs = 0.20 mol dm^{-3}.

bulk solution phase. The following equation, derived to calculate the polyelectrolyte phase volume, V_p, just uses the structural parameter of the linear polyion, b (Å),

$$\frac{V_p}{n_p} = 41.1\left(\frac{7.14}{b} - 1\right)b^3 \quad (cm^3 \ monomol^{-1}) \tag{52}$$

The V_D/n_p values calculated with Eq. (52) are plotted against the $1/b$ values (Fig. 48). The plots show a maximum at $1/b = 0.2$ and diminish when the $1/b$ value decreases to 0.14. The transition at $1/b = 0.2$, observed in the Ca^{2+}-DxS binding system, is consistent with prediction based on Manning's purely electrostatic binding model. It should be recalled that such a predicted change in the physicochemical properties of ion binding in salt-free linear polyelectrolyte systems, i.e., ^{23}Na NMR measurements of Na$^+$ salts of DxS and CmDx (Fig. 6) has been observed. However, such a transition has not been observed in the acid-dissociation property of the weak-acid polyion, CmDx system, even though DxS and CmDx are composed of the same backbone. One possible explanation for the discrepancy found in the plots at lower α/b values in CmDx and DxS systems is the difference in the closest approach by the counterions investigated to the ionic groups fixed in the polymers. For example, in the

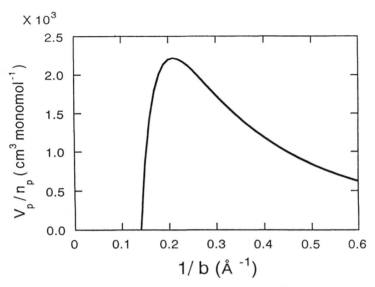

Figure 48 Manning's V_p/n_p as a function of $1/b$ (Å^{-1}).

case of proton (hydronium ion) binding to the carboxylate group of the CmDx molecule, the proton is located just beside the oxygen atom of the carboxylate ion; the site-binding equilibrium is influenced by the nearest neighbors. On the other hand, in the case of Ca^{2+} or La^{3+} ion binding to DxS, the multivalent metal ions are located away from the polyion surface; the territorial-binding equilibrium is influenced by long-range electrostatic interactions including Debye-Hückel type interaction. It should be noted once again that the V_D/n_p values determined for DxS by use of Ca^{2+} and La^{3+} ion-binding equilibria as probes are consistent with each other in spite of the difference in the hydrated ionic radii of the Ca^{2+} and La^{3+} ions. This ensures the presence of a well-defined volume element around the polymer skeleton different from the bulk solution phase.

B. Ca^{2+}-Ion-Binding Equilibria of Heparin

In the preceding discussion of the relationship of the Donnan potential and the Donnan phase volume terms to the structural parameter of a linear polyion, α/b (Å^{-1}), it has been shown that the electrostatic effect on the binding equilibria of ionic polysaccharides can be predicted quantitatively by use of the universal curves shown in Figs. 13 and 14. Also, by separating the overall binding equilibria into two processes, i.e., (1) the concentration of counterion in the polyion domain, territorial binding, and (2) the complexation of counterions in the polyion domain by the repeated ionic groups of the polyion backbone, site binding, the macroscopic binding constant, K_M^0, can be expressed with Eq. (34). Since it has already been shown that the two terms, the Donnan phase volume and the Donnan potential evaluated for the carboxylate and sulfate dextrans are dependent solely on the linear charge separation along the polyion backbone, it is of interest to examine their applicability for the prediction of metal ion-binding equilibria of other ionic polysaccharides containing carboxylate and/or sulfate groups.

Heparin, whose structure is shown in Fig. 49 is a good candidate for examining this application. This mixed acid polysaccharide is a well-known blood anticoagulant. Its complexation of divalent metal ions such as Ca^{2+} ions have extensively been studied in order to understand the role of such complexation on its physicochemical properties. Several different structures have been proposed for the Ca^{2+}-heparin complexes; chelated complexes with two neighboring carboxylate groups coordinated simultaneously among them [68]; it has also been reported that one carboxylate group and the neighboring sulfate group participate in bidentate ligand complex formation [69]. It is apparent from Fig. 49 that heparin is a highly charged polyion and that the polyelectrolyte effect on the Ca^{2+} ion-binding equilibria can be expected to be of the greatest importance. The binding isotherms determined with excess NaCl parallel each other (Fig. 50) as with other polyion systems. The log K_{Ca}^0 values determined by extrapolation of the log K_{Ca} versus θ_{Ca} plots to zero are expressed as a function of log Cs in this figure. These plots are linear with slopes quite close to –2 in value, just as they were for the Ca^{2+}-CmDx(Fig. 25) and the Ca^{2+}-DxS (Fig. 45) systems. The averaged DS value of heparin, estimated to be 11/6, corresponds to a "b" value of 2.73 Å. The Donnan potential term together with the Donnan phase volume term, are taken from the universal curves obtained with CmDx (Figs. 12 – 14) at a specified Cs value. The K_{Ca}^0 values can then be computed with the equation

$$K_{Ca}^0 = G10^{2\Delta pK}(1 + (\beta_S)_D f_S(C_A)_D + (\beta_C)_D f_C(C_A)_D)(C_A)_D^{-1}$$

$$= G10^{2\Delta pK}\left(\frac{V_D}{n_p} + (\beta_C)_D f_C\right) \tag{53}$$

where G stands for the activity coefficient quotient, $\{\gamma_{Ca}/(\gamma_{Ca})_D\}$ $\{(\gamma_{Na})_D/\gamma_{Na}\}^Z$, and the activity coefficient values in the Donnan phase are calculated as described earlier at I_D values determined with Eq. (29). The $(\beta_S)_D$ and $(\beta_C)_D$ terms represent microscopic stability constants for the

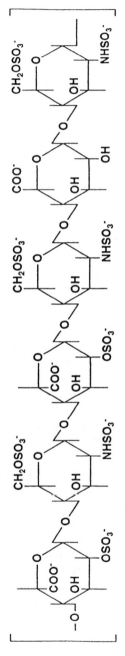

Figure 49 Structure of heparin.

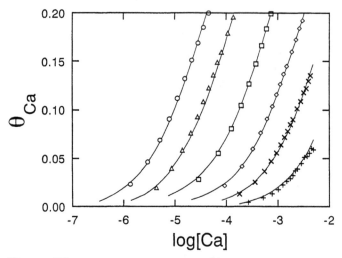

Figure 50   Binding isotherms for Ca^{2+}-heparin ($\alpha = 1$) at different NaCl concentration levels. (○) Cs = 0.01 mol dm^{-3}; (△) Cs = 0.02 mol dm^{-3}; (□) Cs = 0.05 mol dm^{-3}; (◇) Cs = 0.10 mol dm^{-3}; (×) Cs = 0.20 mol dm^{-3}; (+) Cs = 0.50 mol dm^{-3}.

monodentate complexes of Ca^{2+} ion with sulfate and carboxylate groups of the polyion, respectively. The fractions of total ionic groups fixed to the heparin molecule that are sulfate and carboxylate, respectively, are identified by f_S and f_C ($f_C = 3/11$). In this computation, it is assumed that a Ca^{2+} ion binding with sulfate groups is purely electrostatic, i.e., $(\beta_S)_D = 0$, whereas Ca^{2+} ions form *one-to-one complexes with carboxylate groups*, with a $(\beta_C)_D$ value estimated to be 4 mol dm^{-3}, as previously determined for the Ca^{2+}-CmDx binding system. The solid line shown in Fig. 51 corresponds to the plots of the calculated log K_{Ca}^0 value against log Cs. All the experimental plots lie quite close to the calculated curve to show that the carboxylate groups associated with the heparin molecule to form only monodentate complexes with Ca^{2+} ion. With the computational procedures mentioned above, it has been revealed that the metal ion-binding equilibria of these mixed ionic polysaccharides continue to be predictable by taking into account (1) the average linear charge separation on the polyions, (2) the quantity of each ionic groups, and (3) the specific interaction terms ($(\beta_1)_D$ values) with the respective sites.

V.   PHOSPHATE POLYIONS

The counterion-binding properties of polyions with phosphate groups are of special interest because of the biological importance of the role that

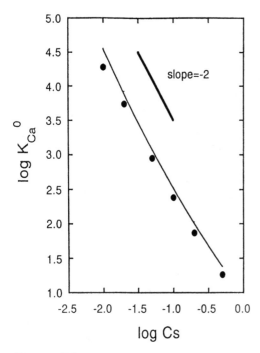

Figure 51   Relationship between log K_{Ca}^0 and log Cs for heparin. Solid line refers to the plots calculated assuming that $(\beta_S)_D = 0$ and $(\beta_C)_D = 4$.

the phosphate group plays in polynucleotides. Among the phosphate polyions, the DNA molecule is known as a "fat" polyion because of double helical structure. The inorganic long-chain polyphosphate ion, PP, is a representative of "thin" polyions because its skeleton is composed only of the phosphate groups [34]; PP is known as one of the most hydrophilic polyions. Its molecule is composed of two weak-acid end phosphate groups and a large number of strong-acid middle phosphate groups. The main contribution to the countercation binding to PP anion is from the middle phosphate groups. The average distance between the neighboring negative charges on the PP molecule is estimated to be ca. 2.8 Å, by using the well-defined bond length of the P-O-P linkage. It has already been shown that the binding equilibria of PP anion to uni- and multivalent metal cations are greatly dependent on the added salt concentration level [70,71]. The plots of log K_M^0 against log Cs for these systems are linear with slopes quite close in value to –Z, where Z is the charge of the metal ion investigated [72]. In this section, the PP-counterion-binding equilibria observed with several combinations of M^{Z+} ions and supporting cations are presented in order to examine the general features of this linear polyions.

The binding of Na^+ ion to PP ($\bar{n}$ = 150) anions has been investigated by using a Na^+ ion-selective glass electrode in the presence of an excess of tetramethylammonium chloride (TMACl). The binding isotherms, just like those compiled in the other linear polyion-counterion-binding systems, parallel each other with an increase in the added salt concentration moving the curves downward as shown in Fig 52. In Fig. 53, the log K_{Na}^0 values determined are plotted against the log Cs values. The straight line, whose slope is close to –1, where 1 corresponds to the valence of Na^+ ion, as expected. In the same figure the log K_{Na}^0 values obtained with oligophosphate ions, such as cyclic polyphosphate ions ($P_nO_{3n}^{n-}$), cP_n (n = 3, 4, 6, and 8) are plotted versus log Cs, in order to show the transition from simple electrolyte to polyelectrolyte behavior [27]. As can be seen in Table 6, the d(log K_{Na}^0)/d(log C_s) value of the oligophosphate ion varies gradually to reach –1 with an increase in the number of the phosphate units which constitute the oligophosphate molecule. It is of sizable interest to note that the plots of the cP_8 system approach those of the PP system(Fig. 54). The chain-length dependence of counterion binding equilibria in polyphosphate systems has also been studied using Ca^{2+} ion as the counterion [27]. The representative Ca^{2+} ion-binding isotherms

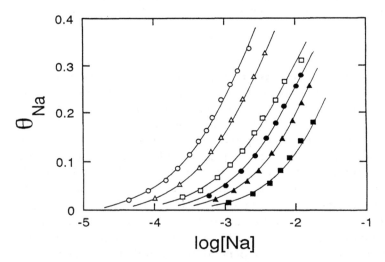

Figure 52 Binding isotherms for Na^+-PP^- ($\bar{n}$ = 150) at different TMACl concentration levels. (O) Cs = 0.01 mol dm^{-3}; (△) Cs = 0.02 mol dm^{-3}; (□) Cs = 0.05 mol dm^{-3}; (●) Cs = 0.10 mol dm^{-3}; (▲) Cs = 0.20 mol dm^{-3}; (■) Cs = 0.50 mol dm^{-3}.

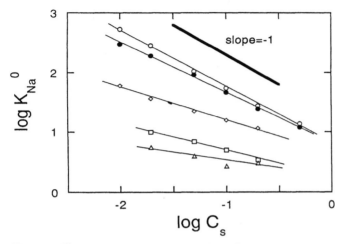

Figure 53 Relationship between log K_{Na}^0 and log Cs for cP_n and PP. ($\triangle$) cP_3; ($\square$) cP_4; ($\diamond$) cP_6; ($\bullet$) cP_8; ($\bigcirc$) PP.

obtained with the linear polyphosphate mixtures, P_n using a Ca^{2+} ion-selective electrode in the presence of 0.1 mol dm^{-3} NaCl are shown in Fig. 55. Additional chain-length-dependent data obtained with oligocyclic phosphate ions are included as well. It is worthwhile noting that the isotherms obtained in the $P_{\bar{n}}$ systems ($\bar{n}$ = 15, 50, 127, and several thousands) agree quite well with each other, showing again that an n value of about 10 is sufficient for the Ca^{2+} ion binding to become independent of a chain length. The binding isotherms of Ca^{2+} ion to PP ($\bar{n}$ = 150) ion obtained at various NaCl concentrations are shown in Fig. 56. A comparison of the vertical displacement of the Na^+ ion-binding (Fig. 52) and Ca^{2+} ion-binding (Fig. 56) isotherms for a particular change in ionic strength

Table 6 $d(\log K_{Na}^0)/d(\log[TMA])$ and $d(\log K_{Ca}^0)/d(\log[Na])$ Values for cP_n (n = 3, 4, 6, and 8) and PP Systems

Polyphosphate	$d(\log K_{Na}^0)/d(\log[TMA])$	$d(\log K_{Ca}^0)/d(\log[Na])$
cP_3	−0.29	−0.70
cP_4	−0.48	−1.21
cP_6	−0.49	−1.70
cP_8	−0.84	−1.98
PP	−0.95	−2.06

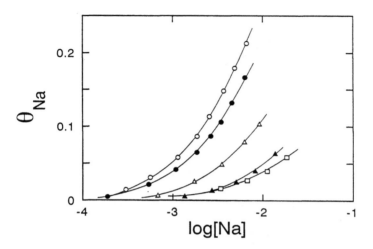

Figure 54 Binding isotherms for Na$^+$-cP$_n$ (n = 3,4,6, and 8) and Na$^+$-PP$^-$ (ñ = 150) in the presence of TMACl. (☐) cP$_3$; (▲) cP$_4$; (△) cP$_6$; (●) cP$_8$; (○) PP.

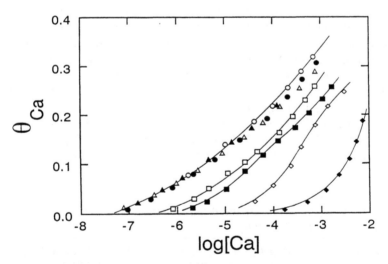

Figure 55 Binding isotherms for Ca^{2+}-cP$_n$ (n = 3,4,6, and 8) and Ca$^+$-PP$^-$ (ñ = 15, 50, 127, and several thousands) in the presence of NaCl (Cs = 0.10 mol dm^{-3}). (◆) cP$_3$; (◇) cP$_4$; (■) cP$_6$; (☐) cP$_8$; (▲) p$_ñ$ (ñ = 15); (△) p$_ñ$ (ñ = 50); (●) p$_ñ$ (ñ = 127); (○) p$_ñ$ (ñ = several thousands).

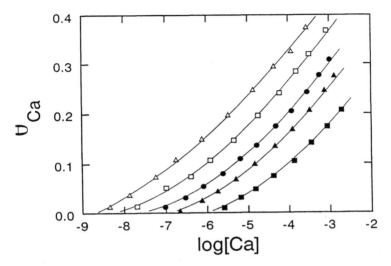

Figure 56 Binding isotherms for Ca$^+$-PP$^-$ (ñ = 150) at different NaCl levels. (Δ) Cs = 0.02 mol dm^{-3}; (□) Cs = 0.05 mol dm^{-3}; (■) Cs = 0.10 mol dm^{-3}; (▲) Cs = 0.20 mol dm^{-3}; (■) Cs = 0.50 mol dm^{-3}.

shows shifts to be approximately a factor of 2 larger in the Ca^{2+}-PP system in accordance with expectation. The log K_{Ca}^0 values determined for both the cP$_n$ and the PP systems are plotted against log C_s (Fig. 57). The linearity in all the cyclic phosphate and long-chain polyphosphate anion systems is quite good. The slopes, d(log K_{Ca}^0)/d(log C_s), and d(log K_{Na}^0)/d(log C_s) found for the straight-line isotherms obtained with the Na$^+$ and Ca^{2+}, cyclic phosphate and long-chain polyphosphate anion systems, were −2 and −1, respectively, once the number of repeating monomer units reached ~8. It is surprising that Gibbs-Donnan concepts are applicable to such small molecules. For example, in the absence of salt it has been determined that about 100 monomer units need to be repeated in the salt-free PP molecule before the linear polyphosphate will exhibit counterion-binding properties that are independent of chain length (Fig. 2). A possible explanation for this effect of salt may be that the added salt greatly suppresses polyion-counterion interaction by shielding. By taking into consideration the electrostatic interaction between a specific mobile ion and a specific fixed ion in linear polyphosphate systems, it may be reasonable to expect that the electrostatic effect due to phosphate unit separated from more 10 neighboring phosphate units, i.e., ca. 30 Å, will be negligible in the presence of an excess of simple salt. A more reason-

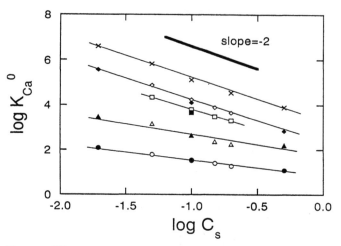

Figure 57   Relationship between log K_{Ca}^0 and log C_s for cP_n and PP. (○) cP_3; (△) cP_4; (□) cP_6; (◇) cP_8; (×) PP. Open symbols; calculated by use of the reported experimental data [74]: closed symbols; this work.

able rationalization may be that shielding of surface charge by the salt may lead to sufficient aggregation of the cyclic linear polyphosphates to endow them with the macromolecular properties observed to prevail.

The complexation equilibria of these oligophosphate ions have conventionally been analyzed by Bjerrum's formation function [73, 74], and the pronounced effect of the supporting electrolyte concentration on the binding equilibria of the oligophosphate ions to metal ions has been attributed to the substantial change in the activity coefficients of the respective chemical species. In order to determine the thermodynamic stability constants, the logarithm of the stability constants of the complexes were plotted against the square roots of the concentration of the supporting electrolyte, i.e., ionic strengths. However, since the present study indicates that the formation of counterion concentrating region around the oligophosphate molecule(s) is responsible for the salt concentration dependence of the binding constants, the ionic strength dependence of the binding equilibrium has been sought by plotting this logarithms of the apparent binding constant against log I. The β_1 values of the one-to-one complexes of the oligophosphate anions so obtained are larger than the intrinsic value by the Donnan potential term associated with the particular salt concentration level. Further precise study is needed for the quantitative expression of the intrinsic binding equilibria through accurate assessment of the Donnan term.

The PP molecule is composed mainly of middle phosphate units. In the H^+ ion form, the group being repeated is a fairly strong acid, with a pK_a value estimated to be about 1. As has examined however, a second mode of interaction of H^+ and PP^- was observed in addition to the usual one-to-one proton complex formed [71]. The H^+-PP^- binding equilibria was studied by measuring the p[H] with a glass electrode in the presence of an excess of TMA Cl. The θ_H versus log[H] plots determined at various concentrations of TMA Cl are shown in Fig. 58. The relationship between the binding isotherms obtained is the same as with other polyelectrolyte systems; however shift to the right hand side with increase in the supporting electrolyte concentration at low salt concentration levels stops when the added salt concentration is higher than 0.5 mol dm^{-3}. This feature known as "wall of site binding" is illustrated in Fig. 10 and is a consequence of salt invasion of the polyion domain until counterion ratios in the two phases reach unity. The protonation behavior of the PP molecule in such high concentrations of supporting electrolyte can be expressed quantitatively by an apparent protonation constant, $(K_H)_{app}$, defined as

$$(K_H)_{app} = \frac{\theta_H}{(1-\theta_H)[H]} \tag{54}$$

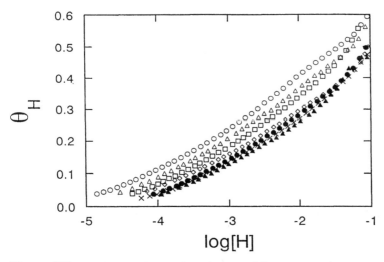

Figure 58 Binding isotherms for H^+-PP^- at different TMACl concentration levels. ($\bigcirc$) Cs = 0.02 mol dm^{-3}; ($\triangle$) Cs = 0.05 mol dm^{-3}; ($\square$) Cs = 0.10 mol dm^{-3}; ($\diamond$) Cs = 0.20 mol dm^{-3}; ($\bullet$) Cs = 0.50 mol dm^{-3}; ($\blacktriangle$) Cs = 1.00 mol dm^{-3}; ($\times$) Cs = 2.00 mol dm^{-3}.

The log $(K_H)_{app}$ value calculated in this way are plotted against θ_H in Fig. 59. It can be seen that the plots are composed of at least two lines that intersect at about $\theta_H = 0.5$. In the high-θ_H regions, i.e., $\theta_H \geq 0.5$, the log $(K_H)_{app}$ values remain constant (=0.6) irrespective of the change in θ_H. This log $(K_H)_{app}$ value is consistent with the log K_H value (ca. 1 [75]) of hypophosphite ion, $PH_2O_2^-$, the monomer analog of PP ion. This result indicates that the one-to-one proton complex formed, HPH_2O_2, is predominant in high-θ_H regions, whereas in low-θ_H regions, i.e., $\theta_H \leq 0.5$, the log $(K_H)_{app}$ value increases drastically with a decrease in θ_H. An additional much stronger binding mode for H^+ ion to the PP middle group(s) is indicated. Additional evidence for this alternate binding mode in the H^+-PP^- system has been provided by measuring the ^{31}P NMR chemical shift change for the PP middle phosphate group (δ_P) upon protonation. As can be seen from Fig. 59, the plots of chemical shift difference $\Delta\delta_P$ (deviation from the shift value determined at full dissociation of PP) against θ_H are composed of two lines, which correspond to the two separate equilibria obtained in the acid-dissociation study. The ^{31}P NMR result together with thermodynamically based evidence for the existence of two H^+-ion complexation paths in the linear polyion show two adjacent phosphate units bind to one H^+ ion simultaneously with the one phosphate group binding being much more weak than the other. The proximity of the adjacent phosphate units of the PP molecules may be responsible for this mode of

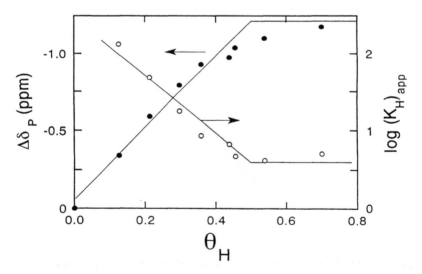

Figure 59 The plots of log $(K_H)_{app}$ and $\Delta\delta_P$ against θ_H. (O) log $(K_H)_{app}$; (●) $\Delta\delta_P$. $Cs = 1$ mol dm^{-3} (TMACl), $Cp = 0.02$ monomol dm^{-3}.

H^+ ion binding to the linear polyion. A similar pattern of protonation has recently been reported for specified polyions with carboxylate groups, i.e., polymers of maleic acid and fumaric acid, whose distance between the adjacent carboxylate groups is as short as ca. 2 Å [76].

When silver ion is substituted for H^+ ion, only a single complex found proved to be predominant when the earlier experimental conditions are duplicated [71]. In Fig. 60 the chemical shift difference, $\Delta\delta_p$ (deviation from the shift value determined in the absence of Ag^+ ion) for the PP middle phosphate group is expressed as a function of θ_{Ag}. The linearity obtained at all ionic strengths is strong evidence for the formation of one-to-one Ag^+ ion complex with the phosphate middle group of the PP molecule. This makes it possible to determine the effective polyelectrolyte phase volume of the PP molecule. As with other M^{Z+}-PP^- systems, binding isotherms of the Ag^+-PP^- system determined in the presence of excess $NaNO_3$ parallel each other. In the case of the $Ag^+/PP^-/Na^+$ (excess) system, the following equation is once again applicable (see Eq. (7)).

$$\frac{(a_{Ag})_D}{a_{Ag}} = \frac{(a_{Na})_D}{a_{Na}} \tag{55}$$

where $(a)_D$ correspond to the activity of a particular ion contained by the Donnan phase developed by the PP molecule. By use of the activity coef-

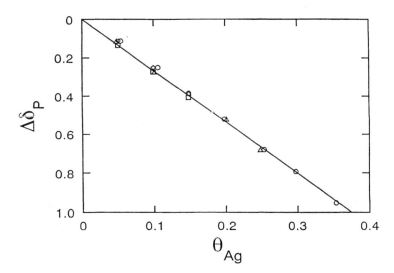

Figure 60 The plots of $\Delta\delta_p$ against θ_{Ag}. (O) Cs = 0.05 mol dm^{-3}; (Δ) Cs = 0.10 mol dm^{-3}; ($\square$) Cs = 0.20 mol dm^{-3}; ($\Diamond$) Cs = 0.50 mol dm^{-3}.

ficient equation, $G \ (=(\gamma_{Na})_D \gamma_{Ag}/\{(\gamma_{Ag})_D \ \gamma_{Na}\})$, the free Ag^+-ion concentration in the Donnan phase, $[Ag]_D$ is formulative as before.

$$\begin{aligned} [Ag]_D &= G[Na]_D[Ag][Na]^{-1} \\ &= \frac{G(1 - \phi_{p,Na} - \theta_{Ag})n_p}{V_D}[Ag][Na]^{-1} \end{aligned} \tag{56}$$

Also the contribution to θ_{Ag} by site-bound Ag^+ ions, $(\theta_{Ag})_{site}/\theta_{Ag}$, can be expressed by use of an intrinsic stability constant, $(\beta_1)_D$, of the monodentate complex, $[Ag(PO_3)]^0$, and the free phosphate unit concentration in the Donnan phase, $[P]_D$.

$$\begin{aligned} \frac{(\theta_{Ag})_{site}}{\theta_{Ag}} &= \frac{[Ag(PO_3)^0]_D}{[Ag]_D + [Ag(PO_3)^0]_D} \\ &= \frac{(\beta_1)_D[Ag]_D[P]_D}{[Ag]_D + (\beta_1)_D[Ag]_D[P]_D} \\ &= \frac{(\beta_1)_D[P]_D}{1 + (\beta_1)_D[P]_D} \end{aligned} \tag{57}$$

Since $(\theta_{Ag})_{site}$ is defined as the difference between the amounts of the total and the free silver ions in the Donnan phase, $(\theta_{Ag})_{site}$ can be expressed with the following equation as

$$\begin{aligned} (\theta_{Ag})_{site} &= \theta_{Ag} - \frac{[Ag]_D V_D}{n_p} \\ &= \theta_{Ag} - G(1 - \phi_{p,Na} - \theta_{Ag})[Ag][Na]^{-1} \end{aligned} \tag{58}$$

By combining Eqs. (57) and (58) we obtain Eq. (59)

$$(\beta_1)_D[P]_D = \theta_{Ag}[Na]G^{-1}(1 - \phi_{p,Na} - \theta_{Ag})^{-1}[Ag]^{-1} - 1 \tag{59}$$

With the presence of trace-level Ag^+ ion, i.e., extrapolating the θ_{Ag} value to zero, $[P]_D$ is equal to n_p/V_D. By use of the $K_{Ag}^0 \ (=\theta_{Ag}/[Ag])$ value, Eq. (59) can be rearranged to Eq. (60).

$$(\beta_1)_D\left(\frac{n_p}{V_D}\right) = K_{Ag}^0[Na]G^{-1}(1 - \phi_{p,Na})^{-1} - 1 \tag{60}$$

Table 7 Constancy of $(\beta_1)_D$ (n_p/V_p) Values for $Ag^+/PP^-/Na^+$ System

[Na] (mol dm^{-3})	0.02	0.05	0.10	0.20	0.50
log K_{Ag}^0	2.25	1.92	1.63	1.33	0.96
$(\beta_1)_D$ (n_p/V_p)	3.45	4.20	4.34	4.35	4.59

As a rough estimate, we may assume that $G = 1$ and that the practical osmotic coefficient of the PP anion is 0.2. The values of $(\beta_1)_D(n_p/V_D)$ calculated by use of the K_{Ag}^0 values determined under various sodium nitrate concentrations are listed in Table 7. The constancy of the product, $(\beta_1)_D(n_p/V_D)$ can be assumed attributable to the constancy of both the $(\beta_1)_D$ and the V_D/n_p values irrespective of the bulk salt concentration level. By assigning a $(\beta_1)_D$ value of 4 mol^{-1} dm^3, the V_D/n_p value is estimated to be ca. 1 dm^3 monomol^{-1}. This quantity is consistent with the V_p/n_p value of 1.4 dm^3 monomol^{-1} obtained from Eq. (52) derived with Manning's two-variable theory when the b value of the PP molecule is assumed to be 2.8 Å.

VI. CONCLUSIONS

The high sensitivity of the metal ion-binding equilibria of negatively charged polymer systems to the added uni-univalent salt concentration levels of their solutions has been interpreted in a unified manner by treating the distribution of counterions between the counterion concentrating region formed around the polyion skeleton and the bulk solution as an *ion-exchange reaction*. The introduction of a volume term for the counterion concentrating region (polyion domain) permits definition of the averaged concentrations of the ionic species in the vicinity of the polyion skeleton, that are needed to express the binding equilibria. The Gibbs-Donnan concept has been proved to be applicable to the analysis of the binding equilibria of not only the polyion gel (ion exchanger), where the boundary between the two regions is well defined, but also to the binding equilibria of linear polyions and even oligoion systems, even though no distinct boundary between the two regions is observable in the water-soluble ionic molecular assemblies. Based on this concept, the apparent (overall) binding equilibria are handled stepwise. First, the activity of the free metal ion in the polyion domain is related to the activity of the free metal ion in the bulk solution phase by the averaged potential, i.e., the Donnan potential between the two regions. Second, the intrinsic binding equilibria of a metal ion to fixed negative charges are expressed as direct binding equilibria in the polyion domain. A rationale for this treatment is based on the fact that for any polyion system, the apparent binding constant, $\log K_A^0$, is simply related to the concentration of the supporting electrolyte, $\log [B]$, by a factor of Z, the charge of an A ion (Eq. (11)), i.e., the fraction of the free A ion in the polyion domain is not affected by substantial change in $[B]$.

The polyion domain volume can be computed by use of the acid-dissociation equilibria of weak-acid polyelectrolyte and the multivalent metal ion binding equilibria of strong-acid polyelectrolyte, both in the presence of an excess of Na^+ salt. The volume computed is primarily related to the solvent uptake of tightly cross-linked polyion gel. In contrast to the polyion gel systems, the boundary between the polyion domain and bulk solution is not directly accessible in the case of water-soluble linear polyelectrolyte systems. Electroneutrality is not achieved in the linear polyion systems. A fraction of the counterions trapped by the electrostatic potential formed in the vicinity of the polymer skeleton escapes at the interface due to thermal motion. The fraction of the counterion release to the bulk solution is equatable to the practical osmotic coefficient, and has been used to account for such loss in the evaluation of the Donnan phase volume in the case of linear polyion systems.

Results of the counterion binding equilibrium analysis of DxS and CmDx which are composed of the common dextran chain, reveal that both the Donnan potential term and the Donnan phase volume term can be expressed simply as a function of the linear charge separation of the polyion and the added salt concentration. Their magnitude is not affected by the nature of the ionic groups of the polyion whether they are carboxylate or sulfate. This makes it possible to predict the extent of the electrostatic effect on the binding equilibrium of certain polyion-counterion combinations, once the structural parameter of the polyion, the charge of the counterion, and the added salt concentration are known. The volume of ionic dextrans with sufficiently high charge density (b = 2.5 Å) is insensitive to the change in the added salt concentration. The computed Donnan phase volume is consistent with the polyelectrolyte phase volume calculated by Manning's two-variable theory.

The agreement between the nonideality terms resolved for CmDx and DxS samples whose DS values are higher than 0.5 is lost when the DS values of the samples are lower than this critical value. The sizable decrease in the phase volume of the DxS polyion when DS has been reduced to lower than 0.5 leads to questions with respect to the mode of ion binding at the lower charge density. Any change should be consistent with the ^{23}Na NMR peak width change with the linear charge separation of the DxS and CmDx samples in salt-free systems. The continuity and discontinuity found in the profiles of the electrostatic effect–charge density relationship are related to the location of the counterions investigated around the polyion surface. The nonideality terms of the acid-dissociation properties of weak-acid polyions reflect the electrostatic potential at the reaction site (carboxylate group) and can be used to estimate directly the degree of the electrostatic effect on the complexation equilibria of the weak-acid polyions.

The ionic atmosphere in the polyion domain is considered to correspond to a homogeneous concentrated electrolyte solution composed of fixed ions, counterions, and co-ions. By defining the standard state of each phase to be the same, valid insights with respect to the measured properties of the various species in the Donnan phase are facilitated. Agreement of the values of the intrinsic binding constants of the monodentate complexes of CmDx and its gel analogs determined with this approach, with those of the corresponding metal complexes of their monomer analogs, i.e., acetate complexes demonstrate the value of this approach.

It has been revealed by the present systematic work on the metal ion-binding equilibria in negatively charged polymers that the Gibbs-Donnan concept can be successfully applied to the thermodynamic expression of the counterion equilibria of polyions of various dimensions, i.e., polyion gels, linear polyions, and even oligoions. This concept can be applied to any polyions irrespective of their shapes, appearances, the nature of the fixed ions, and the degrees of the cross-linking in the polyion gel systems. Also, in spite of the present limited examples of the counterions such as protons and metal ions, this concept is expected to be applicable to the evaluation of the electrostatic effect in the studies of polyion-ionic organic molecule, such as surfactant interactions.

ACKNOWLEDGMENTS

The present author wishes to express his deep gratitude to Professor Jacob A. Marinsky for his continuing encouragement and valuable suggestions throughout this work. He also thanks Dr. Erik Högfeldt and Dr. Mamoun Muhammed in KTH (Stockholm) and Dr. J.C.T. Kwak in Dalhousie University (Halifax) for their interest in this work and many useful discussions. He wishes to acknowledge financial support from the Swedish Natural Science Research Council (NFR), the Natural Sciences and Engineering Research Council of Canada (NSERC), Grant-in-Aid for Scientific Research from Ministry of Education, Science and Culture of Japan (Nos. 04804046 and 05804036), and the Takeda Science Foundation (Japan).

LIST OF SYMBOLS

Quantities and symbols with subscript D and g denote the Donnan phase and the gel phase, respectively.

α	Degree of dissociation of a polyacid
a_i	Activity of species i
b	Linear charge separation on a polyion backbone
β_i	Microscopic stability constant of MA_i-type complex
$(\beta_1)_{app}$	Apparent stability constant of MA-type complex
C_A	Total concentration of A^{Z+} ions (mol dm^{-3})
Cp	Total concentration of polymer ionic groups (monomol dm^{-3})
C_s	Concentration of simple salt
DS	Degree of substitution of ionic dextrans
δ_A	Chemical shift of nucleus A (ppm)
$\Delta\delta_A$	Chemical shift difference of nucleus A (ppm)
D_M^0	Distribution ratio of a trace amount of metal ion between a ion-exchanger and solution
ΔpK	Difference between pK_{app} and pK_0
e	Unit electric charge
f_C	Carboxylate group fraction of total ionic groups of heparin
f_S	Sulfate group fraction of total ionic groups of heparin
ϕ_p	Practical osmotic coefficient
γ_i	Activity coefficient of species i
G	Activity coefficient quotient
[i]	Concentration of species i
I	Ionic strength
K_{app}	Apparent acid dissociation constant of a polyacid
K_{app}^*	Apparent acid dissociation constant of a polyion gel defined by use of H^+-ion concentration in the gel phase
K_0	Intrinsic acid dissociation constant of a polyacid
K_A	Apparent binding constant of A^{Z+} ion to a polyion (monomer basis)
K_A^0	Intrinsic binding constant of A^{Z+} ion to a polyion (monomer basis)
k_f	Territorially bound fraction of total amount of bound A^{Z+} ion
n	Degree of polymerization of polyphosphate (number of monomer units per polymer molecule)
$\tilde{n}$	Average degree of polymerization of polyphosphate
n_i	Total amount of species i (mole or monomole)
θ_A	Average number of bound A^{Z+} ions per ionic group
V_D	Donnan phase volume (dm^3)
V_p	Polyelectrolyte phase volume (cm^3)

$W_{1/2}$ Peak width at half peak height
Z   Charge of an A^{Z+} ion

REFERENCES

1. J. A. Marinsky, *Coord. Chem. Rev.*, *19*: 125 (1976).
2. J. Demirglan, F. Kishk, R. Yag, and J. A. Marinsky, *J. Phys. Chem.*, *83*: 2743 (1979).
3. J. A. Marinsky, F. G. Lin, and K. Chung, *J. Phys. Chem.*, *87*: 3139 (1983).
4. J. A. Marinsky, *J. Phys. Chem.*, *89*: 5294 (1985).
5. J. A. Marinsky, R. Baldwin, and M. M. Reddy, *J. Phys. Chem.*, *89*: 5303 (1985).
6. J. A. Marinsky, T. Miyajima, E. Högfeldt, and M. Muhammed, *React. Polym.*, *11*: 279 (1989).
7. J. A. Marinsky, T. Miyajima, E. Högfeldt, and M. Muhammed, *React. Polym.*, *11*: 291 (1989).
8. J. A. Marinsky and M. M. Reddy, *J. Phys. Chem.*, *95*: 10208 (1991).
9. J. A. Marinsky, *J. Phys. Chem.*, *96*: 6484 (1992).
10. F. Helfferich, *Ion Exchange*, McGraw-Hill, New York, 1962, Chap. 5.
11. H. P. Gregor, *J. Am. Chem. Soc.*, *73*: 642 (1951).
12. E. Gluckauf, *Proc. Roy., Ser. A.*, *214*: 207 (1952).
13. I. Michaeli and A. Katchalsky, *J. Polym. Sci.*, *23*: 683 (1957).
14. F. Oosawa, *Polyelectrolytes*, Marcel Dekker, New York, 1971.
15. J. A. Marinsky, T. Miyajima, E. Högfeldt, F. G. Lin, and M. Muhammed, *Properties of Ionic Polymers-Natural and Synthetic* (L. Salmen and M. Htun, eds.), Swedish Pulp and Paper Research Institute, Stockholm, Sweden, 1991, p.7.
16. T. Miyajima, K. Ishida, M. Katsuki, and J. A. Marinsky, in *New Developments in Ion Exchange* (M. Abe, T. Katayama, and T. Suzuki, eds.), Kodansha-Elsevier, 1991, p. 497.
17. M. Rinaudo, in *Polyelectrolytes*, (E. Selegny, ed.), Reidel, Dordrecht, Holland, 1974, vol. 1, p. 157.
18. G. S. Manning, *J. Chem. Phys.*, *51*: 924 (1969).
19. G. S. Manning, *Q. Rev. Biophys.*, *2*: 179 (1978).
20. G. S. Manning, *Biophys. Chem.*, *7*: 95 (1977).
21. M. Nagasawa, M. Izumi, and I. Kagawa, *J. Polym. Sci.*, *37*: 375 (1959).
22. M. Nagasawa and I. Kagawa, *J. Polym. Sci.*, *25*: 61(1957).
23. W. Kern, *Makromol. Chemie.*, *2*: 279 (1948).
24. H. S. Kielman, J. M. A. M. Van der Hoeven, and J. C. Leyte, *Biophys. Chem.*, *4*: 103 (1976).
25. H. S. Kielman and J. C. T. Leyte, *J. Phys. Chem.*, *77*: 1593 (1973).
26. G. S. Manning, *J. Chem. Phys.*, *62*: 748 (1975).
27. T. Miyajima and R. Kakehashi, *Phosphorous Res. Bull.*, *3*: 37,43 (1993).
28. T. Miyajima, K. Yamauchi, and S. Ohashi, *J. Liq. Chromatogr.*, *4*: 1891 (1981).
29. H. Maki and T. Miyajima, in preparation.

30. H. Noguchi, K. Gekko, and S. Makino, *Macromolecules, 6*: 438 (1973).
31. A. Katchalsky, Z. Alexandrowicz, and O. Kedem, in *Chemical Physics of Ionic Solutions* (B. E. Conway and R. G. Barradas, eds.), Wiley, New York, 1966, p. 295.
32. A. Katchalsky and Z. Alexandrowicz, *J. Polym. Sci., A1*: 2093 (1963).
33. L. Kotin and M. Nagasawa, *J. Chem. Phys., 36*: 873 (1962).
34. G. S. Manning, *Acc. Chem. Res., 12*: 443 (1979)
35. C. F. Anderson, M. T. Record, Jr., and P. A. Hart, *Biophys. Chem., 7*: 301 (1978).
36. C. W. R. Mulder, J. D. Bleijer, and J. C. Leyte, *Chem. Phys. Lett., 69*: 354 (1980).
37. H. Gustafsson, B. Lindman, and T. Bull, *J. Am. Chem. Soc., 100*: 4655 (1978).
38. B. Lindman, in *NMR of Newly Accessible Nuclei*, (P. Laszlo, ed.), Academic Press, 1983, vol. 1, p. 193.
39. H. Gustafsson, G. Siegel, B. Lindman, and L. Fransson, *FEBS Lett., 86*: 127 (1978).
40. J. A. Marinsky, N. Imai, and M. C. Lin, *Isr. J. Chem., 11*: 601 (1973).
41. W. M. Anspach and J. A. Marinsky, *J. Phys. Chem., 79*: 433 (1975).
42. T. Miyajima K. Yoshida, Y. Kanegae, H. Tohfuku, and J. A. Marinsky, *React. Polym., 15*: 55 (1991).
43. T. Miyajima, Y. Kanegae, K. Yoshida, M. Katsuki, and Y. Naitoh, *Sci. Total Environ., 117/118*: 129 (1992).
44. N. Imai and J. A. Marinsky, *Macromolecules, 13*: 275 (1980).
45. F. J. C. Rossoti and H. Rossotti, *The Determination of Stability Constants*, McGraw-Hill, New York, 1961, p. 324.
46. J. Mattai and J. C. T. Kwak, *J. Phys. Chem., 88*: 2625 (1984).
47. Y. M. Joshi and J. C. T. Kwak, *Biophys. Chem., 13*: 65 (1981).
48. J. Mattai and J. C. T. Kwak, *Biophys. Chem., 14*: 55 (1981).
49. J. Mattai and J. C. T. Kwak, *Biophys. Chem., 31*: 295 (1988).
50. J. Mattai and J. C. T. Kwak, *J. Phys. Chem., 86*: 1026 (1982).
51. J. Mattai and J. C. T. Kwak, *Macromolecules 19*: 1663 (1986).
52. J. Schubert, E. R. Russel, and L. T. Myers, *J. Biol. Chem., 185*: 387 (1950).
53. C. Travers and J. A. Marinsky, *J. Polym. Sci., Symp. Ser. No. 47*: 285 (1974).
54. J. A. Marinsky, in *Ion Exchange and Solvent Extraction—A Series of Advances*, (J. A. Marinsky and Y. Marcus, eds.), Marcel Dekker, New York, 1973, vol. 4, Chap. 5.
55. J. A. Marinsky, A. Wolf, and K. Bunzl, *Talanta, 27*: 461 (1980).
56. J. A. Davies, R. O. James, and J. O. Leckie, *J. Colloid Interface Sci., 63*: 480 (1978).
57. G. Gran, *Analyst, 77*: 661 (1951).
58. J. E. Powell, R. S. Kolat, and G. S. Paul, *Inorg. Chem., 3*: 518 (1964).
59. K. Gekko and H. Noguchi, *Biopolymers, 14*: 2555 (1975).
60. P. Slota and J. A. Marinsky, in *Ions in Polymers*, (A. Eisenberg, ed.), Advances in Chemistry Series, vol. 187, American Chemical Society, Washington, DC, 1980, p. 311.

61. H. Oshima and T. Miyajima, *Colloid Polym. Sci., 272*: 803 (1994).
62. S. Alegret, M. Escalas, and J. A. Marinsky, *Talanta, 31*: 683 (1984).
63. S. Sasaki, T. Miyajima, and H. Maeda, *Macromolecules, 25*: 3599 (1992).
64. R. A. Robinson and R. H. Stokes, *Electrolyte Solutions*, Butterworths, London, 1959, Appendix 8,10.
65. A. E. Martell and R. M. Smith (Eds.), *Critical Stability Constants*, Vol. 3, Plenum Press, New York, 1977.
66. N. R. Joseph, *J. Biol. Chem., 164*: 529 (1946).
67. H. Waki and Y. Tokunaga, *J. Liq. Chromatogr., 5(suppl. 1)*: 105 (1982).
68. J. Boyd, F. B. Williamson, and P. Gettins, *J. Mol. Biol., 137*: 175 (1980).
69. J. N. Liang, B. Chakrabarti, L. Ayotte, and A. S. Perlin, *Carbohydr. Res., 106*: 101 (1982).
70. T. Miyajima, J. C. T. Kwak, and S. Ohashi, *Phosphorous and Sulfur, 30*: 581 (1987).
71. T. Miyajima, M. Wada, M. Tokutomi, and K. Inoue, in *Ordering and Organization in Ionic Solutions*, (N. Ise and I. Sogami, eds.), World Scientific, 1988, p. 131.
72. W. Wieker, A. Grossman, and E. Thilo, *Z. Anorg. Allgem. Chem., 307*: 42 (1960).
73. G. Kura and S. Ohashi, *J. Inorg. Nucl. Chem., 38*: 1151 (1976).
74. G. Kura, S. Ohashi, and S. Kura, *J. Inorg. Nucl. Chem., 36*: 1605 (1974).
75. *Stability Constants of Metal Ion Complexes*, Part A, *Inorganic Ligands*, compiled by E. Högfeldt, Pergamon Press, 1982.
76. T. Kitano, S. Kawaguchi, K. Ito, and A. Minakata, *Macromolecules, 20*: 1598 (1987); S. Kawaguchi, T. Kitano, and K. Ito, *Macromolecules, 25*: 1294 (1992).

8

Ion-Exchange Equilibria of Amino Acids

Zuyi Tao

Lanzhou University, Lanzhou, People's Republic of China

I. INTRODUCTION

Both natural and synthesized amino acids are complex mixtures of substances that resemble each other closely. Their initial solutions may contain tens, hundreds, or even thousands of components, many of which are closely related to the desired product in terms of their physical-chemical properties. The separation of an individual or very pure amino acid, a very complicated problem, is amenable to solution through the use of ion-exchange technology. The separation and quantitative analysis of mixtures of amino acids, one of the early outstanding contributions of ion exchange, followed closely the prior contribution of ion exchange to the separation of the rare earths. The classic study of the application of ion exchange to the separation of amino acids by Moore and Stein [1] was published in 1951; it led to the award of the Nobel Prize for chemistry in 1972.

Ion-exchange chromatography continues to be a popular method for the preparation and analysis of amino acids. Even though many techniques available in the rapidly expanding field of high-performance liquid chromatography have been developed, the high capacity of ion-exchange columns makes them ideally suited for extending their use from analytical scale to preparative scale chromatography [2–4].

Since the amino acids are the chemical units from which proteins are formed, the ion-exchange equilibria of amino acids will help provide insight with respect to the ion-exchange behavior of proteins [2,3].

For four decades, ion-exchange equilibria of amino acids have been the subject of numerous experimental and theoretical investigations. However, their ion-exchange equilibria are not nearly as well understood as the ion-exchange equilibria of mineral ions. In spite of the research that has been conducted, there is no unified physiochemical description of the ion-exchange equilibria of amino acids. Detailed mechanisms of interaction of amino acids with ion exchangers are unavailable. The theory of ion-exchange equilibria of amino acids is still in the formative stage.

II. PHYSICAL-CHEMICAL PROPERTIES OF AMINO ACID SOLUTIONS

A. Acid-Base Properties of Amino Acids

Amino acids constitute a particularly important class of bifunctional compounds; the two functional groups in an amino acid are, respectively, basic and acidic, the compounds are amphoteric, and in fact exist as zwitterions or inner salts. For example, glycine, the simplest amino acid, exists mostly in the zwitterion form shown first, rather than as aminoacetic acid

$$H_3^+NCH_2CO_2^- \rightleftharpoons H_2NCH_2CO_2H$$

glycine   glycine
zwitterion form   amino acetic acid form

Some 20 amino acids normally occur in proteins. They are α-amino acids and have the general formula $R—CH(NH_3^+)COO^-$. They exist in the zwitterion form, both in aqueous solution and in solid crystals. In written accounts they are usually shown as $R—CH(NH_2)COOH$, even though alternative zwitterionic forms are available for some. The R groups are different in different amino acids. Table 1 lists the common acids together with the formulas of their R groups [4]. The amino acids are classifiable under three categories, neutral, acidic and basic. The acidic amino acids have an extra carboxyl group in their R groups and the basic amino acids have an extra amino group, while the neutral amino acids have neither.

Most of the amino acids are only sparingly soluble in water, as a consequence of the strong intermolecular electrostatic attraction of zwitterions in the crystal lattice. Exceptions are GLY, ALA, PRO, LYS and ARG, which are quite soluble in water. The water solubility of amino acids is given in Table 2 [5]. All except ASP, GLU and TYR exceed a solubility of 1 g per 100 cm^3 at 25°C.

Table 1 Common Amino Acid Constituents of Protein

R = H   GLY,   Glycine
   CH_3 ALA,   Alanine
   $CH_2(CH_3)_2$ VAL,   Valine
   $CH_2CH(CH_3)_2$ LEU, Leucine
   $CH(CH_3)C_2H_5$ ILE,   Isoleucine
   CH_2OH SER,   Serine
   $CH(OH)CH_3$ THR,   Threonine
   CH_2SH CYS,   Cysteine
   $CH_2CH_2SCH_3$ MET,   Methionine
   $CH_2C_6H_5$ PHE,   Phenylalanine
   $CH_2C_6H_4OH$ TYR   Tyrosine

$$\begin{array}{l} CH_2-CH_2 \\ \ \ | \qquad \quad | \\ CH_2 \ \ CHCOOH \qquad PRO. \quad Proline \\ \quad \diagdown NH \diagup \end{array}$$

$$\begin{array}{l} HOCH-CH_2 \\ \ \ \ \ | \qquad \quad | \\ \quad CH_2 \ \ CH-COOH \qquad HYP, \quad Hydroxyproline \\ \quad \ \ \diagdown NH \diagup \end{array}$$

Acidic amino acids
  R = CH_2COOH ASP,   Aspartic acid
       CH_2CH_2COOH GLU,   Glutamic acid
Basic amino acids
  R = $(CH_2)_4NH_2$ LYS,   Lysine
      $(CH_2)_3C(NH)_2NH_2$ ARG,   Arginine

$$\begin{array}{l} \qquad \ CH-N \\ CH_2C \quad \ \ || \qquad HIS, \ \ Histidine \\ \quad \ \ \diagdown NH-CH \end{array}$$

$$\begin{array}{l} \quad \ \diagup CH_2C \\ CH \qquad \qquad \diagup \\ \quad \diagdown CH \end{array} \quad TRY, \ Tryptophan$$

In acidic solution, the amino acid is completely protonated and exists as the conjugate acid

$$R-CH(NH_3^+)COO^- + H^+ \rightleftharpoons R-CH(NH_3^+)COOH \qquad (1)$$

The dissociation equilibria of the α-carboxyl group and the α-amino group are

$$R-CH(NH_3^+)COOH \rightleftharpoons H^+ + R-CH(NH_3^+)COO^- \qquad (2)$$

$$K_{\alpha\text{-}CO_2H} = \frac{[H^+][R-CH(NH_3^+)COO^-]}{[R-CH(NH_3^+)COOH]} \qquad (3)$$

$$R-CH(NH_3^+)COO^- \rightleftharpoons H^+ + R-CH(NH_2)COO^- \qquad (4)$$

Table 2 Some Physicochemical Properties

		Water solubility (g/100 g, 25°C)	pK_{-CO_2H}	$pK_{\alpha-NH_3^+}$	pK side chain	Partial molal volumes (cm³ mol⁻¹, 25°C)	Molal activity coefficients in water, 25°C		
							m = 0.2	m = 0.3	m = 0.5
1	ALA	16.5	2.34	9.69		60.46	0.993	0.991	0.989
2	ARG	15.0	2.17	9.04	12.84	127.34			
3	ASP	0.5	1.88	9.60	3.65	73.83			
4	CYS	V.Sol	1.71	8.18	10.28	73.44			
5	GLU	0.84	2.16	9.67	4.32	85.98	0.962	0.947	0.917
6	GLY	25	2.34	9.60		43.25			
7	HIS	7.59	1.82	9.17	6.00	98.79			
8	ILE	4.12	2.36	9.68		107.72			
9	LEU	2.3	2.36	9.60		107.75			
10	LYS	V.Sol	2.18	9.12	10.53				
11	MET	3.5	2.28	9.21		105.35			
12	PHE	2.97	1.83	9.13		121.48			
13	PRO	162.3	1.99	10.6		82.23	1.013	1.019	1.034
14	SER	5.0	2.21	9.15		60.62	0.963	0.941	0.901
15	THR	20.5	2.71	9.62			0.983	0.976	0.964
16	TRY	1.14	2.38	9.39		143.91			
17	TYR	0.05	2.20	9.11	10.07	123.6			
18	VAL	8.85	2.32	9.62		90.78	1.016	1.025	1.048

$$K_{\alpha\text{-}NH_3} = \frac{[H^+][R\text{—}CH(NH_2)COO^-]}{[R\text{—}CH(NH_3^+)COOH]} \qquad (5)$$

When the hydrochloride of an amino acid has been half-neutralized by adding base, $[R\text{—}CH(NH_3^+)COO^-H] = [R\text{—}CH(NH_3^+)COO^-]$; after one equivalent of base has been added, the chief species in solution is the zwitterionic form of the amino acid itself. The pH of the solution at this point is simply the pH of a solution of the amino acid in pure water; it is called the isoelectric point (pI). Addition of a further half-equivalent of base corresponds to half-neutralization of the $R\text{—}CH(NH_3^+)COO^-$ and $[R\text{—}CH(NH_2)COO^-] = [R\text{—}CH(NH_3^+)COO^-]$. The $pK_{\alpha\text{-}Co_2H}$ of the α-carboxyl group varies from 1.71 in CYS to 2.71 in THR, while the $pK_{\alpha\text{-}NH_3}$ of the α-amino groups extends from 8.18 in CYS to 9.69 in ALA. The $pK_{\alpha\text{-}CO_2H}$ for the α-CO_2H group, the $pK_{\alpha\text{-}NH_3}$ for the α-amino group and the pK_R for the side chain are listed in Table 2 [5]. Since the amino acids can lose protons to become anions, and can gain protons to become cations, they can be adsorbed on both anion and cation exchangers.

The aliphatic members of the amino acids exhibit no absorption in the ultraviolet region above 220 nm, while the aromatic amino acids, HIS, PHE, TRY and TYR, show characteristic maxima above 250 nm. The zwitteronic character of the α-amino acids shows up clearly in their infrared spectra. No absorption due to the normal NH stretching frequency at 3300–3500 cm^{-1} is observed, indicating the absence of an NH_2 group. Instead, a few of the peaks near 3070 cm^{-1}, 1600 cm^{-1}, 1500 cm^{-1} due to the NH_3^+ are seen (except in PRO, the NH_3^+ group absorbing at ~2900 cm^{-1}). The carbonyl absorption of the unionized carboxyl group at 1700–1750 cm^{-1} is likewise replaced by carboxylate ion absorption at 1560–1600 cm^{-1} [6].

B. Partial Molar Volumes of Amino Acids in Water

The composition dependence of the total volume of a solution at constant temperature and pressure is expressed in terms of the partial molar volumes of the solute and the solvent. Since we are concerned with solvation properties, the quantities which we need to discuss are the partial molar volumes in infinite dilution of the solute so that solute-solute interactions make no contribution. In practice, partial molar volumes are obtained indirectly from precise density measurements. The partial molar volumes at infinite dilution of the amino acids are compiled in Table 2 [7]. It is apparent from these data that an approximately linear correlation exists between the partial molar volume and the number of carbon atoms in the backbone. The data indicate volume contributions from the polar head group (NH_3^+, CO_2^-) and from the CH_2 group and to be about

28 cm^{-3} mol^{-1} and 16 cm^3 mol^{-1}, respectively. The partial molar volumes of the amino acids are larger than those of small mineral ions.

C. Molal Activity Coefficients of Amino Acids in Water

The chemical potential, μ is the partial Gibbs free energy, and the μ of solute in a solution can be expressed in terms of concentration, in an ideal solution

$$\mu = \mu° + RT \ln m \tag{6}$$

where $\mu°$ corresponds to the chemical potential of solute in some standard state, R is the gas constant, T represents the absolute temperature, and m is the molality of solute. Deviations from ideal behavior are allowed for by substituting activity a for m, where a = $m\gamma$, giving

$$\mu = \mu° + RT \ln m\gamma \tag{7}$$

where γ is the molal activity coefficient. The extent of deviation of a real solution at a specific concentration from an ideal solution is expressed by the activity coefficient. The molal activity coefficients of some amino acids in water at 25°C are presented in Table 2 [7].

III. MECHANISMS OF AMINO ACID UPTAKE BY ION EXCHANGERS

A. Factors Influencing the Ion-Exchange Affinity of Amino Acids

Whereas the ion-exchange affinity of various mineral ions to ion exchangers is reasonably well understood, the ion-exchange affinity of various amino acid ions is not because of their more complex nature.

Access to better understanding of the various factors leading to the affinity of various metal ions to an ion exchanger has been at least partially implemented through the development of a model that provides the capability for anticipation of ion-exchange selectivity coefficients measured as a function of the initial concentration level of pairs of ions with the same charge and with different charges over a sizable range of concentration ratios. This Gibbs-Donnan-based model was developed as described below [8].

The chemical potential, μ, of any electroneutral component i is defined by

$$\mu_i = \left(\frac{\partial G}{\partial n_i} \right)_{P, T, n_j, n_k \cdots} \tag{8}$$

where G = free energy, n_i = number of moles of component i, P = pressure, T = absolute temperature, and subscripts, j, k . . . refer to the other electroneutral components of a particular system. The pressure dependence of component i's chemical potential can be shown to reduce to V_j, its partial molar volume, $(\partial V/\partial n_i)_{P,\ T,\ n_j,\ n_k\ldots}$ where V corresponds to volume. By assuming that partial molar volumes are essentially independent of composition and pressure, and that the chemical potential in isotherm systems can be split into two additive terms, one depending only on composition and the other only on pressure, the chemical potential, μ_i, in a solution of molality m and under a pressure P is

$$\mu_i(P,m) = \mu_i(P^\circ,m) + (P - P^\circ)V_i \tag{9}$$

where P° is the standard pressure of 1 atmosphere. The activity, a_i, of component i is defined by

$$\mu_i(P,m) = \mu_i^\circ(P) + RT \ln a_i \tag{10}$$

where μ_i° = chemical potential of component i in the standard state. By combining Eqs. (9) and (10)

$$\mu_i(P,m) = \mu_i(P^\circ) + RT \ln a_i + (P - P^\circ)V_i \tag{11}$$

With such representation of the chemical potential of electroneutral components the equilibrium distribution of diffusible components such as NX, MX, and H_2O between a crosslinked, polyelectrolyte gel phase and an external solution phase can be represented by equating the chemical potential of each diffusible component in the two phases as follows:

$$\bar{\mu}_{NX}(P,m) = \bar{\mu}_{NX}^\circ(P^\circ) + RT \ln \bar{a}_{NX} + (\bar{P} - \bar{P}^\circ)V_{NX} = \mu_{NX}(P,m)$$
$$= \mu_{NX}^\circ(P^\circ) + RT \ln a_{NX} + (P - P^\circ)V_{NX} \tag{12a}$$

$$\bar{\mu}_{MX}(P,m) = \bar{\mu}_{MX}^\circ(P^\circ) + RT \ln \bar{a}_{MX} + (\bar{P} - \bar{P}^\circ)V_{MX} = \mu_{MX}(P,m)$$
$$= \mu_{MX}^\circ(P^\circ) + RT \ln a_{MX} + (P - P^\circ)V_{MX} \tag{12b}$$

$$\bar{\mu}_{H_2O}(P,m) = \bar{\mu}_{H_2O}^\circ(P^\circ) + RT \ln \bar{a}_{H_2O} + (\bar{P} - \bar{P}^\circ)V_{H_2O} = \mu_{H_2O}(P,m)$$
$$= \mu_{H_2O}^\circ(P^\circ) + RT \ln a_{H_2O} + (P - P^\circ)V_{H_2O}$$

(12c)

In these equations the bar is used to identify each component's association with the gel phase. By choosing the standard state to be the same in both phases ($\bar{\mu}^\circ = \mu^\circ$ and $\bar{P}^\circ = P^\circ$) and by equating $\bar{P} - P$ to the "swelling pressure," π, the following convenient relationships are obtained:

$$\ln a_{NX} = \ln \bar{a}_{NX} + \frac{\pi}{RT} V_{NX} \tag{13a}$$

$$\ln a_{MX} = \ln \bar{a}_{MX} + \frac{\pi}{RT} V_{MX} \tag{13b}$$

$$\ln a_{H_2O} = \ln \bar{a}_{H_2O} + \frac{\pi}{RT} V_{H_2O} \tag{13c}$$

Up until this point components have been treated as if they are accessible in an isolated state in order to satisfy Gibbsian thermodynamics. However, when the three components of the two-phase system under scrutiny reach their equilibrium distribution the total concentration of M^+ and N^+ greatly exceeds the concentration of X^- in the gel phase. The extra electric charge and the resultant electrical force on the ions affect their equilibrium distribution and have to be taken into account. In correcting for this aspect thermodynamic rigor is necessarily lost in the process which consists of redefining equilibrium as that condition in which the electrochemical potential, E, of the separate ionic species are equal in both phases:

$$E_{M^+} = \bar{E}_{M^+} \tag{14a}$$

$$E_{N^+} = \bar{E}_{N^+} \tag{14b}$$

$$E_{X^+} = \bar{E}_{X^+} \tag{14c}$$

The electrochemical potential differs from the chemical potential by a term which depends on the electrical potential, σ, of the phase

$$E_i = \mu_i + Z_i F\sigma \tag{15}$$

Here Z_i represents the electrochemical valence of the ith ion and F is the Faraday constant. By combining Eqs. (14), (15), and (11), the Donnan potential, $\bar{\sigma} - \sigma$, is resolved with Eq. (16). Equation (16), applicable to all mobile ion species present, is

$$\bar{\sigma} - \sigma = \frac{1}{z_i F} (RT \ln \frac{a_i}{\bar{a}_i} - \pi V_i) \tag{16}$$

Representation of the equilibrium distribution of N^+, M^+, and X^- with this equation leads finally to Eq. (17) presented next; the $\bar{\sigma} - \sigma$ terms cancel and

$$\log\frac{a_{M^+}\bar{a}_{N^+}}{\bar{a}_{M^+}a_{N^+}} = \frac{\pi}{2.3RT}(V_M - V_N) \tag{17}$$

The activity ratio defined above is not directly measurable and Eq. (17) has been rearranged, as shown below, to provide a term that is

$$\log\frac{\bar{m}_N m_M}{m_N \bar{m}_M} = \log {}^N_M K_{Ex} = \frac{\pi}{2.3RT}(V_M - V_N) + \log\frac{\bar{\gamma}_{M^+}}{\bar{\gamma}_{N^+}} - 2\log\frac{\gamma^\pm_{MX}}{\gamma^\pm_{NX}} \tag{18}$$

In this equation K_{Ex} is the experimentally determined selectivity coefficient, m is the molality of each ion, $\bar{\gamma}$ is the activity coefficient of the ion in the gel phase and $\gamma^\pm$ is the mean molal activity coefficient of the electrolyte in the external solution phase.

Use of Eq. (18) permits identification of the $\log(\bar{\gamma}_{M^+}/\bar{\gamma}_{N^+})$ term as the one contributing most importantly to differences in affinity of pairs of univalent metal ions, M^+ and N^+, for a cation-exchange resin. For example, for the equilibrium distribution of Li^+ and Na^+ ions between dilute solutions of Li^+ and Na^+ chloride ($m_{LiCl} + m_{NaCl} = 0.010$ m) and the much more highly concentrated Dowex-50 (8% cross-linked by weight with divinylbenzene) phase ($\bar{m}_{Li^+} + \bar{m}_{Na^+} \cong 4.5$ m) the $p(V_{Li^+} - V_{Na^+})/2.3$ RT and the $2\log(\gamma^\pm_{LiCl}/\gamma^\pm_{NaCl})$ term yield a small sum (≤ 0.04) while the value of $\log(\bar{\gamma}_{Li^+}/\bar{\gamma}_{Na^+})$ approaches ~0.35 once correction for interaction between the two metal ions is made. Correlation between experiment and computation of the Gibbs-Donnan-based terms is strongly supportive of the model.

The difference in affinity between a pair of cations, C^{+2} and D^+, for a cation-exchange resin, when one of the pair is multivalent, is once again predictable with Eq. (19). Whereas the

$$\log\frac{\bar{m}_C(m_D)^2}{(\bar{m}_D)^2 m_C} = \frac{\pi}{2.3RT}(2V_D - V_C) + \log\frac{(\bar{\gamma}_{D^+})^2}{\bar{\gamma}_{C^{2+}}} - \log\frac{(\gamma^\pm_{DX})^4}{(\gamma^\pm_{CX_2})^3} \tag{19}$$

and solution activity coefficient ratio terms continue to be small enough to neglect when the molality of the mixed electrolyte solution is low, the $(\bar{\gamma}_{D^+})^2/\bar{\gamma}_{C^{2+}}$ term, is always larger than unity and increases in value with the ion-exchanger phase molality, which increases in response to increase in the degree of ion-exchanger cross-linking. The divalent ion is always preferred by the resin phase [9].

The net charge of the mineral ion thus has a strong effect on ion-exchange equilibria [9–12].

As mentioned above the R groups of acidic or basic amino acids have one extra carboxyl or amino acid group at most and the maximum net charge of amino acids in acidic or basic solution is 2. Charge thus plays a role in the interaction of amino acids with ion exchangers. Its contribution is smaller than with mineral ions, however, because of the greater separation of charge affected by the larger amino acids. On the contrary, in most ion exchange systems of mineral ions, the product of osmotic water pressure in the ion exchanger by the difference between the partial molal volumes for a pair of mineral ions is sufficiently small to neglect. The partial molal volumes of most amino acids are much larger than those of small mineral ions, and the product has a considerable influence on selectivity coefficient of amino acids. Other factors, less amenable to assessment, play a part in the uptake of amino acids by the ion exchanger.

The α-COO$^-$ or α-NH$_3^+$ of amino acids which exist as zwitterions or inner salts with zero net charge can be exchanged with anion or cation on the ion exchanger, in spite of the electrostatic repulsion of adjacent α-NH$_3^+$ or α-COO$^-$, as well. The contribution to the selectivity of amino acids of this factor as well as the contributions of London forces [13] and hydrophobic interactions between the amino acids [14] and the matrix of the ion exchanger whose consequence is that the hydrocarbon groups tend to coagulate or to be compressed onto the surface of the ion exchanger remains unresolved. The systematic investigation of sorption of the ions of quaternary ammonium bases by Kressman and Kitchener [13] has indicated that nonionic interactions have a large effect on the selective exchange of organic compounds, the styrene-type ion-exchange resins usually preferring counterions with aromatic groups to those with aliphatic groups, and larger counterions to the smaller ones. Recently, Saunders et al. [14] also pointed out that the hydrophobic interactions may be more significant in the ion-exchange equilibria of aromatic amino acids. However, a systematic investigation aiming at the important role of the nonionic interactions in sorption of the amino acids on the ion exchange has not yet been reported.

The anticipation of the equilibrium distribution of pairs of amino acids with the Gibbs-Donnan-based Eq. (18) is thus not easily realizable, because of the adsorption-enhanced tendencies (London forces, hydrophobic interactions, zwitterion sorption) preventing reliable resolutions of the $\log \bar{\gamma}_{\text{amino acid}_1} / \bar{\gamma}_{\text{amino acid}_2}$ term, and the $\pi \, (V_{\text{amino acid}_1} - V_{\text{amino acid}_2}) /$ (2.3 RT) term.

B. Mechanism of Amino Acid Uptake by Ion Exchanger at Low Solution Concentrations

Ion exchange is a reversible and stoichiometric process with every ion removed from solution replaced by an equivalent amount of another ionic species of the same charge from the ion exchanger. The test of this mass-action-based law for uptake of amino acid and release of counterion from the ion exchanger can be carried out in both static and dynamic experiments. Samsonov [15] carried out the test of this equivalence law by comparing the uptake of GLY with the release of hydrogen ion from the strong-acid ion-exchange resin in dynamic experiments. It was found that the pH at the exit of the column did not change, and the equivalence law appeared to be violated. This researcher and co-workers have carried out tests of the equivalence law by measuring the uptake of GLY or ALA and the release of sodium ion from the strong-acid ion-exchange resin or chloride ion from the strong-base ion-exchange resin used in static experiments [16]. A known amount of the ion-exchange resin in the Na^+ or Cl^- form was equilibrated with a solution of known concentration and pH value. The concentrations of amino acid, Na^+ or Cl^- at equilibrium were analyzed. The results are presented in Fig. 1. As shown in Fig. 1, the amounts of GLY adsorbed are always different from the amount of Na^+ or Cl^- released. For cation-exchange resin, at pH>3 the amount of GLY

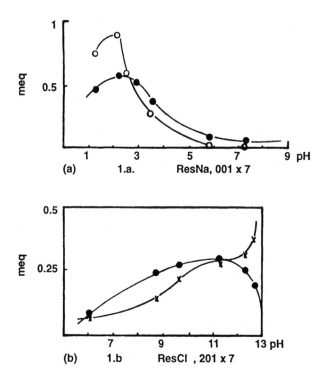

Figure 1 The test of equivalence law for uptake of GLY and release of counterion. (a) ResNa, 001 × 7 (b) ResCl[1], 201 × 7. ●-GLY, O-Na^+, ×-Cl^-.

adsorbed exceeds the amount of Na^+ ion released; for anion-exchange resin, at pH<11 the amount of GLY adsorbed exceeds the amount of Cl^- ion released. The direction of nonequivalence observed was reversed when the amount of GLY adsorbed by cation resin at pH < 3 and by anion resin at pH > 11 was exceeded by the release of counterion. These results are apparently due to use of the ion exchange resins in the salt forms rather than in the H^+ or OH^- form at solution pH values of <3 or >11. This approach favored by the author for testing the equivalence law because of the amphoteric properties of the amino acids, results in the extra release of counterions by their exchange with H^+ or OH^- ions.

The fact that diffusible components of a solution, e.g., simple electrolyte, amino acids, when contacted with ion exchanger permeate the resin phase until the chemical potential of each component is the same in the two phases promotes the mechanism for sorption of the exchanger by electrolyte and amino acids without the release of counterion [17]. To determine the magnitude of such ion-exchanger invasion by an amino acid in order to assess the potential for disturbance of the equivalence test that is under discussion the uptake of GLY by a styrene-divinylbenzene copolymer resin was examined. It was determined that when the concentration of GLY solution is as high as $2.0 \ mol \cdot L^{-1}$ and pH = pI, the amount of GLY adsorbed by this copolymer is sufficiently small to neglect. On the basis of this result sorption of amino acids by ion exchanger was eliminated on a meaningful source of nonequivalence.

Talibudeen [17] and Greenland et al. [18] have demonstrated that cation-exchange and proton-transfer mechanisms both play an important role in the adsorption of amino acids on clays in the hydrogen form; additional forces, including those attributable to polar and van der Waals interactions, supplement these mechanisms. The process of proton transfer can be represented by the equation

$$R—CH(NH_3^+) \ COO^- + Clay\overline{H}^+ \rightleftharpoons ClayRCH(NH_3^+)COOH \qquad (20)$$

The NH_3^+ group of the zwitterion ejects the free hydrogen ion in the clay phase and the hydrogen ion is transferred to the carboxylic group to preserve the electroneutrality of the ion-exchanger phase. Obviously, this explains the apparent violation of the equivalence law. It should be noted that as the distance between the positive and negative charges increases, the contribution of this mechanism is reduced. As mentioned above, the absence of equivalence has been observed in the sorption of amino acids on strong-acid cation-exchange resin in the Na^+ form and strong-base anion-exchange resin in the Cl^- form as well. Eventually the ion-transfer mechanism described by Eq. (20) is encountered in these systems to explain the nonequivalent exchange observed once again.

In such experiments the initial free H^+ ion contained by the initial solution contacted with the cation exchanger must displace Na^+ ion in the resin phase until equilibrium is reached. At this point the ratio of Na^+ and H^+ activity ratios in the resin and solution phases, i.e., the Donnan potential term, are equal. Once this process begins the increasing presence of H^+ ion in the resin phase leads to the ion-transfer process described in Eq. (20) and nonequivalence of ion transfer is the false impression.

With the anion exchanger initially in the Cl^- ion form similar accommodation of the exchange of OH^- ion in solution for Cl^- ion results in the elevation of OH^- ion concentration in the anion exchanger. The process of ion transfer in this system is represented by Eq. (21).

$$R—CH(NH_3^+)COO^- + ResOH^- \rightleftharpoons ResCH(NH_2)COO^- + H_2O \quad (21)$$

where Res represents the ion-exchange resin.

The amount of GLY adsorbed on strong-acid cation- and strong-base anion-exchange resins at different pH values has been compared [16]. The results are shown in Fig. 2 where it is observed that the amounts of GLY adsorbed on the H^+ or OH^- resin forms are, at most of the pH values examined, greater than those on Na^+ or Cl^- form. The dependence on pH values of the amounts adsorbed onto the cation-exchange resin is opposite to that observed with the anion-exchange resin as expected from the opposing nature of the dissociation equilibria, the ion-exchange reactions, and the ion-transfer mechanism of amino acids.

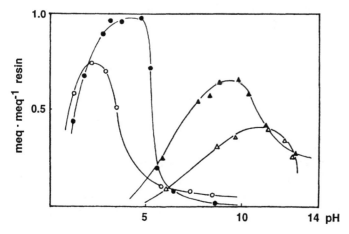

Figure 2 The uptake of GLY by 001 × 7 resin in H^+ and Na^+ forms and by 201 × 7 resin in Cl^- and OH^- forms as a function of pH. O-ResNa, ●-ResH, △-ResCl, ▲-ResOH.

Comparing the cation-exchange reaction (Eq. (21)) with the ion-transfer mechanism

$$R—CH(NH_3^+)COOH + ResH^+ \rightleftharpoons ResR—CH(NH_3^+)COOH + H^+ \quad (22)$$

(Eq. (20)), we find that the same product, $ResR—Ch(NH_3^+)COOH$, is formed. In order to substantiate this observation, the IR spectra of GLY adsorbed on the strong-acid cation-exchange resin in the H^+-ion form were measured [16]. It was found that IR spectra of all samples prepared from aqueous solutions of GLY at a concentration level of 0.1 mol $\cdot$ L^{-1} and at pH values corresponding to 1.42, 4.88 and 5.64 exhibit the band at 1750–1752 cm^{-1} that is assigned to the unionized carboxyl and the band at 1600 cm^{-1} that is assigned to the antisymmetric vibration of carboxylate ion. In spite of the presence of different predominant species in solutions and different mechanism encountered at different pH values, the IR spectra show that the amino acids adsorbed on ion exchanger are dissociated.

The IR spectra of GLY adsorbed on the strong-acid cation-exchange resin in the Na^+-ion form were also measured. It was found that the IR spectrum of sample prepared from aqueous solution at pH = 1.48 also exhibit the bands at 1752 and 1600 cm^{-1}, while the spectra of samples prepared from aqueous solutions of GLY concentrations 0.1 and 1.0 mol $\cdot$ L^{-1} at pH = pI only exhibit the band assigned to the carboxylate ion, and do not exhibit the band assigned to the un-ionized carboxyl. The ion-transfer mechanism is thus supported by the IR spectra.

C. Mechanism of Amino Acid Uptake by Ion Exchanger at High Solution Concentration Levels

The superequivalent sorption of organic electrolytes on ion exchangers has been observed many times. Greenland et al. [19] observed the superequivalent sorption of GLY and of its peptide on montmorillonite in the H^+ ion form. The superequivalent sorption of GLY on a sulfonated cation-exchange resin in H^+-ion form and on a microdisperse anion exchange resin in the Cl$^-$ form [20, 21] was also examined. Very different mechanisms have been offered in an attempt to explain this effect. Muravev and Obrezkov [21], for example, have suggested that this effect can be attributed to the aminecarboxylate interaction between the adsorbed amino acid molecules and the formation of a second layer of amino acid molecules. This mechanism has been explained [21] in the following way;

$$Res\ ^-RCH(NH_3^+)COO\ ^-H^+ + RCH(NH^+)COO\ ^- \rightleftharpoons \quad (23)$$
$$(Res^-RCH(NH_3^+)COO\ ^-)^-RCH(NH_3^+)COO\ ^-H^+$$

In this equation, the carboxylate groups of the first layer of amino acid molecules are presumed to act as the modified functional groups of the ion exchanger by assuming the configuration

$$(Res\ ^-RCH(NH_3^+)COO\ ^-)^-$$

The adsorption isotherms of GLY and ALA on a strong-acid cation-exchange resin in the H^+- or Na^+-ion forms were determined point by point from a series of adsorption measurements affected once equilibrium with solutions of various concentrations was reached. The pH of each series of measurements was kept constant. The amount of amino acid adsorbed on the resin is plotted as a function of the equilibrium concentration in the solution, and is expressed by the mmoles of amino acid per milliequivalent of resin functionality, D (mmole/meq resin) in Figs. 3 and 4. [22] The D value represents the average number of amino acid molecules adsorbed per functional group of resin. When $D > 1$, super-equivalent sorption is occurring; when $D < 1$, it is not possible to deduce whether or not superequivalent sorption is occurring. As can be seen from Figs. 3 and 4 [22,23], the isotherms develop a Langmuir-like pattern with a well-defined plateau, and the maximum D values of GLY and ALA on resin in the H^+-ion form approach 2 in value. The maximum value for GLY on resin in the Na^+ form is 1.3

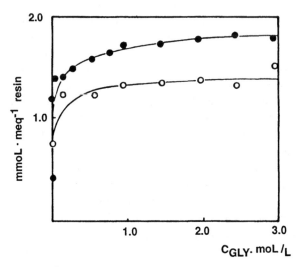

Figure 3   The isotherms of GLY by 001 × 7 resin in H^+ and Na^+ forms. ●-ResH, pH = pI, ○-ResNa, pH = 2.0.

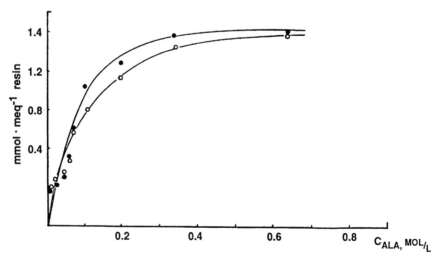

Figure 4 The isotherms of ALA by 001 × 7 resin in H⁺ form. ●-pH = 4.87, ○-pH = 1.90.

In order to substantiate the mechanism postulated, the IR spectra of GLY adsorbed on the strong-acid cation-exchange resin originally in the H⁺-ion form were measured [22,23]. The samples with various amounts of GLY adsorbed were prepared, and the dependence of the ratio of the integrated peak area of the bands at 1599–1600 cm⁻¹ and 1754–1749 cm⁻¹ on the D values was investigated. It was found that when D was <1, the ratios were independent of the D values, while when D was >1, the ratios increased with each increase of the D values. The mechanism projected thus seems to be supported by this result. As shown in Eq. (23), when D > 1 and the formation of a second layer of molecules occurs, the carboxyls of the first layer of amino acid molecules are certain to be dissociated so that they may serve as the new exchange sites.

The mechanism for superequivalent sorption of GLY on the strong-base anion-exchange resin must follow a parallel pattern. In this instance the amine-carboxylate modified anion exchanger functionality, $(Res^+RCHNH^+COO^-)^+$, acts as the new exchange site. This estimate is also supported by the IR spectra of the adsorbed GLY, the integrated peak area ratios of the bands assigned to the NH_3^+ and the benzene ring increasing with increasing D values when D > 1 [24].

IV. EFFECTIVE DISTRIBUTION COEFFICIENT AND AVERAGE SEPARATION FACTOR

A. Effective Distribution Coefficient

Measurement of the distribution of a particular solute between an ion exchanger and its solution can be very informative and useful. Such studies of amino acids have, as a consequence, been the subject of investigation from time to time. In these studies the amino acid concentration includes all positive, negative, and zwitterion species [12].

A number of years ago Seno and Yamabe [25] derived equations that examined the amounts of adsorbed amino acid on an ion exchanger in the presence of competing ions. More recently Helfferich [26] derived equations describing the effective distribution coefficients of amino acids on a strong-acid (cation) and strong-base (anion) exchanger as a function of pH, amino acid concentration and added electrolyte.

The Helfferich equations describing the effective distribution coefficients of amino acids on a strong-acid cation and a strong-base anion exchanger were derived using the following simplifying assumptions [26]:

1. The fixed ionogenic groups are completely ionized at all times.
2. Co-ions do not exist in the ion-exchanger phase.
3. The ion exchanger is homogeneous.
4. The dissociation constant of the carboxyl or amine group, the separation factor of the cation or anion of the amino acid and the distribution coefficient of zwitterion with zero net charge are considered to be approximately constant.
5. Electroneutrality condition in the ion exchanger is preserved.
6. The total concentrations of amino acid in all its species are given as $\{AA\}$ in solution and $\{\overline{AA}\}$ in ion exchanger respectively.
7. Acidic and basic amino acids in principle form two neutral species,

$$R_{CO_2H}—CH(NH_3^+)COO^- \quad \text{and} \quad R_{CO_2}^-—CH(NH_3^+)COOH$$
$$R_{NH_3}—CH(NH_3^+)COO^- \quad \text{and} \quad R^+{}_{NH_3}—CH(NH_2)COO^-$$

only the first of each pair is considered because in most cases their concentrations are much larger.

Various amino acid, ion-exchanger systems are considered in the Helfferich development of these equations. In the first the neutral amino acid can exist as

$R—CH(NH_3^+)COO^-$	neutral species HA
$R—CH(NH_3^+)COOH$	cation H_2A^+
$R—CH(NH_2)COO^-$	anion A^-

The cation and anion separation factors in Helfferich's development are

$$\alpha_{H_2A} = \frac{[\overline{H_2A^+}][M^+]}{[H_2A^+[\overline{M^+}]} \tag{24}$$

$$\alpha_H = \frac{[\overline{H^+}][M^+]}{[H^+][\overline{M^+}]} \tag{25}$$

$$\alpha_A = \frac{[\overline{A^-}][Cl^-]}{[A^-][\overline{Cl^-}]} \tag{26}$$

$$\alpha_{OH} = \frac{[\overline{OH^-}][Cl^-]}{[OH^-][\overline{Cl^-}]} \tag{27}$$

The distribution coefficient of neutral species is

$$D_{HA} = \frac{[\overline{HA}]}{[HA]} \tag{28}$$

The effective distribution coefficient on cation and anion exchanger are

$$D_+ = \frac{D_{HA}K_{\alpha\text{-}CO_2H}}{K_{\alpha\text{-}CO_2H} + [H^+]}$$
$$+ \frac{\alpha_{H_2A}[\overline{X}][H^+]}{(\alpha_H[H^+] + [M^+])(K_{\alpha\text{-}CO_2H} + [H^+]) + \alpha_{H_2A}\{AA\}[H^+]} \tag{29}$$

$$D_- = \frac{D_{HA}K_{\alpha\text{-}NH_3^+}}{K_{\alpha\text{-}NH_3^+} + [OH^-]}$$
$$+ \frac{\alpha_A[\overline{X}][OH^-]}{(\alpha_{OH}[OH^-] + [Cl^-])(K_{\alpha\text{-}NH_3^+} + [OH^-]) + \alpha_A\{AA\}[OH^-]} \tag{30}$$

where $[\overline{X}]$ is the ion-exchange capacity and K corresponds to the dissociation constant of the various amino acid functionalities.

In the second, the acidic amino acid can exist in acidic solution as

R—CH(NH$_3^+$)COO$^-$ neutral species H$_2$A
R—CH(NH$_3^+$)COOH   cation H$_3$A$^+$

The separation factor and the distribution coefficient are given by

$$\alpha_{H_3A} = \frac{\overline{[H_3A^+]}[M^+]}{[H_3A^+]\overline{[M^+]}} \tag{31}$$

$$D_{H_2A} = \frac{\overline{[H_2A]}}{[H_2A]} \tag{32}$$

The effective distribution coefficient on a cation exchanger is

$$D_+ = \frac{[H^+]}{Q} D_{H_2A} K_{\alpha\text{-}CO_2H^+} + \frac{\alpha_{H_3A}[\overline{X}][H^+]^2}{(\alpha_H[H^+] + [M^+])Q + \alpha_{H_3A}\{AA\}[H^+]^2} \tag{33}$$

where $Q = [H^+]^2 + K_{\alpha\text{-}CO_2H}[H^+] + K_{\alpha\text{-}CO_2H}K_{R\text{—}CO_2H}$.

Next the acidic amino acid can exist in basic solution as

R—CH(NH$_3^+$)COO$^-$ neutral species H$_2$A
R—CH(NH$_2$)COO$^-$ univalent anion HA$^-$
R$^-$—CH(NH$_2$)COO$^-$ bivalent anion A^{2-}

The separation factor for HA$^-$ and the selectivity coefficient for A^{2-} are given respectively by

$$\alpha_{HA} = \frac{\overline{[HA^-]}[Cl^-]}{[HA^-]\overline{[Cl^-]}^2} \tag{34}$$

$$K_A = \frac{\overline{[A^{2-}]}[Cl^-]^2}{[HA^-]\overline{[Cl^-]}^2} \tag{35}$$

The effective distribution coefficient on anion exchanger is

$$D_- = \frac{[\overline{OH^-}]}{\alpha_{OH}^2[OH^-](K_{\alpha\text{-}NH_3^+} + [OH^-])}(\alpha_{OH}\alpha_{HA}K_{\alpha\text{-}NH_3^+} + K_A[\overline{OH^-}]) \tag{36}$$

where

$$[\overline{OH^-}] = \left(\frac{\alpha_{OH}^2 (A' + B' + C')^2}{2E'} + \frac{\alpha_{OH} A'[\overline{X}]}{E'} \right)^{1/2} - \frac{\alpha_{OH}(A' + B' + C')}{2E'}$$

$$A' = \alpha_{OH}[OH^-](K_{\alpha\text{-}NH_3^+} + [OH^-])$$

$$B' = [Cl^-](K_{\alpha\text{-}NH_3^+} + [OH^-])$$

$$C' = \alpha_{HA} K_{\alpha\text{-}NH_3^+} \{AA\}$$

$$E' = 2K_A \{AA\}$$

With the basic amino acid in acidic solution the following species can exist:

R$^+$—CH(NH$_3^+$)COO$^-$ univalent cation HA$^+$
R$^+$—CH(NH$_3^+$)COOH   bivalent cation H$_2$A^{2+}

The separation factor for HA$^+$ and the selectivity coefficient for H$_2$A^{2+} are, respectively,

$$\alpha'_{HA} = \frac{[\overline{HA^+}][M^+]}{[HA^+][\overline{M^+}]} \tag{37}$$

$$K_{H_2A} = \frac{[\overline{H_2A^{2+}}][M^+]^2}{[H_2A^{2+}][\overline{M^+}]^2} \tag{38}$$

The effective distribution coefficient on cation exchanger is

$$D^+ = \frac{[\overline{H^+}]}{\alpha_H^2[H^+](K_{\alpha\text{-}CO_2H} + [H^+])} (\alpha_H \alpha'_{HA} K_{\alpha\text{-}CO_2H} + K_{H_2A}[\overline{H^+}]) \tag{39}$$

where

$$[\overline{H^+}] = \left(\frac{\alpha_H (A + B + C)^2}{2E} + \frac{\alpha_H A[\overline{X}]}{E} \right)^{1/2} - \frac{\alpha_H (A + B + C)}{2E}$$

$$A = \alpha_H[H^+](K_{\alpha\text{-}CO_2H} + [H^+])$$

$$B = [M^+](K_{\alpha\text{-}CO_2H} + [H^+])$$

$$C = \alpha'_{HA} K_{\alpha\text{-}CO_2H} \{AA\}$$

Finally, the basic amino acid can exist in basic solution as

$RCH(NH_3^+)COO^-$ neutral species HA
$RCH(NH_2)COO^-$ univalent anion A$^-$

The distribution coefficient for HA and the separation factor for A$^-$ are

$$D'_{HA} = \frac{\overline{[HA]}}{[HA]} \tag{40}$$

$$\alpha'_A = \frac{\overline{[A^-]}[Cl^-]}{[A]\overline{[Cl^-]}} \tag{41}$$

The effective distribution coefficient on anion exchanger is

$$D_- = \frac{[OH^-]}{Q'} D'_{HA} K_{\alpha\text{-}NH_3^+}$$
$$+ \frac{\alpha'_A [\overline{X}][OH^-]^2}{(\alpha_H[OH^-]+[Cl^-])Q' + \alpha'_A\{AA\}[OH^-]^2} \tag{42}$$

where

$$Q' = [OH^-]^2 + K_{\alpha\text{-}NH_3^+}[OH^-] + K_{\alpha\text{-}NH_3^+} K_{R-NH_3^+}$$

B. Average Separation Factor

Up to this point the equations developed for the evaluation of separation factors in the various amino acid, ion-exchanger systems are based on their standard definition, the concentration ratio, at equilibrium, of a particular ionic species of the amino acid and the counterion it exchanges within the resin divided by their concentration ratio in solution. Since such information may be sought for the design of analytical or larger scale separation process for amino acids the neglect in both phases of all other amino acids species for the evaluation of α should be avoided. The separation factor so modified is identified as the average separation factor [27,28]. The earlier computations of α have been modified to permit evaluation of the average separation factors. The equations developed for this purpose are presented below [27,28]. Since ion-exchange chromatography of amino acids is almost exclusively performed on strong-acid ion-exchange resins (sulfonated polystyrene-divinylbenzene copolymers) the development of the average separation factors of amino acids by this resin with respect to a common ion M^+ as a function of pH is presented as follows [29,30].

1. As mentioned above, the neutral amino acid can exist as neutral species HA and cation H_2A^+ in acidic solutions. The distribution coefficient of HA is replaced by HA of α_{HA}, its separation factor. In spite of zero net charge and low affinity, α_{HA} can be written as

$$\alpha_{HA} = \frac{[\overline{HA}][M^+]}{[HA][\overline{M^+}]} \tag{43}$$

Indeed, comparison of Eq. (43) with Eq. (28), shows that there is essentially no difference between D_{HA} and α_{HA} at a given experimental condition.

The average separation factor is then

$$\alpha_t = \frac{([\overline{H_2A^+}]+[\overline{HA}])[M^+]}{([H_2A^+]+[HA])[\overline{M^+}]} \tag{44}$$

$$\alpha_t = \frac{\alpha_{H_2A} + \alpha_{HA}10^{pH-pK_{\alpha\text{-}CO_2H}}}{1+10^{pH-pK_{\alpha\text{-}CO_2H}}} \tag{45}$$

2. The acidic amino acid can exist in acidic solutions as H_2A and H_3A^+, and the D_{H_2O} is once again replaced by the separation factor

$$\alpha_{H_2A} = \frac{[\overline{H_2A}][M^+]}{[H_2A][\overline{M^+}]} \tag{46}$$

Then the average separation factor is

$$\alpha_t = \frac{\alpha_{H_3A} + \alpha_{H_2A}10^{pH-pK_{\alpha\text{-}CO_2H}}}{1+10^{pH-pK_{\alpha\text{-}CO_2H}}} \tag{47}$$

3. The basic amino acid can exist in acidic solution as HA^+, H_2A^{2+} and small amounts of $R—CH(NH_3^+)COO^-(HA_\alpha)$ and $R^+—CH(NH_2)COO^-$ (HA_R); the separation factors for HA_α and HA_R are written as

$$\alpha_{HA_\alpha} = \frac{[\overline{HA_\alpha}][M^+]}{[HA_\alpha][\overline{M^+}]} \tag{48}$$

$$\alpha_{HA_R} = \frac{[\overline{HA_R}][M^+]}{[HA_R][\overline{M^+}]}$$

(49)

Then one obtains

$$\alpha_t = \frac{K_{H_2A}[\overline{M}]/[M^+] + (\alpha'_{HA} + \alpha_{HA_\alpha} + \alpha_{HA_R})10^{pH-pK_{\alpha \cdot CO_2H}}}{1 + 10^{pH-pK_{\alpha \cdot CO_2H}}}$$

(50)

The equations formulated for predicting separation and average separation factors affected by contacting amino acid solutions with cation and anion exchangers in the M^+ and Cl^- forms are approximate in nature. Though the concentrations of charged and uncharged species in the solution phase for this purpose can be calculated and the selectivity coefficient can be estimated from the Donnan potential term and the activity coefficients of the amino acid species, the best values of these equilibrium parameters are obtained from regression analyses of average separation factors [28] and effective distribution coefficients at various experimental conditions. More elaborate equations which include a knowledgeable treatment of the Donnan potential contribution as well as correction for nonideality would seem to be needed to approach their accuracy.

V. THERMODYNAMIC FUNCTIONS

A. Early Efforts

Adsorption isotherms were obtained for four amino acids in an investigation of their interaction with calcium montmorillonite and sodium and calcium illite. Linear isotherms were obtained in the study of their adsorption by the calcium clay. These isotherms were described in terms of a constant partition of solute between the solution and the adsorbent Stern layer. Free-energy values were calculated using the van't Hoff relation [16,27].

Samsonov [15] studied the direct sorption of ALA and other dipolar ions by SDV-3 ion exchanger resin at pH = 7. The enthalpy and entropy components of these sorptions were obtained from the isotherm dependence on temperature. It was found that the transformation of the resin from the hydrogen to the amino acid form was accompanied by a rise in the system's entropy. The thermodynamic-based description of the exchange of α-amino acids with hydrogen on three ion-exchange resins at pH $<$ pK$_{\alpha \cdot CO_2}$ were determined as well. However, precise descriptions of the experimental measurements and the calculations program used were not given.

The thermodynamic treatment of the ion-exchange equilibria of amino acids with strong-acid cation-exchange resin in the H^+-ion form has been presented by Russian scholars [30]. The general treatment given by Gaines and Thomas [31] differed from their treatment in that the dissociation equilibria of the amino acids were taken into account. In spite of this the Gaines and Thomas treatment provides less insight with respect to the physical causes for the behavior of ion-exchange, amino acid systems. It is also difficult to apply.

B. Thermodynamic Functions for the Interaction of Strong-Acid Cation-Exchange Resin in the H^+-Ion Form with Amino Acid Cations

The exchange of an amino acid cation, $R—CH(NH_3^+)COOH$, for hydrogen ion in the strong-acid cation-exchange resin at a pH < $pK_{\alpha\text{-}CO_2H}$ can be described with the equation

$$ResH^+ + R—CH(NH_3^+)COOH \underset{K_H'}{\overset{\Delta H_H'}{\rightleftharpoons}} ResR—CH(NH_3^+)COOH + H^+ \quad (51)$$

The concentration-based equilibrium distribution of ions in this exchange reaction is defined by

$$K_H' \; \overline{\frac{[ResR—CH(NH_3^+)COOH][H^+]}{[R—CH(NH_3^+)COOH][\overline{H^+}]}} \quad (52)$$

The K_H' values that were measured for nine amino acids are listed in Table 3.

It can be seen from Table 3 that the concentration-based equilibrium distribution of amino acid cations and H^+-ion between solution and strong-acid cation-exchange resin is small. Its value has been determined to be insignificantly changed by increases in the concentration of the amino acid cations in the ion-exchange resin [15]. In addition, the activity coefficients of some amino acids in the aqueous solutions at 0.2, 0.3, and 0.5, mol L^{-1} are approximately equal to 1 (Table 2). It may be concluded, on this basis, that this ion exchange system can be considered a quasi-ideal system, the quasi-ideality being attributable to the weak electrostatic attraction between large amino acid cations and fixed ionogenic group in the resin and the similarly weak interaction between the large amino acid cation and simple anions in the solution. Since the thermodynamic equilibrium constant is not sizably different from the concentration-based equilibrium constant, the standard free-energy change of the

Table 3 Thermodynamic Functions of Ion Exchange of Amino Acid Cations on Strong-Acid Cation-Exchange Resin in Hydrogen Form

Amino acid	pH	K'_H	$\Delta G'_H$	$\Delta H'_H$	$T\Delta S'_H$
				(kJ/eq)	
L-Phen	1.96–2.01	3.19 ± 0.07	−2.87	2.99	5.86
L-Ser	2.11–2.20	1.52 ± 0.05	−1.05	6.09	7.14
L-Try	1.71–2.15	4.57 ± 0.07	−3.76	3.84	7.60
L-Leu	2.30–2.34	1.16 ± 0.06	−0.37	2.98	3.35
L-Gly	2.22–2.31	0.25 ± 0.02	3.43	5.15	1.72
DL-Thr	2.20–2.32	0.52 ± 0.06	1.62	5.38	3.76
L-Ala	1.92–2.31	0.51 ± 0.02	1.67	3.75	2.08
L-Glu	1.92–2.15	0.55 ± 0.02	1.48	5.27	3.79
L-Arg	2.45–2.49	0.35 ± 0.05	2.60	3.42	0.82

ion-exchange reaction is not noticeably in error when K'_H is substituted for the thermodynamic value in Eq. (53).

$$\Delta G_1^0 = -RT \ln K'_H \tag{53}$$

The proton transfer process can be written as

$$ResH + R-CH(NH_3^+)COO^- \underset{\Delta H_t}{\overset{K_t}{\rightleftharpoons}} ResR-CH(NH_3^+)COOH \tag{54}$$

and can be considered as the sum of the ion-exchange reaction (51) and the protonation reaction

$$R-CH(NH_3^+)COO^- + H^+ \underset{\Delta H_1}{\overset{K_1}{\rightleftharpoons}} R-CH(NH_3^+)COOH \tag{55}$$

Thus, $K_t = K'_H K_1$ and $\Delta H_t = \Delta H'_H + \Delta H_1$. Micro-titration-based calorimetric experiments carried out during the sorption of amino acids on strong-acid cation-exchange resin in aqueous solutions at pH = pI have indicated that $\Delta H_t \approx 0$ [15, 32]. On this basis $\Delta H'_H = -\Delta H_1$. The ΔH_1 values of nine amino acids determined by these microtitration calorimetric experiments [33], are also listed in Table 3.

It is known that

$$\Delta G_1^0 = \Delta H'_H + T \Delta S'_H \tag{56}$$

where $\Delta S'_H$ is the entropy change of the ion-exchange reaction (51) and $T \Delta S'_H$ is the difference between ΔG_1^0 and $\Delta H'_H$. The $T \Delta S'_H$ values obtained are listed in Table 3.

It is hoped that the above ideas and observations will stimulate research of the ion-exchange equilibria of protein.

ACKNOWLEDGMENTS

Some of the research presented here was supported by the National Natural Science Foundation of China. I also wish to acknowledge Professor J. A. Marinsky for his encouragement.

REFERENCES

1. S. Moore and W. H. Stein, *J. Biol. Chem., 192*:663 (1951).
2. S. Yamamoto, K. Nakanishi, and R. Matsuno, *Ion-Exchange Chromatography of Proteins,* Marcel Dekker, New York, 1988, Chaps. 1, 9.
3. Tao Zuyi, *Huaxue Tongbao (Chemistry), 9*:25 (1990).
4. H. F. Walton and R. D. Rocklin, *Ion Exchange in Analytical Chemistry*, CRC Press, Boca Raton, 1990, Chap. 7.
5. R. M. Hardy, in *Chemistry and Biochemistry of the Amino Acids* (G. C. Barrett, ed.), Chapman and Hall, London, 1985, Chap. 2.
6. K. Nakamoto, *Infrared and Raman Spectra of Inorganic and Coordination Compounds,* 4th ed., Wiley, New York, 1986, Part III-8.
7. T. H. Lilley, in *Chemistry and Biochemistry of Amino Acids* (G. C. Barrett, ed.), Chapman and Hall, London, 1985, Chap. 21.
8. J. A. Marinsky, in *Ion Exchange and Solvent Extraction*, (J. A. Marinsky and Y. Marcus, eds.), Marcel Dekker, New York, 1993, pp. 237–334.
9. M. M. Reddy and J. A. Marinsky, *J. Macromol. Sci.-Phys. B5(1)*: B5-138 (1971).
10. J. A. Marinsky, in *Proceedings of the Symposium on Ion Exchange, Transport and Interfacial Properties* (S. Y. Richard, D. B. Richard, eds.), The Electrochemical Society, Pennington, 1981, p. 1.
11. J. A. Marinsky, *J. Phys. Chem., 89*:5294 (1985).
12. F. Helfferich, *Ion Exchange*, McGraw-Hill, New York, 1962, Chap. 5.
13. T. R. E. Kressman and J. H. A. Kitchener, *J. Chem. Soc.,* 1208, 1949.
14. M. S. Saunders, J. B. Vierow, and G. Carta, *AIChE J, 35*:53 (1989).
15. G. V. Samsonov, *Ion-Exchange Sorption and Preparative Chromatography of Biologically Active Molecules*, Consultants Bureau, New York and London, 1986, Chap. 3.
16. Tao Zuyi, Zhang Baolin, and Seng Fenling, *Acta Physico-Chimica Sinica, 8*: 464 (1992).
17. O. Talibudeen, *Trans. Faraday Soc., 51*:582 (1955).
18. D. J. Greenland, R. H. Laby, and J. P. Quirk, *Trans. Faraday Soc., 61*:2013,2024 (1965).
19. D. J. Greenland, R. H. Laby, and J. P. Quirk, *Trans. Faraday Soc., 58*:829 (1962).
20. V. Ya. Vorob'eva, L. V. Naumova, and G. V. Samsonov, *Zh. Fiz. Khim., 55*:1679 (1981).
21. D. H. Murav'ev and O. N. Obrezkov, *Zh. Fiz. Khim., 60*:396 (1986).
22. Zhang Baolin, Seng Fenling, and Tao Zuyi, *Chem. J. Chinese Univ., 13*:840 (1992).

23. Zhang Hui, *Interactions of Amino Acids, Oxidized Glutatione and Bovine Serum Albumin with Cation Ion Exchanger*, M.S. thesis, Lanzhou University, 1990.

24. Shao Tong, *Interactions of Carboxylic Acids, Amino Acids and Oxidized Glutathione with Anion Ion Exchanger*, M.S. thesis, Lanzhou University, 1990.

25. M. Seno and T. Yamabe, *Bull. Chem. Soc. Jpn.*, *33*:1532, 1021 (1960).

26. F. Helfferich, *Reactive Polym.*, *12*:95 (1990).

27. Q. Yu, J. Yang, and N.-H.L. Wang, *Reactive Polym.*, *6*: 33 (1987).

28. N.-H.L. Wang, Q. Yu, and S. U. Kim, *Reactive Polym.*, *11*:261 (1989).

29. E. A. Moelwyn-Hughes, *Physical Chemistry*, Pergamon Press, Oxford, 1961, p. 1023–1024.

30. V. A. Pasechnik, N. N. Nemtsova, and G. V. Samsonov, *Zh. Fiz. Khim.*, *46*:1210 (1972).

31. G. L. Gaines, Jr. and H. C. Thomas, *J. Chem. Phys.*, *21*:714 (1953).

32. Zhang Baolin, *Interactions of Amino Acids with Ion Exchanger and Application of Titration Calorimeter*, M. S. thesis, Lanzhou University, 1989.

33. Zhang Baolin, Wang Wenqing, Li Zhong, and Tao Zuyi, *Chem. J. Chinese Univ.*, *12*:1522 (1991).

9

Ion-Exchange Selectivities of Inorganic Ion Exchangers

Mitsuo Abe

Tsuruoka National College of Technology and Tokyo Institute of Technology, Tsuruoka, Yamagata, Japan

I. INTRODUCTION

Many investigators continue to look for new inorganic ion-exchange materials, whose special properties such as resistance to high temperature and radiation fields can be employed to advantage. The most important groups of inorganic ion exchangers are clays, zeolites, hydrous oxides, insoluble acid salts, heteropolyacids, and hexacyanoferrates. Some of these inorganic ion exchangers exhibit especially high selectivities for certain elements or groups of elements. Zeolites and inorganic ion exchangers that include zirconium phosphate, polybasic acid salt, heteropoly salts and hydrous oxides have been discussed in earlier volumes of this series [1, 2]. In 1982, the book *Inorganic Ion Exchange Materials* was edited by A. Clearfield [3].

This chapter examines representative examples of the selectivities of inorganic exchangers for comparison with those of organic ion-exchange resins. The inorganic ion exchangers, unlike their organic analogs, usually have rigid structures and do not undergo any appreciable dimensional change during the ion-exchange reaction. The rigid structure leads to specific and unusual selectivities. The selectivity of inorganic ion exchangers can be discussed in terms of an ion sieve effect, steric factors, ion size preference, an entropy effect and ion memory preference.

II. QUANTITATIVE DESCRIPTION OF SELECTIVITY OF ION EXCHANGERS

The framework of inorganic ion exchangers possesses positive or negative charges which are compensated by ions of opposite charge called counter-ions. In the ion-exchange process these counter-ions are replaced by ions of the same charge in electrolyte solutions in contact with the ion exchanger preserving the electroneutrality of the exchanger framework. The interstices in the framework are known as pores. The channels or cavities in the framework depend on their inter-connections. Certain materials are capable of exchanging both cations and anions. These are known as amphoteric ion exchangers.

A. Ion-Exchange Ideality

When an exchanger (solid phase) in the H^+-ion form is equilibrated with the n-valent metal ion of an electrolyte solution, the ion exchange reaction can be represented by the expression

$$n\overline{H^+} + M^{n+} \Leftrightarrow \overline{M^{n+}} + nH^+ \tag{1}$$

where the bar refers to the exchanger phase and M^{n+} corresponds to the n-valent metal ion. The thermodynamic equilibrium constant, K, of the above reaction can be defined as

$$K = \frac{m_{H^+}^n \, \overline{E}_{M^{n+}} \, \gamma_{H^+}^n \, \overline{f}_{M^{n+}}}{m_{M^{n+}} (\overline{E}_{H^+})^n \, \gamma_{M^{n+}} (\overline{f}_{H^+})^n} = \frac{(\overline{a}_{M^{n+}})(a_{H^+})^n}{(\overline{a}_{H^+})^n (a_{M^{n+}})} \tag{2}$$

where m_{H^+} and $m_{M^{n+}}$ are the molalities of H^+ and metal ions in the solution, γ_{H^+} and $\gamma_{M^{n+}}$ represent the activity coefficients of the H^+ and M^{n+} ions in solution, $\overline{E}_{H^+}$ and $\overline{E}_{M^{n+}}$ are the concentrations of H^+ and M^{n+} in the exchanger phase, expressed in equivalent fraction terms, and $\overline{f}_{H^+}$ and $\overline{f}_{M^{N+}}$ are the activity coefficients of H^+ and M^{n+} in the exchanger phase, respectively. Such combination of concentration and activity coefficient terms yield the activity-, a-, based resolution of K as shown.

Other representations of the ion-exchange data are by the equilibrium coefficient, K_c,

$$K_c = \frac{\overline{E}_{m^{n+}}}{(\overline{E}_{H^+})^n} \frac{(m_{H^+})^n}{(m_{M^{n+}})} \tag{3}$$

where concentration terms are uncorrected for nonideality in both phases and by the corrected equilibrium coefficient, K_H^M:

$$K_H^M = \frac{(\overline{E}_{M^{n+}})(a_{H^+})^n}{(\overline{E}_{H^+})^n(a_{M^{n+}})} \tag{4}$$

where correction for nonideality in the solution phase alone is included.

When measurements are made to investigate the ion exchange behavior of a cation or anion it is common practice to determine the distribution coefficient, D_i^*, where

$$D_i = \frac{\text{concentration of solute in exchanger}}{\text{concentration of solute in solution}}.$$

In this instance

$$D_{M^{n+}} = \frac{[\overline{M}^{n+}]}{[m_{M^{n+}}]} = K \frac{[\overline{H}^+]^n \gamma_{M^{n+}} \overline{f}_{H^+}^n}{\gamma_{H^+}^n \overline{f}_{M^{n+}}} \cdot \frac{1}{[m_{H^+}]^n} \tag{5}$$

When the exchange reactions are carried out using micro quantities of the metal ion $[m_{H^+}]^n \gg [m_{M^{n+}}]$ and $[\overline{H}^+]^n \gg [M^{n+}]$. Under these conditions, Eq. (5) reduces to

$$\log D_{M^{n+}} = \text{const.} - n \log[m_{H^+}] \tag{6}$$

For an ideally reversible exchange reaction the plot of $\log D_{M^{n+}}$ vs. $\log[m_{H^+}]$ should resolve a straight line with a slope of $-n$, the valency of the exchanged metal ion (Fig. 1). Much attention has been paid as to whether the assumptions of Eq. (6) are applicable for the situation where $[m_{H^+}]^n \gg [m_{M^{n+}}]$ with weak-acid ion exchangers.

The ratio of the distribution coefficient for two different ions, A and B, is defined as their separation factors, $\alpha_{A/B}$

$$\alpha_{A/B} = \frac{D_A}{D_B} \tag{7}$$

This parameter is of practical value when the separation of different ions is considered. In general, if $\alpha_{A/B}$ is > 1, ions A are more selectively re-

---

*IUPAC recommendations for defining D are [7] distribution coefficient, D_g: the ratio of the total (analytical) amount of a solute per gram of dry ion exchanger to its analytical concentration (total amount per cm^3) in the solution; concentration distribution ratio, D_c; the ratio of the total (analytical) concentration of a solute in the ion exchanger to its concentration in the external solution (dimensionless); the concentrations are calculated per cm^3 of the swollen ion exchanger and cm^3 of the external solution; volume distribution coefficient, D_v: the ratio of the total (analytical) concentration of a solute in the ion exchanger calculated per cm^3 of the column or bed volume to its concentration (total amount per cm^3) in the external solution.

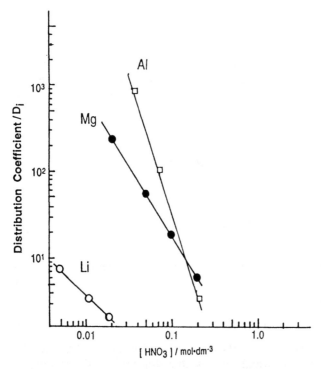

Figure 1 Ion exchange ideality of uni-, di, and trivalent metal ions on cubic antimonic acid (C-SbA). Metal ions: Li$^+$ (O), Mg^{2+} (●) and Al^{3+} (□). Initial concentration of metals; 10^{-3} mol dm^3; weight of C-SbA, 0.10 g; total volume, 10.0 cm^3, temp., 30°C. (From data of Refs. 4, 5, and 6.)

moved from solution than ions B by the ion exchanger. The separation of A from B on an ion-exchange column is facilitated by an increase in the value of $\alpha_{A/B}$. Many studies have been carried out for the determination of D values. This information is useful for the separation of trace and micro amounts of the elements in the fields of analytical chemistry, radiochemistry and environmental chemistry. The selectivity sequences on inorganic ion exchangers have been reported by various laboratories (Table 1). It can be seen from Table 1 that selectivity depends on the following factors:

1. Acid strength of the substances; cation, amphoteric and anion-exchange reactions
2. Concentration of the exchanging ions
3. Crystallinity of the materials
4. Solution media and ionic strength

Table 1 Ion-Exchange Selectivities on Inorganic Ion Exchangers[a]

Oxides and hydrous oxides	Selectivities	Amount	References
PbO	$Fe(CN)_6^{3-} < SO_4^{2-} < I^- < SO_3^{2-} < S_2O_3^- < Cl^- < PO_4^{3-} <$ $Fe(CN)_6^{4-} < CrO_4^{2-} < S^{2-}$	micro	8
ZnO	$NO_3^- < Cl^- << CNS^- < I^- < Fe(CN)_6^{3+} < SO_3^{2-} < S_2O_3^- <$ $SO_3^{2-} < C_2O_4^{2-} < CrO_4^{2-} < Fe(CN)_6^{4-} < S^{2-}$	micro	8
Bi_2O_3	$Fe(CN)_6^{3-} < Fe(CN)_6^{4-} < I^- < SO_4^{2-} < PO_4^{3-} < C_2O_4^{2-} <$ $CrO_4^{2-} < S^{2-}$	micro	8
$H\text{-}Fe_2O_3$ heated at 125°C	$Ba < Co, Ni < Mn(II) < Cd < Zn < Cu < Ag < Pb < Al < Fe(III) < H^+$	trace	9
heated at 170°C	$ClO_4^-, I^- < Br < ClO_3^-, Cl^- < NO_3^-, NO_2^- < CNS^- < BrO_3^- <$ $Fe(CN)_6^{3-} < S_2O_3 < IO_4^-, SO_4^{2-} < SO_3^{2-}, CrO_4^{2-}, Fe(CN)_6^{4-} < CO_3^{2-},$ $F^-, PO_4^{3-} < B_4O_7^{2-}, S^{2-}, AsO_4^{3-}, OH^-$	trace	10
heated at 450°C	$Na < Mg < Ca < Sr < Ba < Tl(I) < Ag, Cd < Mn(II) <$ $Ni < Co < Zn < UO_2(II), Cu(II) < Pb < Hg(I) < Hg(II) < Al, Cr(III) <$ $Th < Fe(II), Fe(III) < H^+$	trace	10
α-Hematite	$VO_3^{4-} < MoO_4^{2-} < Cr_2O_7^{2-}$ (in 0.1 M NaCl)	micro	11
La_2O_3	$Cl^- < Fe(CN)_6^{3-} < SO_4^{2-} < Fe(CN)_6^{4-} < S^{2-} < I^- < MnO_4^- <$ $CrO_4^{2-} < C_2O_4^{2-} < PO_4^{3-}$	micro	8
$H\text{-}SiO_2$	$Li < Na < K < Ag, Mg < Ca\text{-}Sr < Ba, Ni < Cd < Zn < Cu < Co,$ $La < Cr(III) < Al < Sc < UO_2^{2+} < Th < Fe(III) < H^+$	micro	12
	$Rb < K < Na < Li; Sr, Ba < Mg < Be; In < Ga < Al$	micro	13, 14
	Uni-valent Cations < Bi- < Tri	micro	13, 14
	$Na < Ba, Ca < Gd < UO_2(II) < Pu(IV), U(IV) < Zr, Nb$	micro	15
NH_4^+ or Ca^{2+} form	$Ni < Co(II) < Cd < Zn, Ag < Cu(II) < UO_2(II) < Pu(IV), U(IV) <$ Zr, Nb	micro	16
Mg^{2+} form	$Mg < Zn < Ni < Cu(II)$	micro	17
	$SO_4^{2-} < C_2O_4^{2-} < PO_4^{3-}$	micro	18-22,

Table 1 Continued

Oxides and hydrous oxides	Selectivities	Amount	References
H-TiO₂			
heated at 30°C	Na<Rb<Cs<Sr, Co<Ni<Cu, Fe(III)<Cr(III)	trace	23
dried at room temp.	Na<K<Cs<Ca<Sr<Ba, Co<Ni<Mn(II)<Zn<Cd<Cu	micro	24
H₂Ti₄O₉nH₂O	Li<Na<K<Rb<Cs, Mg<Ca<Sr<Ba	micro	25,26
	Ni<Co<Mn(II)<Zn<Cu(II)	micro	27
H-SnO₂	Cs<Na<alkaline earth<Mn(II)<Ni<Fe(II), Co(II)<Zn,	micro	28-30
	UO₂(II)<Cu(II)<Al, Cr(III)		
	I⁻<Br⁻<MnO₄⁻<Cl⁻<<Fe(CN)₆³⁺<Fe(CN)₆⁴⁻<Cr₂O₇²⁻<	micro	2
	SO₄²⁻<<C₂O₄²⁻<<PO₄³⁻		
H-ZrO₂	Cs<Rb<K<Na<Li (in alkaline solution)	micro	31
H-ThO₂			
heated at 400°C	Cs<Rb<Na, Co<Ni<Cu, Fe(III)<Cr(III)	trace	23
H-MnO			
Manganite	Li<Na<K; Ni<Cu<Co	micro	32
δ type	Mg<Ca<Sr<Ba; Ni<Co<Cd, Zn<Mn	trace	32
γ type	Ni<Zn<Co<Cu	trace	33
λ type	Na, K<Rb<Cs<<Li	micro	34
Birnesite	Li<<K<Rb<Cs	micro	33
Cryptomelane	Li<Na<Cs<<K<Rb	micro	35,36
H-Sb₂O₅			
Amorphous	Li<Na<K<Rb<Cs (in HNO₃ solution)	micro	37,38
	Li<Cs<Na<K=Rb (in NH₄NO₃ solution)	micro	37
Glassy	Li<Na<K<Rb<Cs (in HNO₃ solution)	micro	37,38

Cubic	Li<Cs<Na<K=Rb (in NH_4NO_3 solution)	micro	37
	Li<K<Cs<Rb<Na (in HNO_3 solution)	micro	37,38
	Li<K<Rb<Cs<Na (in NH_4NO_3 solution)	micro	37
	Li<K<Cs<Rb<Na	macro	39
	Cs<K<Rb<Na (in HCl, HNO_3, $HClO_4$ soln.)	trace	40
	Cs<Rb<K<Na (in NH_{4aq}, CH_3COOH)	trace	41
	Cs<Li<Rb<NH_4<K<Na; Mg<Ca<Sr<Ba	trace	41
	Cs=Rb<Na; Ra<Ba<Ca<Sr	trace	42
	NH_4^+<$CH_3NH_3^+$<$C_2H_5NH_3^+$<$(CH_3)_2NH_2^+$	micro	43
	Mg<Ba<Ca<Sr; (in HNO_3 solution)	micro	5
	Mg<Ca<Sr<Ba	macro	44
	Ni<Mn(II)<Zn<Co<Cu(II)<<Cd (in HNO_3 solution)	micro	45
	Pt(IV)<Au(III)<Pd(II)<<Hg(I),Hg(II)<Ag (in HNO_3 solution)	micro	46
Monoclinic	K<Rb<Cs<Na<<Li	macro	47
	Na<K<Rb<Cs<<Li	micro	47
H-Nb_2O_5	Li<<**K**<<Cs; Mg<Ca<Sr<Ba;	micro	48
	Mn(II$_2$)=Co<<Y=Ce(III)=Eu<<Sc<Fe<<UO_2(II)	micro	48
H-WO_3			
Amorphous	Mg<Ba, Co<Zn, Cu<Sr<Ca<Ni, Mn(II)	micro	49
Cubic ammonium	Li<<Na<K<Rb<Cs	micro	50
Acid salts	Zirconium based		
Zirconium phosphate			
Amorphous	Li<Na<Ag<NH_4^+<K<Rb<Cs, Mg<Ca<Sr<Ba	0.014 salt	51
	Li<Na<K	macro	52-53
	Ca, B, Sr<<Cs	micro	54
	Co<Cu<Fe(III); Cr(III)<Al<Fe(III)	paper	55
	Ce(III)<Am(III)<Cm(III)<Eu<Cf<U(VI)<Ce(IV)	trace	56
	Na<Sr<Ce(III)<Cs (pH=0-2)	trace	56
	Na<Rb<Cs; Y<Ce(III); Sr<Ce(III)<U(VI)	trace	57

Table 1 Continued

Acid salts	Selectivities	Amount	References
Semicrystalline	Li<K<H<Cs (in MCl + HCl)	macro	58
	Co<Ni<Cn>Zn; Cr(III)<Fe(III)	micro	55
α-crystalline	Cs<Na<Ce(III)<U(VI) (pH = 2)	micro	59
Zirconium arsenate			
Amorphous	Na<K<Cs (pH = 2.3-3.3)	micro	60
Crystalline	K<Na<Li<Rb<Cs (0.1 M MOH + MCl)	macro	61
Zirconium antimonate			
Semicrystalline	Na<K<Rb<Cs<<Li (in HNO$_3$)	micro	62
	Li<Na<K<Rb<Cs (in MOH + HNO$_3$)	macro	62
Amorphous	Li<Cs<Rb<NH4<K<Na (pH = 3)	micro	63
Zirconium tungstate	Li<Na<K<Rb<Cs (Zr/W = 0.44)	micro	64
Zirconium tellurate	Na<<K<Rb<Cs; Sr<Ca<Ba		65
Zirconium oxalate			
Gel-like	Cs=Rb=K<Na		66
Crystalline	Na<Cs<Rb<K		66
Zirconium nolbdate	Ba<Sr<Ca		53
	Titanium based		
Titanium arsenate	Co<Ni<Mn<Zn<Ga<Sr<Cd<Cu<Pb<Ba (AS/Ti = 2)	micro	67
(Amorphous)	Co<Zn<Sr<Ba<Hg(II)<Cd<Ni<Pb (As/Ti = 1.8)	micro	68
	Ca<Sr<Ba; Cu<Mn; (in water, As/Ti = 1.7)	micro	69
	Co<Zn<V(IV)<Ni<Fe(III)	micro	69
Titanium antimonate			
Amorphous	Hg<Sr<Mn; Zn<Ni<Cd<Cu<Ca=Ba<Pb (Sb/Ti = 1.2)	micro	70
	Hg<Ni<Ca<U(VI)<Co<Sr<Pb<Cd<Zn<Ba (Sb/Ti = 1)	micro	71
	Pb<Hg<Ag	micro	72
Semicrystalline	Na<K<Rb<Cs<<Li	macro	73
	Li<Na<K<Rb<Cs (in MOH + HNO$_3$)	micro	73
	Mg<Ca<Sr<Ba	micro	74
	Mn<Ni<Cd<Zn<Co<Cu<Fe(II)<Pb	micro	74

Material	Selectivity order		Ref.
Titanium molybdate	Mg<Sr<Pb; Zn<Cd<Cu<Ca<Ba (Mo/Ti = 0.5-2.0)	micro	75
Titanium tungstate	Sr<Ga<Mg<Pb (W/Ti = 2)	micro	76
Amorphous	Li<Na<K; Mg<Ca<Sr<Ba (W/Ti = 1-2)	micro	77
Titanium selenite	Li<Na<K (pH=3); K<Na<Li (pH=8); K<Na=Li (pH=12)	micro	68
Titanium vanadate	Mg<Ca<Sr<Ba	micro	78
Thorium based			
Thorium phosphate			
Thorium molybdate	Mg, Cu, La<Sr<Hg, Cd<Ni<Ba, Al<Ga<In	micro	79
Cerium based			
Cerium phosphate			
(amorphous)	Na<Ag<Cs	micro	80
(crystalline)	Na<Ag<Cs (TLC); Li<Na<K; Co<Eu(II)<Fe(II) (1MHClO4 TLC)	micro	81
Cerium phosphate-sulfate (Ce:P:S=2:1:1)	Ca<Cs<Ba<Sr<Ag<Na	micro	82
Cerium arsenate	Li<Na<K (0.1M MCl + MOH at low pH)	macro	83
(crystalline)	K<Na<Li (0.1M MCl + MOH at low pH)	macro	83
Cerium antimonate	Cd<Hg(II)	macro	84
Cerium tungstate	Mn(II)<Cd<Ni, Zn<<Cu<Cs<Ag<Tl(III)Co<Tl(I)<Hg(II)	micro	85
Tin(IV) based			
Tin phosphate			
(amorphous)	Mg<Ca<Sr<Ba (pH2.5)	micro	86
	Na<K<Rb<Cs (HNO3)	micro	87
	Ru<Sr<Ce(III)<Y<Nb<Zr<Cs (0.1 M HNO3)	micro	88
	Rb, Ce(IV), Sr, Y, Ce(III)<<Cs<Nb<Zr (0.1 M HNO3)	micro	88
	Ni, Cu(II), Co<<Rb, U(VI), Fe(III)<Fe(II) (0.1M HNO3)	trace	89
	Co<Ni<Zn<Cu	micro	90
(crystalline)	Cs<Li<Na<K (pH=4.0, 0.1M); Cs<K,Na<Li (pH=4.0, 0.1M)	macro	90

Table 1 Continued

Acid salts	Selectivities	Amount	References
Tin arsenate			
(amorphous)	Li<Na<K	macro	86
(crystalline)	Mn<Ni<Co<Zn<Cu (pH=2.5)	micro	81
	Ni<Zn<Co<Cu; Sn/As=0.33)	micro	91
Tin antimonate	Na<K<Rb<Cs<<Li	micro	73
	Cs<Rb<K<Na<Li	macro	92
	Ca<Cu<Mg<Mn<Co<Hg(II), Zn<Sr, Ba<<Cd (pH=1)	micro	91
Tin tungstate	Zn<Cd<Mg<Sr, Mn, Cu<Ni<Pb<Ba<Co (water)	micro	93
Tin selenate	Cs<Na<Li	macro	94
	Ni<Co<Zn<Cu; In<Ga<Al<Y<Ce<La; Mg<Ga<Sr<Ba; 25metalions<<Sr, In, Pr<Fe(III), Pb	micro	68
Chromium based			
Chromium phosphate	Cs<Rb<<K<Na (H-form); Rb<<Cs<K<Na (NH$_4$ form)	micro	71
Chromium tripoly-	many polyvalent metal ions<<H<Na<K<Rb<Cs	macro	94
phosphate			
	Co<Zn<Na<Rb<Cs<Ba	micro	95
Other acid salts			
Tantalum phosphate	Na<<K<Rb<Cs (amorphous)	micro	96
Tantalum antimonate	Li<Na<NH$_4$<K<Rb<Cs<Ag<Tl	micro	97
	Mn(II)<Mg<<Zn<Co<Cu<Cd<Ni<Ca<Sr<Pb<Ba	micro	97
	Al<Sc<Y<Eu<Sm<Nd<Pr<Ce<La	micro	97
Phosphoantimonate	Li<Na<K<Rb<Cs	micro	98, 99
(silica gel support)	Li<K, Eu<Rb<<Ca<Sr<Na; Sr<Na<K<Rb<Eu<Cs	micro	100

Heteropoly acid salt

AMP	Na<K<Rb<Cs	trace	101
Hexacyanoferrate			
Ni-hexacyanoferrate	Na<K<Rb<Cs	trace	102

Selectivity patterns of zeolites[103]

Selectivity series	Type of zeolites
Li<Na<K<Rb<Cs	Chabazite, mordenite, norton zelon, erionite, phillipsite, clinoptilolite, Linde Y, ZK-4
Li<Na<Rb<K<Cs	Mordenite, norton zeolon, erionite, phillipsite, clinoptilolite, ZK-4
Li<Na<Rb<Cs<K	Erionite
Li<Cs<Rb<K<Na	Linde 13-X
Cs<Li<Rb<K<Na	Linde 4A
Cs<Rb<Li<K<Na	Ultramarine, analcite, NaPt
Cs<Rb<K<Li<Na	Basic sodalite, ultramarine

aIn some cases insufficient data are available to differentiate between the series.

Generally, amorphous materials exhibit selectivity sequences that are similar to those reached with organic ion-exchange resins. This result is attributed to the fact that both materials have relatively open and elastic structures. The concentration dependence of D_i values is generally much larger for ions with large ionic crystal radii than it is for ions with small ionic crystal radii. However, D values depend on the loading of metal ions in inorganic ion exchangers. Ion-exchange ideality is usually maintained up to relatively high concentrations of the metal ions, but the D values increase with decreasing concentration of metal ions. A typical example of such concentration dependence is provided in Fig. 2 for Cs^+ and Mg^{+2} ions in equilibrium with cubic antimonic acid [5]. Generally, an unique D_i value is obtained with relative higher concentration for organic ion-exchange resins than that for inorganic ion-exchangers. Only when the concentration of the metal ion under investigation is very small ($\leq 10^{-7}$ mol. dm^{-3}) can a unique response of D_i to solution concentration level be D_i expected. This is the case with organic ion exchangers as well. However, it is often difficult to make valid comparisons between data from different laboratories. To circumvent this problem, the initial concentration must be provided along with the total volume of the solution, the weight of exchanger used and the temperature.

The formula of heteropoly salts can, in general, be represented by $H_m XY_{12}O_{40} \cdot nH_2O$ (where H = H, Ag, Tl, NH_4, or $NR_n H_{4-n}$, X = P, As, Si, B, Al, Ge, Ti, or Zr, and Y = Mo or W). Ammonium 12-molybdosphosphate (AMP) is obtained in the form of yellow microcrystalline powder with a composition corresponding to $(NH_4)[P(Mo_3O_{10})_4] \cdot 5H_2O$. Van R. Smit et al. studied its sorption, at 25°, of monovalent cations at trace concentration levels in ammonium nitrate media, pH = 2, by means of batch equilibration and column chromatographic techniques [101]. The selectivity sequence of AMP for Ag^+ and Tl^+ is $Ag^+ < Tl^+$; for alkali metal ions it is $Na^+ < K^+ < Rb^+ < Cs^+$ and parallels the increase in crystallographic ionic radii of the alkali metals.

Selective sorption of Cs^+ was first reported for the hexacyanoferrate(II) of nickel. Its general formula can be represented by $AB[Fe(CN)_6]$ (where A = K, Rb, Cs, etc., B = Ni, Zn, Cu, Co or Cd). The hexacyanoferrate(II) of nickel exhibits the same selectivity sequence for alkali metal ions as the AMP: $Na^+ < K^+ < Rb^+ < Cs^+$ [102]. At low loading, an extremely high selectivity is resolved for Cs. The separation factor ($\alpha_{Cs/Na}$) is found to be larger than 10^3.

Zeolites with a silicon/aluminum ratio of 1, because of their low water content, have the highest possible ion-exchange capacity, when computed on this basis. Linde 4A and Linde 13-X exhibit selectivity patterns characteristic of high-field-strength exchangers despite their open structure and higher water content.

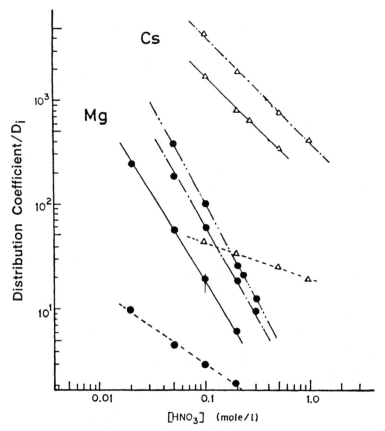

Figure 2   The dependence of distribution coefficient, D_i, values upon the initial concentration of Mg^{2+} and Cs^+ ions in contact with cubic antimonic acid (C-SbA). Initial concentration of metal ions: Mg^{2+} (●), (– –) 10^{-2} M, (—) 10^{-3}M, (—·) 10^{-4} M, (··) 10^{-5} M; Cs^+ (▵), (– –) 10^{-2} M, (—) 10^{-3}M, (—·) 10^{-4} M. (Data from Ref. 5.)

III.   ACID STRENGTH OF INORGANIC ION EXCHANGERS

All inorganic ion exchangers are sparingly soluble, solid compounds in which the binding of exchangeable species is more or less ionic in character. Generally cation uptake by amphoteric inorganic ion exchangers is increased by increasing the pH of the external solution, while anion uptake is increased by decreasing the pH. One of the important methods of characterizing ion-exchangers is pH titration studies to determine acidity or basicity (strong or weak). On the inorganic ion-exchangers, rate of ion-exchange reactions is generally very slow. When the solutions of (MOH + MCl) or (HCl + MCl) with various ratios at a constant ionic strength and temperature are added to the ion-exchanger by batch technique, the pH changes occur. The uptakes of the ions exchanged can be calculated from pH change and/or determination of ions between before and after equilibration for the supernatant solutions. The pH plots of the solution against the volume of the [OH⁻] or [H⁺] used is so-called as pH titration curve.

A. Acid Strength of Hydrous Oxides

The reactive groups of inorganic exchangers in the hydrogen ion form are hydroxide groups. The various reactions they undergo depend on the nature of the hydrous oxide of the element (E) and the pH. For the OH group in E-O-H, typically the following three types of behavior [104] can be expected:

(a) As an ion with spherical symmetry
(b) As OH⁻ pointed toward the central atom with cylindrical symmetry
(c) As polarized with a tetrahedral charge distribution

Case (a), implying random orientation or free rotation, is apparently realized in the high-temperature form of KOH. The transformation from (b) to (c) is to be expected with increasing dissociation of the proton.

The arrangement of sites repeated on the internal surfaces of hydrous oxides can be represented as E-O-H---OH$_2$. Proton transfer may occur in the following ways: (1) rupture of the OH group in EOH, (2) formation of an OH bond in the hydronium ion, and (3) removal of the proton from the electrostatic field of the central metal-oxygen anion. If E has a small radius and is a highly electronegative element, acidic properties will appear. With increase in the size of E and/or with a decrease in its electronegativity, amphoteric behavior will result first and this will be followed by basic behavior. Their exchange reactions can be described as follows:

Dissociation at high pH

$$\overline{\text{E-OH}} \Leftrightarrow \overline{\text{E-O}^-\text{H}^+} \tag{8}$$

leads to uni-univalent cation-exchange reaction

$$\overline{\text{E-O}^-\text{H}^+} + \text{M}^+ \Leftrightarrow \overline{\text{E-O}^-\text{M}^+} + \text{H}^+ \tag{9}$$

Protonation at low pH

$$\overline{\text{E-OH}} + \text{H}^+ \Leftrightarrow \overline{\text{E-OH}_2^+} \tag{10}$$

and simultaneous bonding of an anion X⁻

$$\overline{\text{E-OH}_2^+} + \text{X}^- \Leftrightarrow \overline{\text{E-OH}_2^+\text{X}^-} \tag{11}$$

ION-EXCHANGE SELECTIVITIES

leads to an anion-exchange reaction

$$\overline{\text{E-OH}_2^+\text{X}^-} + \text{Y}^- \Leftrightarrow \overline{\text{E-OH}_2^+\text{Y}^-} + \text{X}^- \tag{12}$$

Amphoteric behavior can be deduced by following the progressive dissociation reaction of the acid formed in reaction (11). From Eqs. (10) and (11),

$$\overline{\text{E-OH}_2^+\text{X}^-} \Leftrightarrow \overline{\text{E}^+\text{OH}^-} + \text{H}^+ + \text{X}^- \qquad K_1 \tag{13}$$

From Eq. (9)

$$\overline{\text{E-O}^-\text{H}^+} + \text{M}^+ \Leftrightarrow \overline{\text{E-O}^-\text{M}^+} + \text{H}^+ \qquad K_2 \tag{14}$$

Then

$$K = K_1 K_2 = \frac{\overline{[\text{E-O}^-\text{M}^+]} \cdot [\text{X}^-][\text{H}^2]^2}{\overline{[\text{E-OH}_2^+\text{X}^-]}[\text{M}^+]} \tag{15}$$

When the concentrations in the ion exchanger of $\overline{\text{E-O}^-\text{M}^+}$ and $\overline{\text{E-OH}_2^+\text{X}^-}$ are equal, the equiadsorption point (IAP), pH_{IAP}, can be defined by

$$\text{pH}_{IAP} = \frac{1}{2}\left\{(\text{pK}_1 + \text{pK}_2) + \log\frac{[\text{M}^+]}{[\text{X}^-]}\right\} \tag{16}$$

With its determination carried out at constant ionic strength the pH_{IAP} value, so obtained, provides an assessment of the acidity of the amphoteric ion exchanger. This pH, relatable to the average value of pK_1 and pK_2 with Eq. (16), is very close to the value of the isoelectric point (IEP). Parks [105] has pointed out that the relationship between the IEP of a solid surface and the valency-effective ionic radii, when corrected for crystal field effects, coordination, hydration, and other factors, is quite good. He also has indicated that the broad probable IEP range characteristic of a cation oxidation state may be selected from the data in Table 2 as shown below. It is known that the IEP for amphoteric oxides is affected by the presence of impurities, crystallinity and the chemical species under investigation.

Table 2 Isoelectric Points of Metal Oxides

M_2O		IEP>	pH 11.5
MO	8.5<	IEP>	pH 12.5
M_2O_3	6.5<	IEP>	pH 10.4
MO_2	0<	IEP>	pH 7.5
M_2O_5, MO_3	0.5<	IEP	

Source: From Ref. 105.

1. Hydrous Oxides of Trivalent Metals [106]

Hydrous aluminum oxide is used frequently in chromatographic columns. Hydrous oxides of Ga, In, and the lanthanides may also be employed as inorganic ion exchangers. However, substantive research of these oxides has not been performed.

a. Aluminum Various forms of the hydrous oxides of aluminum are found in natural and synthetic materials such as gibbsite (hydrargillite), bayerite and norstrandite; the oxide hydroxide; AlOOH, as diaspore and beomite. These crystalline oxides are sparingly soluble, solid compounds. However, amorphous aluminum oxide is soluble in weak acid and alkaline solution. Three types of alumina are commercially available: neutral alumina (pH 6.9–7.1), basic alumina (pH 10–10.5), and acidic alumina (3.5–4.5). Most chromatographic alumina is a mixture of γ-alumina with a small amount of hydrous aluminum oxides (and perhaps a small quantity of sodium carbonate that remains from the manufacturing and activation process) [107].

b. Bismuth Anion exchange response of the hydrous bismuth oxide, $Bi(OH)_3$ or $Bi_2O_3 \cdot 3H_2O$, during neutralization with acid is similar to that observed with strongly basic quaternary ammonium ion exchange resins to document its strong base properties (Fig. 3) [108]. The adsorption capacity is 3.0–3.8 meq/g [109]. The strong preference of the hydrous oxide for chloride ion reflects the fact that the amorphous material readily forms BiOCl [110]. The oxychloride is transformed to α- and/or γ-Bi_2O_3 by regeneration with 2 M (M = mol/dm^3) NaOH solution [110].

c. Iron(III) Various types of hydrous iron(III) oxides are known to exist with different crystalline forms that depend on the method of preparation. Three of them are amorphous α-FeOOH (geothite), β-FeOOH, and γ-FeOOH (lepidocrocite). Amorphous iron(III) oxide with the composition $Fe_2O_3 \cdot 2H_2O$, exhibits typically amphoteric ion-exchange behavior during its neutralization with acids and bases, while only anion-exchange behavior is encountered with β-FeOOH (Fig. 4) [108]. Both oxides are practically insoluble in the solution above pH 1.6, whereas they are transformed to α-Fe_2O_3 by immersion in strong alkaline solution for an extended period.

2. Hydrous Oxides of Tetravalent Metals

The oxides of metals, such as Ti, Zr, Ce(IV) and Th exhibit amphoteric ion-exchange behavior (Fig. 5). The decrease in acid strength of their oxides, Ti > Zr > Ce > Th, is paralleled by the decrease in the effective crystal ionic radii of these metals. The pH_{IAP} of the oxides follows the opposite pattern, Ti < Zr < Ce < Th. It has been believed until recently that silica gel shows no anion-exchange character, even in acid solution, but anion

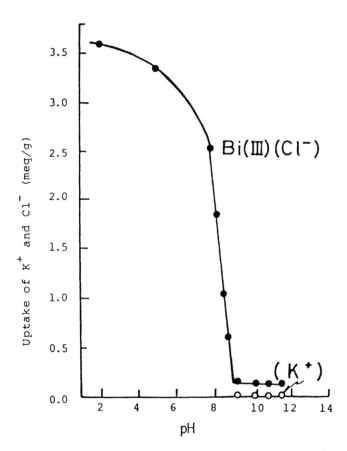

Figure 3 Uptake of K⁺ and Cl⁻ by hydrous bismuth oxide. Exchanger, 0.25 g; total volume, 12.5 cm³; ionic strength, 0.1 M (HCl + KCl) or (KOH + KCl), temp., 30°C. (From M. Abe and T. Ito, *Nippon Kagaku Zasshi*, 86:817 (1965). With permission.)

exchange has been observed at a pH of 3 [112]. The isoelectric point (IEP) of silica gel is generally accepted to be about 2 [105, 113]. When polyvalent anions are introduced to amphoteric hydrous oxides the oxides exhibit cation-exchange behavior. Hydrous manganese dioxide shows mainly cation-exchange properties and is a relatively weak acid. This behavior is characteristic of even those manganese oxides which are known to have endured a number of different modifications. The IEP of hydrous manganese (IV) oxides has been reported to change, depending on the modifications: Mn(II)-manganite (1.8±0.5) < δ-type (2.8 ± 0.3)<α-type (4.5 ± 0.5) < γ-type (5.5 ± 0.2) < β-type (7.3 ± 0.2) [114].

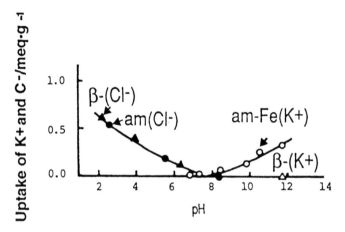

Figure 4   Uptake of K$^+$ and Cl$^-$ by hydrous iron (III) oxides. amFe, amorphous; β-Fe, β-FeOOH; exchanger, 0.25 g; total volume, 12.5 cm^3; ionic strength, 0.1 M (HCl + KCl) or (KOH + KCl), temp., 30°C. (From M. Abe and T. Ito, *Nippon Kagaku Zasshi*, 86:817 (1965). With permission.)

3.   Hydrous Oxides of Pentavalent Metals [111]

The hydrous oxides of Sb, Ta and Nb exhibit cation-exchange behavior. Their uptakes of potassium ion are increased by increasing the pH of the solution in contact with them. Examination of their ion-exchange properties as a function of pH during their neutralization has shown that the

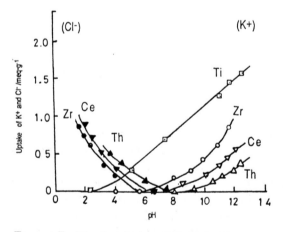

Figure 5   Uptake of K$^+$ and Cl$^-$ by hydrous oxides of Ti, Zr, Ce(IV) and Th. Exchanger, 0.25 g; total volume, 12.5 cm^3; ionic strength, 0.1 M (HCl + KCl) or (KOH + KCl), temp., 30°C. (From M. Abe and T. Ito, *Nippon Kagaku Zasshi*, 86:1295 (1965). With permission.)

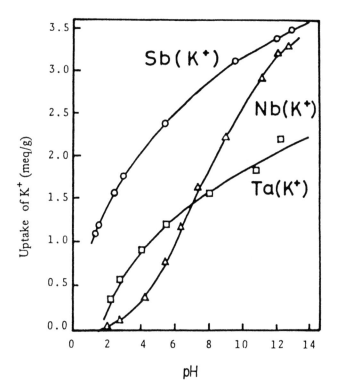

Figure 6 Uptake of K^+ by hydrous oxides of Nb, Ta, and Sb(V). Exchanger, 0.25 g; total volume, 12.5 cm^3; ionic strength, 0.1 M (HCl + KCl) or (KOH + KCl), temp., 30°C. (From M. Abe and T. Ito, *Nippon Kagaku Zasshi, 86*:1295 (1965). With permission.)

acid strength of these amorphous oxides is increased in the order; Sb > Ta > Nb (Fig. 6). The acid strength of crystalline hydrous oxides of pentavalent metals is found to be almost the same as that of amorphous oxides.

B. Classification of Acid Strength

Abe and Ito have reported the uptake of K^+ and Cl^- as a function of pH on various hydrous oxides [111]. When a neutral solution of potassium chloride is added to different hydrous oxides, the pH of the supernatant solution is either decreased by K^+ exchange or increased by Cl^- exchange. Amphoteric hydrous oxides affect a particular pH, depending on the nature of their oxides (Fig. 7) [106]. If we consider the polarizing power of a cation M^{n+} to be proportional to Z/r^2, where Z is its ionic charge and r

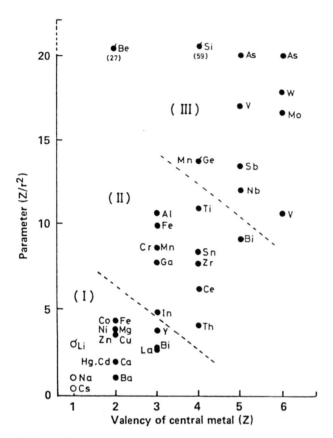

Figure 7 Correlation between the parameter (Z/r^2) and valency of central metal. Values of r are taken from Ref. 115, assuming that the metals have a co-ordination number (CN) of 6 (●) except Li, Be, Si, and Mn which have CN4 (◓). (From M. Abe, in *Inorganic Ion Exchange Materials* (A. Clearfield, ed.), CRC Press, Chap. 6, 1982. With permission.)

is the effective crystal ionic radius, various hydrous oxides of polyvalent metals can be classified as anion exchangers(I), amphoteric exchangers(II) and cation exchangers(III) by plotting Z/r^2 versus Z as shown in Fig. 7. The soluble hydroxides are also included in this figure.

The presence of additional oxygen attached to M increases the acid strength of the compound. Thus, in the series of M oxides the approximate order of increasing acid strength and increasing cation-exchange character for the insoluble hydrous oxides [111] can be represented as $MO < M_2O_3 < MO_2 < M_2O_5 < MO_3$.

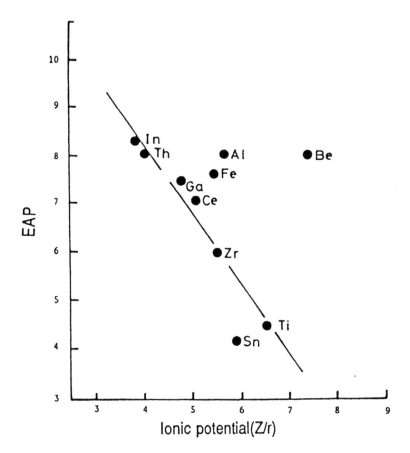

Figure 8 Correlation between ionic potential and equiadsorption points (EAP) of amphoteric hydrous oxides of tri- and quadrivalent metals. Values of r are taken from Ref. 115, EAP data from Ref. 111. (From M. Abe, in *Inorganic Ion Exchange Materials* (A. Clearfield, ed.), CRC Press, Chap. 6, 1982. With permission.)

The pH_{IAP} of the amphoteric hydrous oxides, when plotted against the effective ionic radii of central metals, decreases with decreasing ionic radii (Fig. 8) [106].

C. Change in Crystallinity and Crystal System

The dissociation properties and the resultant pH values obtained for inorganic ion exchangers are strongly influenced by crystallinities, crystal system, supporting electrolyte, ratio of solution to exchanger, ionic strength and the temperature selected. It is possible, for example, to obtain a measure of the degree of crystallinity of a particular exchanger through the controlled addition of an alkali metal ion to it. It has been known, as well, that amorphous antimonic(V) acid is gradually transformed into crystalline (cubic) material by aging in acidic solution. Novikov et al. have determined how the pH profile resolved during neutralization of a particular hydrous oxide respond to different degrees of aging (Fig. 9) [116]. It can be seen that the monomeric species of $H[Sb(OH)_6]$ exhibit strong-acid behavior when freshly precipitated. In the subsequent structure ordering to the crystalline state the solids retain their acidic properties. Crystallization does not alter the total quantity of hydrogen ion accessible to neutralization. It merely reduces the dissocia-

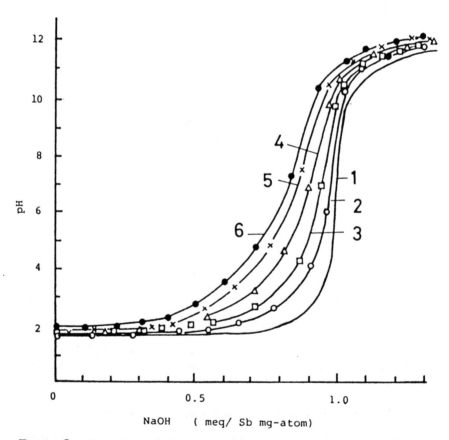

Figure 9 The pH titration curves of precipitated antimonic acid. 1, published titration curve for hydrogen hexahydroxo-antimonate $H[Sb(OH)_6]$; 2, freshly precipitated; 3, after 5 months; 4, after 10 months; 5, after 1.5 years; 6, after 2 years. (From B. G. Novikov, E. A. Mateova, and F. A. Belinskaya, *Zh. Neorg. Khim.,* *20*:1566 (1975). With permission.)

tion tendencies of the acid by its polymerization. Two crystal systems of antimonic acid, cubic [117] and monoclinic [118], have been shown to prevail. The pH affected during the exchange of alkali metal ion for hydrogen ion in the course of neutralization of monoclinic antimonic acid (M-SbA) with MOH has yielded a curve with a single rise in pH response to the addition of the base. This result has shown that M-SbA behaves like a strong monobasic acid exchanger (Fig. 10) [119]. The apparent capac-

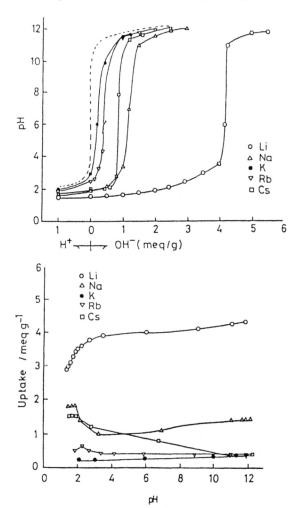

Figure 10 The pH titration curves (top) and dependence of alkali metal ion uptake on pH (bottom). Exchanger, 0.10 g; total volume, 10.0 cm³; temp., 30°C, ionic strength, 0.1 (MNO_3 + HNO_3) or (MNO_3 + MOH), M = Li, Na, K, Rb, and Cs. (From R. Chitrakar and M. Abe, *Solvent Extr. Ion Exch.*,7:721 (1989). With permission.)

ity found to be 4.3 meq/g for lithium ion was less than 0.5 meq/g for potassium ion. The lithium ion can enter the tunnel of the M-SbA crystal. Even though K^+ ion cannot, because of the ion sieve effect, when the potassium chloride solution is added to M-SbA the decreased pH is ascribed to the displacement of surface protons only for k^+.

Similar behavior has been observed with different modifications of Mn(IV) oxides. The cryptomelane type of hydrous manganese dioxide prepared by the reaction of Mn(II) and $KMnO_4$ in sulfuric acid solution exhibits an extremely high selectivity for potassium ions [120]. The spinel type of manganese oxide prepared by the introduction of lithium ion has exhibited an extremely high selectivity for lithium ion [121, 122]. These aspects are considered later.

D. Acid Strength of Insoluble Acid Salts of Polyvalent Metals

Acid salts with the general formula $M(H_nXO_y)_m nH_2O$ (where M = Ti, Zr, Th, Ce, Sn, UO_2^{2+} and X = Si, P, As, Sb, Mo, or W) can usually be prepared in either an amorphous or a layered crystalline state. Hydrous oxides of P, As and Sb are more acidic than those of the tetravalent metals [2, 3]. The acid strength of ion exchangers can be characterized by determining pH response to different metal cations during neutralization at constant ionic strength and a specified temperature. The acid strength of these acid salts increase with increasing amounts of the second acid group. For example, the pH response of tin(IV) antimonates during their neutralization shows that the acid strength increases with increasing Sb/Sn ratios (Fig. 11) [123].

The structure of α-zirconium phosphate (α-ZrP) has been well established by Clearfield and Smith [124]. Figure 12 shows the idealized structure of α-ZrP with one of the zeolitic cavities created by the arrangement of the layers [125]. The free space available to the sides of the layered structure is sufficiently large to allow a spherical ion of 0.263-nm radius to diffuse into the cavity without obstruction [126]. The pH response to metal ion during neutralization of α-ZrP with standard base depends sizably on the degree of crystallinity and the kind of metal ions used (Fig. 13). Amorphous ZrP yields a pH curve with a single slope, whereas with α-ZrP, a sharp change in the slope occurs when about half the protons have been exchanged with Li^+, Na^+, and K^+.

$$Zr(HPO_4) \cdot H_2O(0.76 \text{ nm}) + Na^+$$
$$\rightarrow ZrNaH(PO_4) \cdot 5H_2O(1.18 \text{ nm}) + H^+ \qquad (17)$$

Sodium hydroxide is required to neutralize the acid and allow the ion-exchange reaction to reach completion. There are three phases present

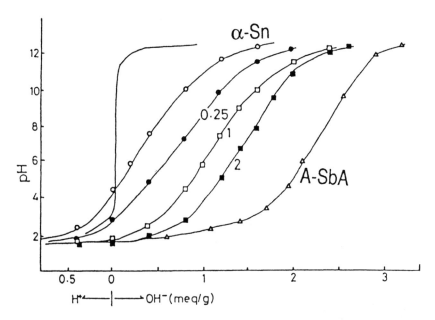

Figure 11 The pH response of tin(IV)antimonate with various compositions during their neutralization with base. α-Sn; α-hydrous tin(IV) oxide, A-SbA; amorphous antimonic(V) acid, numerical number; mol ratio of Sb/Sn, exchanger, 0.25 g; total volume, 12.5 cm^{-3}; temp., 25°C; ionic strength, 0.1 M (KOH + KCl). (From M. Abe and T. Ito, *Kogyo Kagaku Zasshi, 79*:440 (1967). With permission.)

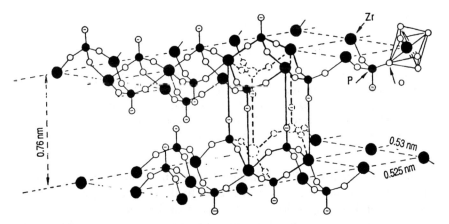

Figure 12 Idealized structure of α-zirconium phosphate (α-ZrP) showing one of the zeolitic cavities created by the arrangement of the layers. (From G. Alberti, in *Study Week on Membranes* (R. Passino, ed.), Pontificiae Acad. Sci. Scripta, Varia, Rome, p. 629, 1976. With permission.)

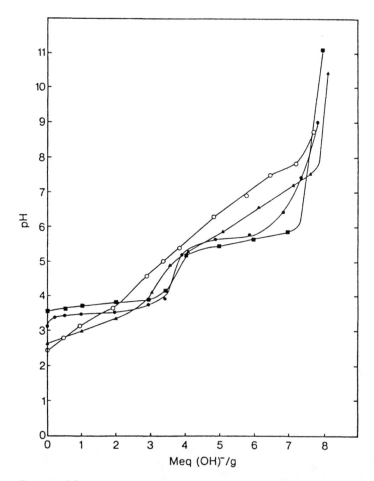

Figure 13 The pH response of α-ZrP of low and intermediate crystallinities during neutralization with standard base. Titrant, 0.1 M NaOh + 0.1 M NaCl; sample designations are 0.8; 48 (0.8 M H_3PO_4; 48-hr reflux time) (○) 2.5; 48 (▲); 4.5; 48 (●), 12; 48 (■). (From A. Clearfield, Å. Oskarsson, and C. Oskarsson, *Ion Exch. Membr., 1:9* (1972). With permission.)

along the plateau; α-ZrP, the half-exchange phase, and the aqueous solution. Three components are necessary to describe the system: the sodium ion and hydrogen ion concentration of the solution and the total amount of the exchanger. Thus, according to the phase rule, the system is invariant at constant temperature and pressure, and the pH titration curves should have zero slopes [127]. The second plateau has a zero slope as well since the reaction is

$$ZrNaH(PO_4) \cdot 5H_2O + Na^+ \rightleftharpoons Zr(NaPO_4) \cdot 3H_2O(0.98 \text{ nm}) + H^+ \quad (18)$$

X-ray study of the ion-exchanged solid shows that partial solid solution occurs, the composition range increasing with decreasing crystallinity. The pH curve resolved for amorphous ZrP, however, exhibits a positive slope to show that the zero degrees of freedom restriction no longer applies to this system.

It should also be noted that the pH curves obtained by reverse reaction with standard acid are displaced from foreward reaction at low sodium content. This is due to the formation of a phase whose composition corresponds to $Zr(HPO_4) \cdot 6H_2O(1.04 \text{ nm})$; it has been referred to as θ-ZrP. No ion-exchange reaction occurs in the Rb^+H^+ and Cs^+/H^+ systems on α-Zrp [128].

E. Acid Strength of Aluminosilicates

Zeolites and clay minerals belong to the group of silicious ion exchangers. Zeolites can be defined as crystalline hydrated aluminosilicates of Group I and II metals (alkali and alkaline earth metals) of the periodic table. Most of the natural ion-exchange minerals are crystalline aluminosilicates with cation-exchange properties. The three-dimensional framework structure of zeolites provide a relatively open window to channels and/or interconnecting cavities affected by tetrahedral SiO_4 and AlO_4 units cross-linked by the sharing of oxygen atoms. Since Al is trivalent, the lattice carries a negative electric charge (one elementary charge per aluminum atom). This charge is balanced by alkali or alkaline earth cations which do not occupy fixed positions but are contained by the solvent associated with the cavities and channels. In contrast to zeolites, clay minerals are layered silicates. They carry their counterions in between the layers of the lattice. They exit in different structural varieties and are primarily cation exchangers. Most zeolites and clay minerals usually are unstable in acidic media and exhibit cation-exchange properties in neutral and weak-base media. A few clay minerals such as kaolinites can simultaneously exchange anions.

The pK_a value of clay minerals usually lies between 8 and 10 [129].* This implies that the protons of the SiOH groups contribute to cation exchange only at very high pH values, thus being responsible for a pH-dependent portion of the cation-exchange capacity. The number of nega-

---

*Editor's note: When careful attention is paid to the effect of counterion concentration levels of the solution phase on the pH measured, the intrinsic pK_a value of the hydroxylated silica, $Si(OH)_4$, repeated in the clay mineral phase can be expected to correspond to the literature value of 9.3 to 9.4 reported for silicic acid.

tively charged sites in the layered structure is related to the capacity. Crystalline silicic acids associated with minerals in their H^+ ion form contain protonated Si-OH groups and the pK_a values measured for this species have ranged from 5.0 to 3.3 [130].[*]

Zeolites with a high silica content have been studied intensively because of their high thermal stability and important catalytic properties when exposed to severe operating conditions. They can be prepared by removing the aluminum (dealumination) from original zeolites, their direct synthesis being very difficult. The dealumination of the zeolites is usually carried out with the following two methods. The starting material is prepared by slow extraction of the Al with e.d.t.a. as described by Kerr [131]. The reaction with $SiCl_4$ for this purpose can be formally described as [132]

$$Na_x(AlO_2)_x(SiO_2)_y + SiCl_4$$
$$\rightarrow NaCl + AlCl_3 + Na_{x-1}(AlO_2)_{x-1}(SiO_2)_{y+1} \qquad (19)$$

The acid strength of the zeolite is known to be increased by increasing the Si/Al ratio. The total negative charge density of the framework is reduced in the dealuminated zeolites, but the effective charge of an individual site is strengthened. These effects may change the cation distribution among the available sites. The stability in acid media and the acid strength are also increased by removing aluminum from the aluminosilicate structure. The cation-exchange capacity is thus inversely proportional to the silicon/aluminum ratio.

IV. ION-EXCHANGE PROPERTIES OF INORGANIC ION EXCHANGERS

The solvent content of organic ion exchangers depends on the exchanging ion. Inorganic ion exchangers with their much more rigid three-dimensional frameworks undergo much less shrinkage or expansion in parallel situations. These inorganic ion exchangers will show different properties from organic ion-exchange resins by experiencing ion sieve and steric effects.

A. Ion Sieve Effect

In order to replace protons in the acid form of inorganic ion exchangers, the cation present in the external solution must diffuse through the win-

---

[*]Editor's note: Once again this variability arises from neglect of the Donnan potential term.

dows connecting the cavities. The water molecules of the hydrated ions are exchanged frequently with bulk water molecules in the solution. When the size of the window is smaller than the diameter of the hydrated counterions, a part or all of the water molecules of their hydration shell must be lost to allow the cation to pass through the window. If the cations can pass the inorganic ion exchanger pores only after having lost water molecules coordinated to them in solution, the distinct kinetic effect observed varies with the hydration energies of the various ions. If the counterions have larger crystal ionic radii than this opening of the window, ion sieve effects can prevail.

Most zeolites have three-dimensional structures with cavities connected to channels that exhibit ion sieve properties. If the pore volume of the cavity is large enough, the hydration stripped cations exchanged can be rehydrated in the cavity to simulate their hydration in the solution. The large cations (Rb^+, Cs^+, organic cations) cannot enter the channels or windows of the cavities of zeolites with small pores. Table 3 lists the pore openings of some hydrated zeolites [133].

The ions Li^+, Na^+, K^+ and Ag^+ diffuse readily into the zeolite structure of sodalite while Ti^+, Rb^+ and Cs^+ diffuse with great difficulty. This result is consistent with the 0.26-nm diameter accessibility of the channels in sodalite [134]. The Ca^{2+}/Na-A zeolite exchange is often compared with

Table 3 Pore Openings of Hydrated Zeolites

Zeolite	Pore opening (nm)
A	0.42
Analcite	0.26
Chabazite	0.37 × 0.42
Clinoptilolite	0.40 × 0.55
	0.41 × 0.47
Erionite	0.36 × 0.52
Linde L	0.71
Linde W	0.42 × 0.44
Mordenite	0.67 × 0.70
	0.29 × 0.57
Offretite	0.36 × 0.52
Philipsite	0.42 × 0.44
	0.28 × 0.48
X	0.74
Y	0.74

Source: From Ref. 133.

the Mg^{2+}/Na-A zeolite exchange, even though the ionic crystal radius of Ca^{2+} ion is larger than that of the Mg^{2+} ion. Because of the higher charge density of Mg^{2+}, the water ligands are bound more strongly to Mg^{2+} than Ca^{2+}. The influence of hydration is especially noticeable for La^{3+}. At room temperature no exchange of La^{3+} ion for Na^{2+} ion is observed. This ion-exchange reaction may occur at elevated temperatures because the hydration sphere is less tightly bound, and the ligand water molecules are exchanged more frequently than at room temperature [135]. The ion sieve effect can be used to evaluate the window size of structurally uncharacterized inorganic ion exchangers. A series of cations differing in size is tested in exchange experiments for this purpose. The van der Waals dimension for alkyl ammonium cations available over a wide range of values [136] has been of use to such a program. When cations are smaller than 0.6 nm in two directions, they can pass through the window of the C-SbA (Table 4) structure without any resistance and exchange with the cations initially present. For the cations extending less than 0.6 nm in only one direction, the cation may pass its window with some distortion. However, a cation larger than 0.6 nm in all three directions cannot pass the window, even if the cavity has enough room for the large cations. Cations such as $(CH_3)_4N^+$ and $(C_2H_5)_4N^+$ can only be adsorbed on the surface of the C-SbA exchanger and its window size, as a consequence can be estimated to be about 0.6 nm [104,137,138].

Table 4 The van der Waals Dimensions of Organic Ions and Their Uptake on C-SbA

Ion	x direction length (nm)	y direction width (nm)	z direction height (nm)	Uptake value (meq g^{-1})
NH_4^+	0.286	0.286	0.286	1.35
$CH_3NH_3^+$	0.49_1	0.40_0	0.40_0	0.54
$C_2H_5NH_3^+$	0.59_0	0.48_8	0.40_0	0.31
$(CH_3)_2NH_2^+$	0.64_2	0.42_8	0.40_0	0.18
$n\text{-}C_3H_7NH_3^+$	0.72_6	0.48_8	0.40_0	—
$n\text{-}C_4H_9NH_3^+$	0.84_2	0.48_8	0.40_0	0.10
$iso\text{-}C_3H_7NH_3^+$	0.65_2	$.055_6$	0.47_7	0.20
$tert\text{-}C_4H_9NH_3^+$	0.65_2	0.61_8	0.56_8	—
$(CH_3)_3NH^+$	0.64_2	0.61_0	0.41_7	0.10
$(CH_3)_4N^+$	0.64_2	0.61_0	0.62_2	0.04
$(C_3H_5)_4N^+$	—	—	—	0.04

Source: From Ref. 137.

The α-zirconium phosphate has a layered structure in which each layer consists of a plane of zirconium atoms coordinated octahedrally to oxygen. The free space in the sides of the layered structure is large enough to allow a spherical ion of 0.263 nm diameter to diffuse the cavity without any obstruction and is accessible to Li^+, Na^+ and K^+ [139]. However, the size of the window is smaller than the ionic spheres of Rb (0.296 nm) and Cs (0.338 nm) and an ion sieve effect is encountered with these ions.

Acid salts of Group IV phosphates and arsenates have layered structures with interlayer distances large enough to accommodate a number of ions. The size of the opening to the cavities of the acid salts with the same layered structure is dependent upon the particular composition. Table 5 lists the correlation between ion sieve effects and the interlayer distance of various acid salts [140].

When a small amount of Na^+ is added to solution used to examine the exchange of Cs^+ and H^+-ion with an α-zirconium phosphate ion exchanger, the Na^+-ion exchanged can expand interlayer distance. The complete exchange of H^+ for Cs^+ that may occur in the exchanger can be attributed to catalysis by the Na^+ ion. When tertiary ammonium ions are introduced in the layer of the α-zirconium phosphate ion exchanger, the interlayer distance expansion that results is proportional to the length of the carbon chain of the amine.

Table 5 Ion Sieve Effect for Alkali Metal Exhibited by Layered Acid Salts

Exchanger	$d^*(nm)^a$	Exchangable Ions
$Zr(HPO_4)_2 \cdot 2H_2O$ (γ-ZrP)	1.220	Na-Cs
$Th(HPO_4)_2 \cdot 3H_2O$	1.147	Na-Cs
$Ce(HPO_4)_2 \cdot H_2O$	1.095	Na-Cs
$Zr(HPO_4)_2$ (β-ZrP)	0.940	Na-Cs
$Ce(HAsO_4)_2 \cdot H_2O$	0.910	Li, Na, K (partly)
$Zr(HAsO_4)_2 \cdot H_2O$	0.780	Li, Na, K, Cs (partly)
$Sn(HAsO_4)_2 \cdot H_2O$	0.777	Li, Na, K and Cs (partly)
$Ti(HAsO_4)_2 \cdot 2H_2O$	0.777	Li, Na, K
$Sn(HPO_4)_2 \cdot H_2O$	0.776	Li, Na, K, Cs (partly)
$Zr(HPO_4)_2 \cdot H_2O$ (α-ZrP)	0.760	Li, Na, K
$Ti(HPO_4)_2 \cdot H_2O$	0.756	Li, Na
$Th(HAsO_4)_2 \cdot H_2O$	0.705	Li

$^a d^*$ is interlayer distance.
Source: From Ref. 140.

B. Ion-Exchange Isotherms

At a given temperature, the ion-exchange reaction affected at equilibrium on a particular exchanger can be characterized by the isotherm. The results of selectivity measurements in ion-exchange investigations are usually presented as isotherms; the mole or equivalent fraction of one of the competing ions in the exchanger phase is plotted against its mole or equivalent fraction in solution at a given temperature. Such a graphical representation is illustrated in Fig. 14 (left). In the case where complete exchange is not obtained a number of investigators have normalized the isotherms. This normalization procedure involves setting the maximum limit of exchange, $(\overline{X}_M)_{max}$, equal to unity and then multiplying each point of the exchange isotherm by the normalization function, $f_N = 1(\overline{X}_M)_{max}$. However, such treatment of the thermodynamic data has been questioned in recent publications [135] and their presentation in this manner is controversial. The results obtained do not provide a meaningful comparison between ions which are exchanged to different extents. The standard procedure needs to be based on a theoretical capacity made accessible through empirical formulas for the exchangers. The exchange isotherm has been classified by five different types [141] of behavior (curves in Fig. 14a–e for the uni-univalent exchange reaction) [138].

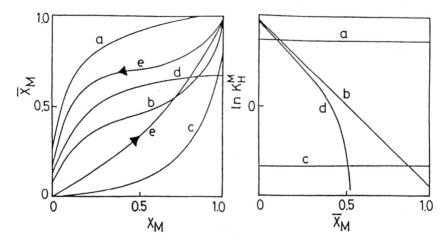

Figure 14 The ion-exchange isotherms for uni-univalent ion-exchange reactions (left) and the ln K_H^M dependence (right) of various types of reaction on $\overline{X}_{M^+}$ (a) Langmuir type, (b) selectivity reversal, (c) anti-Langmuir type, (d) incomplete exchange, (e) hysteresis loop. (From M. Abe, in *Ion Exchange Processes: Advances and Applications* (A. Dyer, M. J. Hudson, and P. A. Williams, eds.), The Royal Society of Chemistry, Science Park, Cambridge, UK, p. 199, 1993. With permission.)

Each curve can be classified into the following five types of ion-exchange behavior:

(a) Langmuir type; higher selectivity for the entering cation over the entire range of exchanger composition.
(b) Selectivity reversal; entering cations show a selectivity reversal with increasing equivalent fraction in the exchanger.
(c) Anti-Langmuir type; the displaced cations are selectively held over the entire exchanger composition range.
(d) Incomplete exchange; exchange of ions does not go to completion although the entering cation is initially preferred, and the degree of exchange reaches a limiting value lower than unity.
(e) Hysteresis loop; hysteresis effects may result from the formation of two exchanger phases.

It is often difficult to make valid comparisons between data from different laboratories. The ion exchange isotherms obtained with inorganic ion exchangers are altered by various factors: origin of the product, method of preparation, determination of $\overline{X}_M = 1$, solution media, ionic strength, temperature, etc.

1. Various Types of Ion-Exchange Isotherms

The various types of ion-exchange isotherms are summarized in Tables 6.1–6.3. Through their examination the isotherm type developed with various ion exchangers for a particular ion-exchange reaction, e.g.,

$$n\overline{Na^+} + M^{n+} \Leftrightarrow \overline{M^{n+}} + nNa^+ \tag{20}$$

can be determined. This reaction is expressed as M/Na in Table 6.

a. Hydrous Oxides The research that has been published up to 1994 for hydrous metal oxides is very limited. The types of isotherms resolved are listed in Table 6.1. Systematic studies of cubic antimonic acid (C-SbA) have been performed by a research group of Abe [106].

b. Acid Salts Ion-exchange behavior of α-ZrP is determined by two features of the structure, the size of the opening into the cavities, and the weak forces holding the layers together. As mentioned earlier, the largest opening in the cavity will allow a spherical cation with a diameter of 0.263 nm to diffuse unobstructed into the cavity. Thus, Li^+, Na^+, and K^+ can exchange in acid solution, while Rb^+ and Cs^+ cannot [149]. During neutralization with NaOH solution, the interlayer spacing of α-ZrP increases from 0.756 to 1.18 nm in the half-exchanged phase. The half exchanged α-ZrP can then exchange with cesium ions [150,151]. The complete replacement of sodium ions by hydrogen ions, upon reversing the above neutralization reaction yields θ-ZrP, a highly hydrated phase. In

Table 6.1 Type of Isotherm for the Various Hydrous Metal Oxides

Hydrous oxides	Exchange cations	References (type of isotherm*)
Almina[a]	I/NO_3	142(b)
	SO_4/NO_3	142(a)
	TcO_4/NO_3	142(c)
	NO_3/TcO_4	142(e)irreversible
Ferrox[b]	I/NO_3	142(b)
	SO_4/NO_3	142(c)
	TcO_4/NO_3	142(c)
	NO_3/TcO_4	142(c)irreversible
Zirox[b]	I/NO_3	142(b)
	SO_4/NO_3	142(a)
	TcO_4/NO_3	142(c)
	NO_3/TcO_4	142(c) partial irreversible
Zirconia	Li/K	143(a)
	Na/K	143(a)
	Cl/NO_3	144(b), 145
	SO_4/NO_3	145
Cubic antimonic acid	Li/H	39(b)
	Na/H	39(c), 40(c), 146(c)
	K/H	39(c)
	Rb/H	39(c)
	Cs/H	39(c), 45(c)
	Mg/H	147(b), 146(b)
	Ca/H	147(c), 146(c)
	Sr/H	147(c), 146(c)
	Ba/H	147(c), 148(c)
	Mn/Ni	44(c)
	Ni/H	44(b)
	Co/H	44(b)
	Zn/H	44(b)
	Cu/H	44(b)
	Cd/H	44(a-$CdOH^+$ exchange)
	Pb/H	44(a-$PbOH^+$ exchange)
	NH_4/H	137(c)
	$C_nH_mNH_x/H$	137(b)

[a]Supplied as 63–150 μm granules by Fision, Loughborough, UK.
[b]Supplied as 0.1–0.5 mm granules by Recherche Applique du Nord, Hautmont, France.
*Type of isotherm refers to Fig. 14.

the back-titration reaction, the fully exchanged phase of $Zr(NaPO_4)_2 \cdot 3H_2O$ is converted back to the half-exchanged phase in what appears to be a reversible exchange reaction. However, the replacement of sodium ions by protons occurs at a lower pH than the forward reaction. Thus, the forward and reverse reactions do not follow the same path. Torracca has determined the isotherms for the Na^+/K^+. Li^+/Na^+ and Li^+/K^+ exchange reactions. In each case the forward and reverse paths do not coincide exhibiting appreciable differences [152].

Similar tendencies are observed for the crystalline acid salts with layered structures. These types of ion-exchange isotherms are summarized in Table 6.2

Table 6.2 Type of Isotherm Developed by the Various Acid Salts

Type of acid salts	Exchange cations	References (type of isotherm*)
Amorphous ZrP	Li/H	153(c)
	Na/H	153(c)
	K/H	153(c)
	Rb/H	153(c)
	Cs/H	153(c)
	Cs/K	154(a)
	Cs/Rb	154(a)
	Rb/H	154(c)
heated (25-100°C)	UO_2^{2+}/H	155(b)
(175,250°C)	UO_2^{2+}/H	155(a)
Semicryst. ZrP		
(25-100°C)	UO_2^{2+}/H	155(b)
(175,250°C)	UO_2^{2+}/H	155(a)
Cryst. Zrp (50-250°C)	UO_2^{2+}/H	155(b)
α-Zrp	K/Li	156(e)
	K/Na	157(e), 158(e)
Fibrous CeP	Pb/H	159(c)
Tin antimonate	Li/H	92(c)
	Na/H	92(c)
	K/H	92(c)
	Rb/H	92(c)
	Cs/H	92(c)
Titanium antimonate	Li/H	160(c)
	Na/H	160(c)
γ-ZrPPa	Ba/H	161(a)

c. Zeolites Lack of knowledge with respect to the ion-exchange properties of zeolites as well as their limited availability was responsible for the delay in their wide use in industry until the 1960s. During the last few years, however, considerable ion-exchange equilibria data have been obtained in industrial laboratories for more than 100 different zeolite types. Unfortunately, only few of the results obtained have been published. These publications cover mostly the investigations of the commercially important zeolite A, X, Y, Z, chabazite, clinoptilolite and mordenite. The types of isotherms obtained with these zeolites are summarized in Table 3.

Table 3   Types of Exchange Isotherms Resolved in Different Zeolites for Different Ion-Exchange Reactions

Type of zeolite	Exchange cations	References (type of isotherm)
A	Li/Na	162(b), 163(b)
	K/Na	162(c), 163(c)
	Rb/Na	162, 163(c)
	Cs/Na	162(d)
	NH_4/Na	163(c)
	Ag/Na	164(a, two sites), 165(a)
	Tl/Na	165(a)
	Mg/Na	166(d), 167(c), 168, 171(c)
	Ca/Na	162(a), 165(c), 166(a), 169(c)
	Ba/Na	165(d)
	Mn/Na	168(c)
	Ce/Na	170
	NH_4/K	163(a)
	Cs/Rb	163(c)
	Rb/NH_4	163(c)
	NH_4/Cs	163(c)
Analcite	K/Na	172(e)
	Ag/Na	172(a)
	Tl/Na	172(d)
	Ca/Na	173(b)
Chabazite	Li/Na	174(b)
	K/Na	174(a)
	Cs/Na	170
	NH_4/Na	174(a), 175
	Ca/Na	174(c)
	Sr/Na	174(c)
	Pb/Na	174(c)
	Sr/Cs	170

Table 3 Continued

Type of zeolite	Exchange cations	References (type of isotherm)
Clinoptilolite	Cs/Na	170
	NH_4/Na	176, 177(b)
	Na/NH_4	136(a)
	Ca/Na	170
	Sr/Na	170
	Cs/K	170
	NH_4/K	170
	Cs/Rb	170
	Cs/NH_4	170
	Cs/Ca	170
	Mg/NH_4	176
	Ca/NH_4	176
	Sr/Cs	170
	Sr/Ca	170, 178(c)
Erionite	Li/Na	179(b)
	K/Na	179(a)
	Rb/Na	179(a)
	Cs/Na	179(a)
	Ca/Na	178(b), 179(d)
	Sr/Na	178(b), 179(d)
	Ba/Na	179(d)
	Sr/Ca	178(c)
Hector clinoptilolite	Sr/Na	178(c)
Linde L	K/Na	180(b)
	NH_4/Na	180(b)
	Cs/K	180(b)
	Ba/K	180(b)
Linde W	NH_4/Na	181
Mordenite	Na/H	182
	Li/Na	182, 183(b)
	K/Na	182
	Cs/Na	170, 182
	Ag/Na	182
	Mg/Na	182
	Ca/Na	170, 182
	Sr/Na	170, 182
	Ba/Na	182(b)
	Mn/Na	182
	Co/Na	182, 183(d)
	Ni/Na	182
	La/Na	184

(continued)

Table 3 Continued

Type of zeolite	Exchange cations	References (type of isotherm)
	Cs/K	170
	Cs/Rb	170
	Cs/NH$_4$	170
	Cs/Ca	170
	Sr/Cs	170
	Sr/Ca Mordenite	170
	Offr./Erionite, K/Na	185
	Ca/Na	185
	K/NH$_4$	185
Phillipsite	Na/Li	177(a)
	K/Na	177(b), 186
	Rb/Na	177(c)
	Cs/Na	177(a)
	Ca/Na	177(b), 178(b)
	Ba/Na	177(a)
	Sr/Ca	178(c)
Stilbite	K/Na	187
	Cs/Na	187
	Cs/K	187
	Sr/Cs	187
X	Li/Na	188(b), 189
	K/Na	166(c), 188(c), 189
	Rb/Na	188(d), 189
	Cs/Na	188(d), 189
	Ag/Na	190
	Tl/Na	189, 190(d)
	Mg/Na	191(d)
	Ca/Na	188(a), 192(d)
	Sr/Na	188(c), 192(c), 193(a)
	Co/Na	194(c)
	Ni/Na	194(d)
	Cu/Na	194(d & c)
	Zn/Na	194(d & c)
	Y/Na	193(c)
	Ce/Na	193(c), 189
	La/Na	189
	Ce/Sr	193(c)
Y	Li/Na	189
	K/Na	189
	Rb/Na	189
	Cs/Na	189, 195
	Tl/Na	189
	NH$_4$/Na	189

Table 3 Continued

Type of zeolite	Exchange cations	References (type of isotherm*)
	Ca/Na	189, 192(d)
	Sr/Na	192(d), 195
	Ba/Na	192(d), 196(d)
	Mn/Na	191(d)
	Co/Na	194(c)
	Ni/Na	194(d)
	Cu/Na	191(d), 194(d & c)
	Zn/Na	191(d), 194(d & c)
	Cd/Na	191(d), 197
	Pb/Na	172($PbOH^+$ exchange)
	Y/Na	191(d)
	Ce/Na	195
	La/Na	189
	NH_4/K	198(a)
	Co/K	197
	Ni/K	197
	Zn/K	197
	Cd/K	197
	Rb/NH_4	198(a)
	NH_4/Cs	198(a)
	Li/NH_4	198(b)
	Ca/NH_4	198(c)
	Sr/NH_4	198(a)
	Ba/NH_4	198(a)
	Mn/NH_4	198(c)
Z	Li/Na	199(e)
	K/Na	199(e)

2. Determination of Thermodynamic Ion-Exchange Equilibrium Constants

The ion-exchange isotherms resolved at various laboratories differ because of differences in the definition of the total exchange capacity that is employed (see Sec. IV.B). The contribution of the ratio of the activity coefficients in the solution can be neglected for uni-univalent exchange reaction on inorganic ion-exchangers. The neglect of the temperature dependence of $\gamma_{H^+}/\gamma_{M^+}$ does not give a serious deviation in the values of thermodynamic equilibrium constant in dilute solution and a narrow temperature range. However, the values of the activity coefficients cannot be neglected in the uni-multivalent exchange reaction for the compilation of selectivity coefficients. The values of ionic activity coefficient ratios

$(\gamma_{H^+})^n/(\gamma_{M^{n+}})$ in solution can be calculated with the following equation [200]:

$$\log\frac{(\gamma_{H^+})^n}{\gamma_{M^{n+}}} = \log\frac{(\gamma_{\pm HNO_3})^{2+n}}{(\gamma_{\pm M(NO_3)_n})^{n+1}} = \frac{2S\sqrt{I}}{1+1.5\sqrt{I}} \tag{21}$$

where $S = 1.8252 \times 10^6 \sqrt{\rho/\varepsilon^3 T^3}$

and ρ and ε are the density and dielectric constant of water, respectively. The I is the ionic strength, and T is the absolute temperature of the system.

The thermodynamic equilibrium constant can be evaluated by using the simplified form of the Gaines-Thomas equation [201], assuming that the change in water content in the exchanger is negligible. The equation, with K_H^M defined as the corrected selectivity coefficient by Eq. (4), follows:

$$\ln K = -(n_M - n_H) + \int_0^1 \ln K_H^M \, d\overline{X}_{M^{N+}} \tag{22}$$

When the plot of $\ln K_H^M$ versus $\overline{X}_{M^{n+}}$, known as the Kielland plot, is linear, $\ln K_H^M$ is given by

$$\ln K_H^M = 4.606C\overline{X}_{M^{n+}} + (\ln K_H^M)_{\overline{X}_{M^{n+}} \to 0} \tag{23}$$

where C is the Kielland coefficient [202] and $(\ln K_H^M)_{\overline{X}_{M^{n+}} \to 0}$ is the value of $\ln K_H^M$ obtained by extrapolation of $\overline{X}_{M^{n+}}$ to zero. The slope of the line in the Kielland plot is 4.606C. Kielland plots corresponding to ion-exchange isotherms are illustrated in Fig. 14 (right). When small cations in the exchanger are exchanged with large cations, ion exchange becomes progressively more difficult with increasing amounts of the large cations in the solid phase because of the steric effect. The ion-exchange isotherms and Kielland plots thus are of the type (b) or (d), respectively, in Fig. 14. Such behavior arises in the rigid structure of an inorganic ion exchanger which undergoes relatively little swelling when there is a large difference in the size of the two exchanging cations. The change in the free energy ($\Delta G°$) of the ion-exchange reaction, calculated per cation equivalent, is as follows:

$$\Delta G° = -(RT/n_{M^{n+}}n_{H^+})\ln K \tag{24}$$

The equilibrium constant can also be determined through studies of the dissociation of the exchanger in its H^+-form during neutralization with standard base (pH titration) by the method of Argersinger et al. [203]. However, numerical values of thermodynamic functions include the difference in hydration between the two cations in the solid and solution phases and the change in water activity cannot be ignored. The changes in enthalpy ($\Delta H°$) and entropy ($\Delta S°$) are calculated in the same way as they are for standard thermodynamic reactions; $\Delta G° = \Delta H° - T \Delta S°$

3. Steric Effect

If there is not enough available space for ingoing large ions within the cavities in the exchangers, their exchange becomes increasingly difficult in the course of the ion-exchange reaction. The absolute values of the Kielland coefficient, $|C|$, indicate the extent of the steric effect and increase generally with increasing difference in the crystal ionic radii of the two exchanging cations. Incomplete exchange can also arise with large ions from lack of interstitial space even when no ion sieve effect occurs. When the Kielland plot is linear site homogeneity characterizes in the exchangers.

Alkyl ammonium ions are readily available in a wide range of different diameters. Table 6 and Fig. 15 are presented next to provide typical

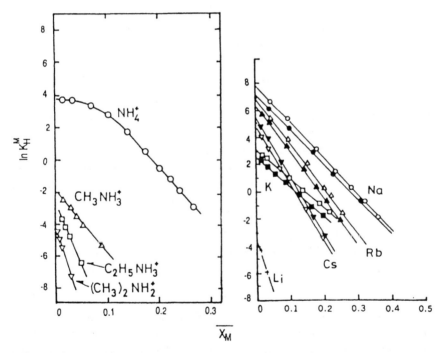

Figure 15 Kielland plots for NH_4^+/H^+ and R_4N^+/H^+ systems (left) and alkali metal ions/H^+ (right) on cubic antimonic acid (C-SbA). Exchanger, 0.10 g; total volume, 10.0 cm^3; temp., 30°C; ionic strength, 0.1 ($MNO_3 + HNO_3$) or (MNO_3 + MOH), M = Li, Na, K, Rb, and Cs. (Left, from M. Abe, K. Yoshigasaki, and T. Sugiura, *J. Inorg. Nucl. Chem.*, 42:1753 (1980); right, M. Abe, *J. Inorg. Nucl. Chem.*, 41:85 (1979) With permission.)

examples of ion-exchange isotherms and Kielland plots obtained for the NH_4^+/H^+, the R_4N^+/H^+, and the alkali metal ion systems equilibrated with cubic antimonic acid (C-SbA) [137]. The ions having a size smaller than 0.6 nm for one or two dimensions can pass through the window of the cavity. The Kielland plot, quite linear for most of the systems studied, are not for the NH_4^+/H^+ system. With the NH_4^+/H^+ system, the slope of the Kielland plot is small initially and then becomes steeper with increase in the equivalent fraction of NH_4^+ in the solution phase. This behavior indicates that the exchange reaction does not go to completion even though the entering cation is initially preferred. The absolute values of Kielland coefficients increase in the order $NH_4^+ < CH_3NH_3^+ < C_2H_5NH_3^+ < (CH_3)_2NH_2^+$, paralleling the increasing order of their van der Waals dimensions. Uptake of $(CH_3)_4N^+$ and $(C_2H_5)_4N^+$ is very small, indicating that adsorption may occur only on the surface of the exchanger because of the ion sieve effect. A similar sequence is found for the organic cation/H^+ systems on clinoptilorite [136].

The selectivity sequence depends very much on the effect of steric factors as the reaction proceeds. For the alkali metal ions/H^+ system on the C-SbA, the selectivity sequences is a function of $\overline{X}_M$. The order of increase is $Li^+ < K^+ < Cs^+ < Rb^+ < Na^+$ for $\overline{X}_M = 0.0-0.1$, $Li^+ < Cs^+ < K^+ < Rb^+ < Na^+$ for $\overline{X}_M = 0.1-0.32$, and $Li^+ < Cs^+ < Rb^+ < K^+ < Na^+$ for $\overline{X}_M = 0.32-1.0$ (Fig. 15 (right)). The high selectivity coefficient for Na^+ is attributable to highly negative values of $(\Delta H°)$ and minimal contributions of $(\Delta S°)$ at trace level concentrations of sodium ion [39].

Amorphous zirconium phosphate also exhibits selectivity dependence on loading for the alkali metal ions/H^+ system. The order $Li^+ < Na^+ < K^+ < Rb^+ < Cs^+$ for $\overline{X}_M = 0.0-0.19$ is reversed to $Cs^+ < Rb^+ < K^+ < Na^+ < Li^+$ when $\overline{X}_M$ reaches 0.62. Seven additional selectivity sequences were found for $\overline{X}_M = 0.19-0.62$ (Fig. 16) [153].

The magnitude of the Kielland coefficient depends not only on the size of the ionic radii, but also on the size of the hydrated ions. The free energy of the ion-exchange reaction can be represented by

$$\Delta G = (\Delta G_e^B - \Delta G_e^A) - (\Delta G_h^B - \Delta G_h^A) \tag{25}$$

The first term on the right is attributable to the difference in G_e, resulting from the electrostatic interaction of two cations with the electrastatic field of the negative ion-exchanger site. The second term represents the difference in the hydration energy, G_h, of the two cations. Small cations having a high charge density are associated with larger hydration shells. These hydrated cations are subject to sizable steric hindrance if the hydration energy of the ions is larger than the energy of removal of a part

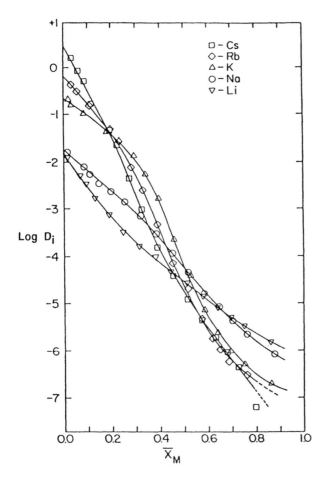

Figure 16 log K_c as a function of the equivalent fraction of alkali metal cation in the Am-ZrP exchanger 0.5; 48 (the same designation as Fig. 13). (From L. Kullberg and A. Clearfield, *J. Phys. Chem.*, *85*:1578 (1981). With permission.)

or all of the water molecules bound to the ions by electrostatic interaction. For example, the water molecules of the ion hydrates are bonded more strongly to Li^+ than to Na^+ on C-SbA. As a consequence the hydrated Li^+ undergoes more steric interference than the hydrated Na^+ ion. Similar steric interference tendencies are observed for the Mg^{2+} and Al^{3+}/H^+ exchange systems. Thus, minimum steric effects and maximum selectivity coefficients are observed for the exchanging ions having effective ionic radii of about 0.10 nm (Fig. 17) [138].

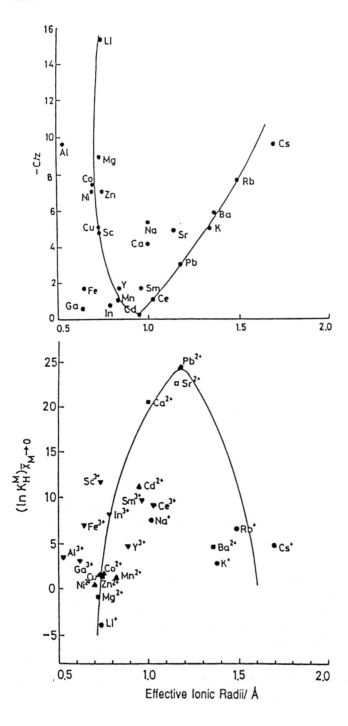

As shown by Sherry and Walton, the ion-exchange reaction can be separated into the following two reactions [165]:

$$H^+(gas) + M^+(aq) \Leftrightarrow H^+(aq) + M^+(gas) - \Delta Y^{\circ}_{hyd} \tag{26}$$

$$M^+(gas) + \overline{H^+} \Leftrightarrow \overline{M^+} + H^+(gas) \ \Delta Y^{\circ}_{ex} \tag{27}$$

where Y represents thermodynamic functions such as G, H, and S. The numerical values of $\Delta Y^{\circ} = \Delta Y^{\circ}_{ex} - \Delta Y^{\circ}_{hyd}$ include the difference in the particular thermodynamic function related to the hydration (ΔY°_{hyd}) of the ions in aqueous solution and (ΔY°_{ex}) the exchanger. The values of (ΔY°_{hyd}) can be found in the Rosseinsky tables [204]. If the numerical value of $\Delta Y^{\circ}_{hyd} - \Delta Y^{\circ}_{ex}$ is equal to zero, the cations behave the same in both the solution and the exchanger. The numerical value of ΔY°, when the entropy of hydration of the ions is the thermodynamic function of interest, takes the form $\Delta S^{\circ}_{ex} - \Delta S^{\circ}_{hyd}$. The magnitude of ΔS°_{ex} may be enhanced relative to ΔS°_{hyd} by the removal of some or all of the water molecules from the hydration shell of the ions. The selectivity is governed by the magnitude of the ΔS° term, when the contribution of the $T\Delta S^{\circ}_{ex}$ term is larger than that of the ΔH°_{ex} term.

C. Selectivities of Exchangers with Different Crystallinities

It has been shown that the selectivity of organic ion-exchange resins for alkali metal ions increases with decreased loading of these ions and with increase in their degree of their cross-linking. The inorganic oxides have rigid structures and can, by analogy, be regarded as highly cross-linked materials.

The affinity series of the inorganic exchangers for alkali metal ions vary depending on both the crystalline form of the exchanger and the nature of the sorption media. Most of the hydrous oxides can be obtained as amorphous structures. At low loading the amorphous hydrous oxides of Al, Si, Ti, Zr, Ce(IV) and Sb(V) exhibit the usual selectivity sequence $Li^+ < Na^+ < K^+ < Rb^+ < Cs^+$. This selectivity sequence is observed with the

---

Figure 17 Plots equivalent of Kielland coefficients (top) and $(\ln K^M_H)_{\overline{x}_M \to 0}$ (bottom) values versus effective crystal ionic radii (EIR) for C-SbA. The values of EIR are taken from Ref. 115. Exchanger, 0.10 g; total volume, 10.0 cm^{-3}; temp., 30°C; ionic strength, 0.1 ($MNO_3 + HNO_3$) or ($MNO_3 + MOH$), M = Li, Na, K, Rb, and Cs. (From M. Abe, in *Ion Exchange Processes: Advances and Applications* (A. Dyer, M. J. Hudson, and P. A. Williams, eds.), The Royal Society of Chemistry, Science Park, Cambridge, UK, p. 199, 1993. With permission.)

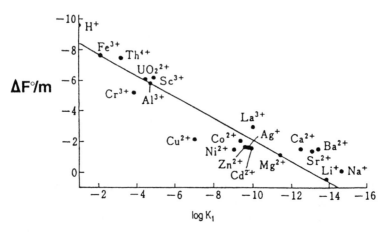

Figure 18 Plot of $\Delta F°/m$ vs. log K_1 for hydrous oxides of silica. (From L. L. Dugger, J. H. Stanton, B. N. Irby, B. L. McConnell, W. W. Cummings, R. W. Maatman, *J. Phys. Chem.*, 68:757 (1964).)

strong acid organic ion exchange resins. Dugger et al. has shown that $\Delta F°$ values for ion-exchange reaction of 20 metal ion/H^+ systems relate linearly to the logarithm of the first hydrolysis constant of these metal ions (Fig. 18) [205]. In alkaline media, the selectivity sequence is reversed in the order of $Cs^+ < Rb^+ < K^+ < Na^+ < Li^+$ for the hydrous oxides of Zr and Sn.

Change in the selectivity patterns of transition metal ion/H^+ systems has been encountered with the amorphous and anatase types of hydrous titanium oxides with different crystallinities [24]. Potassium titanate, $K_2O \cdot nTiO_2$ (n = 2–4), in particular, exhibits a layered structure. Fibrous titanic acid, $H_2Ti_4O_9 \cdot nH_2O$, is obtained by acid treatment of fibrous $K_2Ti_4O_9 \cdot nH_2O$ and shows higher selectivity for K, Rb and Cs than the amorphous titanic acid [206].

Four types of hydrous antimony oxide (antimonic acid), the amorphous (A-SbA), the glassy (G-SbA), the cubic (C-SbA), and the monoclinic (M-SbA) are known so far [138]. Both the A-SbA and G-SbA affect the selectivity sequence $Li^+ < Na^+ < K^+ < Rb^+ < Cs^+$, while the selectivity sequence of C-SbA is unusual with $Li^+ \ll K^+ < Cs^+ < Rb^+ \ll Na^+$ for micro amounts in acid media (Fig. 19). The degree of crystallinity of α-ZrP strongly influences its ion-exchange behavior as mentioned earlier. The pH versus base added plots for α-ZrP with different crystallinity are shown in Fig. 13. It is seen that each increase in acid concentration at a fixed reflux time is reflected in the shape of the curves. The titration curves with the most well-defined plateaus were obtained with the most highly crystalline samples [126].

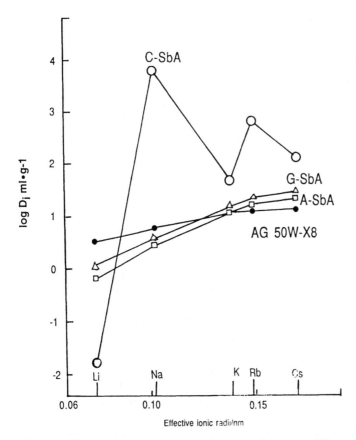

Figure 19 The log D_i values for alkali metal ions on different antimonic acids and AG 50 W-X8. Exchanger, 0.25 g, total vol., 25.0 cm^3; alkali metal ions, 10^{-4} M; HNO$_3$, 1 M. C-SbA; cubic, A-SbA; amorphous; G-SbA; glassy.

D. Ion Memory Effect

The retention of high selectivity for a particular cation after exchanging with other cations has been sought by endowing the crystalline structure of the inorganic ion exchanger with a fixed and well-defined structure. Examples in the literature which exhibit this characteristic and endow the ion-exchange memory sought, as a result, have been reported.

For example, Vol'Khin et al. [121] have prepared manganese oxide in spinel type, by exchanging proton for lithium ions in manganese oxide containing lithium ions. Similar results were obtained with the λ-type MnO$_2$ [34,122]. These oxides showed an extremely high selectivity for lithium ions. Cryptomelane type manganese dioxide prepared from potassium permanganate and manganous sulfate showed an extremely high selectivity for potassium ions [35]. The crystal structure was observed to remain essentially unchanged after exchanging lithium or potassium ions in these oxides with H^+.

These results suggested that if material can be constructed by keeping ionic bonding for exchangeable ions in framework of coordination bonding, the product will show high selectivity just as to be retained in ion memory after exchanging particular ion with other ion to maintain fundamentally original structures.

E. Lithium Ion Memory Exchangers

A new crystalline antimonic acid $HSbO_3O \cdot 0.12H_2O$ has been prepared from $LiSbO_3$ [47] by the Li^+/H^+ ion-exchange reaction affected with concentrated nitric acid solution. The $LiSbO_3$ was obtained by heating $LiSb(OH)_6$ at 900°C. The $LiSb(OH)_6$ was prepared by the addition of LiOH solution to a Sb(V) chloride solution at 60°C. The x-ray diffraction pattern (XRD) of $HSbO_3O \cdot 0.12H_2O$ has been attributed to a monoclinic cell configuration (space group $P2_1/m$ or P_21). This monoclinic cell of $HSbO_3$(M-SbA) has essentially the same arrangement as the $LiSbO_3$ where the oxygen atoms form a distorted hexagonal close packing configuration with antimony and lithium occupying some of the octahedral holes. The $LiSbO_3$ is also obtained by heating a mixture of Sb_2O_3 and Li_2CO_3.

The pH response of M-SbA to neutralization with standard base is apparently that of a strong monobasic acid for the alkali metal ion/H^+ systems [47]. The order of metal ion uptake is $K^+ < Rb^+ < Cs^+ < Na^+ \ll Li^+$ throughout the pH range studied (Fig. 10). The M-SbA exhibits an extremely high selectivity for lithium ions (ion memory effect). Apparent capacities of Na^+ and Cs^+ at pH 1.6 are higher than those for K or Rb at pH 3. The XRD patterns of the products exchanged with Na^+ and Cs^+ at low pH show the presence of a mixture of cubic and monoclinic antimonic acids. No phase transformation is observed at pH > 5, however. A large ion-exchange capacity of 4.3 meq/g is obtained at pH 11. Thermodynamically based parameters have been resolved for the Li^+/H^+ exchange

Table 7 Thermodynamic Data for Li^+/H^+ Exchange on M-SbA at 298 K

ΔG°_{298} (kJ mol^{-1})	ΔH°_{298} (kJ mol^{-1})	ΔS°_{298} (J(K eq)$^{-1}$)	ΔH°_{hyd} (kJ mol^{-1})	ΔH°_{ex} (kJ mol^{-1})	ΔS°_{hyd} (J(K eq)$^{-1}$)	ΔS°_{ex} (J(K eq)$^{-1}$)
12.25	47.03	117	574.2	621.2	−10.03	107

Source: From Ref. 47.

with M-SbA by applying the simplified Gaines-Thomas equation (Eq. (22)) through use of the pH data compiled during neutralization of the theoretical capacity of 5.7 meq/g assigned to it. The calculated $\Delta Y°$ values are summarized in Table 7. The equilibrium constant for the Li^+/H^+ exchange at 25°C was determined to be 7.09×10^{-3} leading to a $\Delta G°_{298}$ value of 12.25 kJ mol^{-1}. The $\Delta G°_{298}$ value decreases with increase of temperature, indicating that the reaction is favored by increase of temperature. This indicates an entropy-producing process prevails in the Li^+/H^+ exchange system.

The employment of MAS FT-NMR spectroscopy has been very useful for understanding the Li^+ exchange mechanism [208,210]. The original M-SbA in its H^+ form, when exchanged with relatively low quantities of Li^+, produces two peaks with different line widths in the H^+ NMR spectra. Computer simulation shows that the chemical shifts for both spectra are similar, and the half-widths are found to be 17.6 kHz for the broad one and 1.5 kHz for the narrow one, respectively (Fig. 20). The change in the area of the ^1H NMR spectrum is plotted against lithium uptake in Fig. 21. The area of both absorption lines decreases monotonously with increasing uptake of lithium ion. The area of the broad spectrum almost reaches zero in sample E, where lithium ion exchange has almost reached its theoretical capacity of 5.7 meq/g. The area of the narrow spectrum does not reach zero, indicating that about 30% of the water originally in the M-SbA still remains.

Analysis of the relaxation time indicates that a narrow linewidth in the H^+ NMR spectra generally implies that the protons on the surface of M-SbA are mobile, while a wide one indicates that proton motion on the surface of M-SbA is rather restricted. The IR spectra study yield absorption bands at 2800–3200 cm^{-1} to show the presence of free water in addition to the —OH of Sb-OH (2800–3000 cm^{-1}). The peaks attributed to the Sb-OH group decrease markedly with increasing uptake of Li^+. These results indicate that lithium ion exchange is attributed to the dehydration in the M-SbA structure.

The ^7Li NMR of the samples with different Li^+ uptake show simple Gaussian peaks for lower loading and doublet peaks with an additional sharp peak at the center (Fig. 22) at higher loading. The simple Gaussian shape indicates the absence within the lower loading range of second-order quadrupole interaction and a fourfold symmetrical coordination of the lithium ions. The transformation from a single peak to a doublet indicates that the lithium ions are transformed from a fourfold symmetrical site to a three-fold symmetrical electric field when the uptake of lithium ions exceeds half of the theoretical capacity. This is attributed to the electric repulsion between the lithium ions. The fractional lithium content in the $LiSbO_3$ corresponding to these different spectra is esti-

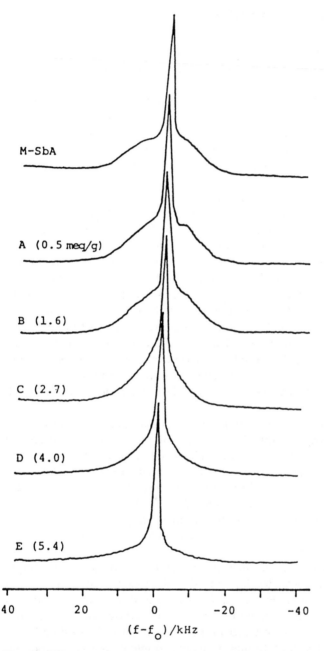

M-SbA

A (0.5 meq/g)

B (1.6)

C (2.7)

D (4.0)

E (5.4)

40   20   0   -20   -40

$(f-f_o)/kHz$

Figure 20 ^1H NMR spectra of M-SbA and M-SbA exchanged with increasing quantities, 0.5 to 5.4 meq/g, of Li^+. The numerical values in parentheses indicate uptake of Li^+ in meq/g, and the ^1H NMR spectra were measured with a high-power wide band FT-NMR spectrometer 90°, pulse 1.0 µs.

430

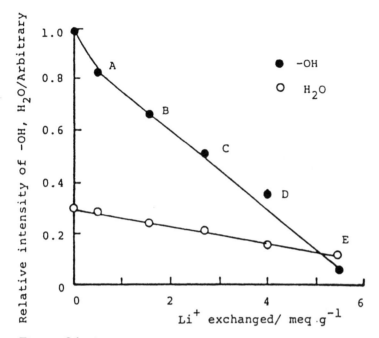

Figure 21 ^{1}H NMR signal intensity of –OH and H_2O vs. uptake of Li^+ on M-SbA.

mated to be (0, 0.16, 0.5, etc.) by Edstrand and Ingri [209]. The schematic view of the lithium ions in the sample exchanged to different extents is illustrated in Fig. 23.

In conclusion, the NMR studies indicate that lithium ion exchange occurs predominantly with protons immobilized in Sb-OH^+. Similar conclusions are reached from the results obtained with Li^+/H^+ systems in contact with cubic niobic acid [210] and cubic tantalic acid [211]. An extremely high selectivity for lithium ions in these ion-memory-based exchangers is due to the sizable electrostatic force between the sites and the lithium ions whose hydration shell has been removed by the exchangers.

M-SbA can be applied for the selective removal of lithium ions from seawater and hydrothermal water [207].

V. CONCLUDING REMARKS

The major objective of this chapter has been to demonstrate the utility of concepts that derive from the interpretation of the selectivities of inorganic ion-exchange materials. It should be realized that even in the simplest cases understanding of the factors and mechanisms underlying ion-

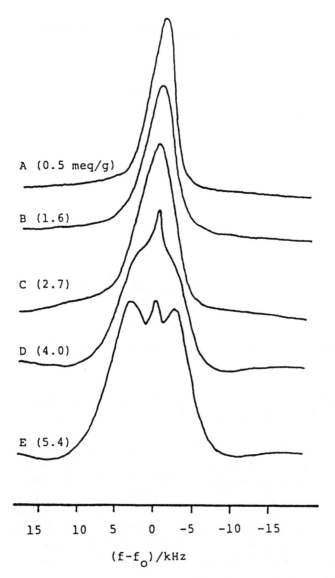

Figure 22 ⁷Li NMR spectra of M-SbA and M-SbA exchanged with different quantities of Li⁺.

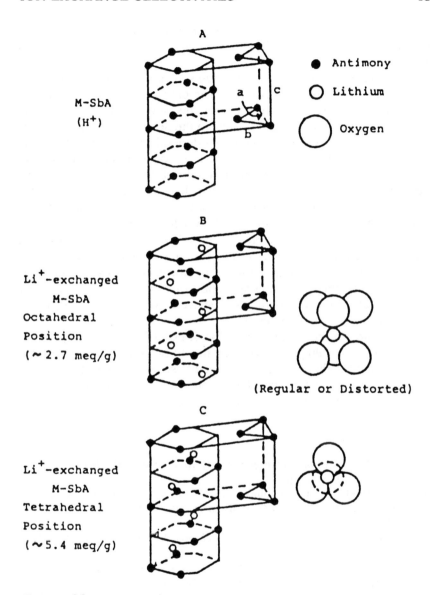

Figure 23 Cation ordering (left) in M-SbA (A) and Li⁺-exchanged M-SbA (B and C) and schematic coordination of lithium ions (right) to oxygen; (B) regular octahedral; (C) regular tetrahedral.

exchange selectivity is, as yet, imperfect. There are many factors that make valid comparisons between data from different laboratories difficult as well.

Many scientists are interested in the separation of the elements for analytical and technological reasons. The emphasis placed on high selectivity in inorganic ion exchanger, for this purpose has led many researchers to restrict their studies solely to the determination of D_i values for the elements. However, the D_i values depend strongly on various factors; the concentration of the elements to be determined, solution media, ratio of the volume of solution to weight of the exchanger, temperature, etc., and this leads to uncertainty in their interpretation. A smaller number of publications in which both forward and reverse reactions are examined to obtain ion-exchange isotherms have appeared.

The characteristics of inorganic ion exchangers depend strongly on the preparative conditions, e.g., starting substances, temperature of preparation, concentration, aging, drying temperature, etc. It is very important to list in detail the experimental conditions employed in preparing samples for obtaining XRD patterns, IR spectra, DTA data, etc., for proper characterization of these materials.

Even though the much more difficult task of providing a quantitative basis for describing selectivity remains to be accomplished, the material contained in this chapter should be of use both in pointing the way toward more rigorous theories of ion exchange and for providing a more solid basis for the application of ion exchange to separation techniques.

REFERENCES

1. H. S. Sherry, in *Ion Exchange Vol. 2* (J. A. Marinsky, ed.), Marcel Dekker, New York, 1969, Chap. 3.
2. A. Clearfield, G. H. Nancollas and R. H. Blessing, in *Ion Exchange and Solvent Extraction Vol. 5* (J. A. Marinsky, ed.), Marcel Dekker, New York, 1973, Chap. 1.
3. A. Clearfield, ed., *Inorganic Ion Exchange Materials*, CRC Press, Boca Raton, FL, 1982.
4. M. Abe, E. A. A. Ichsan, and K. Hayashi, *Anal. Chem., 52*:524 (1980).
5. M. Abe and K. Uno, *Sep. Sci. Technol., 14*:355 (1979).
6. M. Abe, M. Tsuji, and M. Kimura, *Bull. Chem. Soc. Jpn., 54*:130 (1981).
7. Recommendation on ion exchange nomenclature. *Int. Workshop Uniform of Reliable Formulations, Nomenclature and Experimentation for Ion Exchange*, Helsinki, Finland, 1994.
8. E. Hayek and H. Schimann, *Montash. Chem., 88*:686 (1958).
9. A. Lewandowski, W. Czemplik, and A. Dzik, *Zeszyty Nauk. Univ. Poznaiu, Mat. Fix. Chem., No. 3*:3 (1960).

10. A. Lewandowski and S. Tustanowski, *Chem. Anal. (Warsaw)*, *14*:77 (1969).
11. D. Kyriacov, *Surface Sci.*, *8*, 370 (1967).
12. D. L. Dugger, J. H. Stanton, B. N. Irby, B. L. McConnell, W. W. Cummings, and R. M. Maatman, *J. Phys. Chem.*, *68*:757 (1964).
13. L. F. Kirichenko, D. N. Strazhesko, and G. F. Yankovskaya, *Ukr. Khim. Zh.*, *31*:160 (1965).
14. S. K. Rubanik, A. A. Baran, D. N. Strazhesko, and V. V. Strelko, *Teor. Eksp. Khim.*, *5*:361 (1969).
15. S. Ahrland, I. Grenthe, and B. Noren, *Acta Chem. Scand.*, *14*:1059 (1960).
16. A. P. Dushina and V. B. Aleskovskii, *Zh. Neorg. Khim.*, *8*:2194 (1963).
17. M. F. Smirnova, A. P. Doshina, and V. B. Aleskovskii, *Izv. Akad. Nauk. SSSR.*, *4*:248 (1968).
18. V. Veselý and Pekárek, *Talanta*, *19*:219 (1972).
19. R. W. Dalton, J. L. McClanahan and R. W. Maatman, *J. Coll. Sci.*, *17*:207 (1962).
20. H. Ti Tien, *J. Phys. Chem.*, *69*:350 (1965).
21. R. W. Maatman, *J. Phys. Chem.*, *69*:3196 (1965).
22. L. A. Allen and E. Matijevic, *J. Collid Interface Sci.*, *33*:421 (1970).
23. G. Heitner-Wirguin and A. Albu-Yaron, *J. Inorg. Nucl. Chem.*, *28*:2379 (1966).
24. M. Abe, M. Tsuji, S. P. Qureshi, and H. Uchikoshi, *Chromatographia*, *13*:626 (1980).
25. Y. Komatsu, Y. Fujiki, and T. Sasaki, *Bunseki Kagaku*, *31*:E225 (1982).
26. Y. Komatsu, Y. Fujiki, and T. Sasaki, *Bunseki Kagaku*, *32*:E33 (1983).
27. T. Sasaki, Y. Komatsu, and Y. Fujiki, *Sep. Sci. Technol.*, *18*:49 (1983).
28. J. D. Donaldson and M. J. Fuller, *J. Inorg. Nucl. Chem.*, *30*:1083 (1968).
29. J. D. Donaldson and M. J. Fuller, *J. Inorg. Nucl. Chem.*, *32*:1703 (1970).
30. J. D. Donaldson and M. J. Fuller, *J. Inorg. Nucl. Chem.*, *30*:2841 (1968).
31. D. Britz and G. H. Nancollas, *J. Inorg. Nucl. Chem.*, *31*:3861 (1969).
32. D. J. Murray, T. W. Hearly, and D. W. Fuerstenau, *Adv. Chem. Ser.*, *79* (R. F. Gould, ed.), Am. Chem. Soc., Washington, D.C., 1968, p. 74.
33. D. J. Murray, *J. Collid Interface Sci.*, *46*:357 (1974).
34. K. Ooi, Y. Miyai, S. Katoh, H. Maeda and M. Abe, *Bull. Chem. Soc. Jpn.*, *61*:407 (1988).
35. M. Tsuji and M. Abe, *Solv. Extr. Ion Exch.*, *2*:253 (1984).
36. Y. Tanaka, M. Tsuji and M. Abe, *Sep. Sci. Technol.*, *28*:2023 (1993).
37. M. Abe and T. Ito, *Bull. Chem. Soc. Jpn.*, *40*:1013 (1967).
38. M. Abe, *Bull. Chem. Soc. Jpn.*, *42*:2683 (1969).
39. M. Abe, *J. Inorg. Nucl. Chem.*, *41*:85 (1979).
40. I. N. Bourrelly and N. Deschamps, *J. Radioanal. Chem.*, *8*:303 (1971).
41. J. Lefebvre and F. Gaymard, *Compt. Rend.*, *260*:6911 (1965).
42. L. H. Baetsle and D. Huys, *J. Inorg. Nucl. Chem.*, *30*:639 (1968).
43. M. Abe and T. Itoh, *J. Inorg. Nucl. Chem.*, *42*:1641 (1980).
44. M. Abe and K. Sudoh, *J. Inorg. Nucl. Chem.*, *43*:2537 (1981).
45. M. Abe and K. Kasai, *Sep. Sci. Technol.*, *14*:895 (1979).
46. M. Abe and M. Akimoto, *Bull. Chem. Soc. Jpn.*, *53*:121 (1980).

47. R. Chitrakar and M. Abe, *Solv. Extr. Ion Exch.*, 7:721 (1989).

48. Y. Inoue, H. Yamazaki, K. Okada, and K. Morita, *Bull. Chem. Soc. Jpn.*, 58:2955 (1985).

49. A. K. De and K. Chowdhury, *Chromatographia*, 11:586 (1978).

50. A. J. Khan, M. Abe, M. Tsuji and Y. Kanzaki, *Bull. Chem. Soc. Jpn.*, 64:694 (1991).

51. C. B. Amphlett and L. A. McDonald, *J. Inorg. Nucl. Chem.*, 6:220 (1958).

52. E. M. Larsen and D. R. Vissers, *J. Phys. Chem.*, 64:1732 (1960).

53. J. M. Peixoto-Cabeal, *J. Chromatogr.*, 4:86 (1960).

54. G. Alberti and G. Grassian, *J. Chromatogr.*, 4:83 (1960).

55. S. Ahrland, J. Albertsson, and Å. Oskarsson, *J. Inorg. Nucl. Chem.*, 32:2069 (1970).

56. E. P. Horwitz, *J. Inorg. Nucl. Chem.*, 28:1469 (1966).

57. S. Ahrland and J. Albertsson, *Acta Chem. Scand.*, 18:1861 (1964).

58. V. A. Perevozova and E. S. Boichinova, *Zh. Prikl. Khim. (Leningrad)*, 40:2679 (1967); *Chem. Abstr.*, 68: 72668 (1968).

59. J. Albertsson, *Acta Chem. Scand.*, 20:1689 (1966).

60. A. Clearfield, G. D. Smith, and B. Hammond, *J. Inorg. Nucl. Chem.*, 30:277 (1968).

61. E. Michel and A. Weiss, *Z. Naturforsch, B22*:1100 (1967).

62. M. Abe and K. Hayashi, *Solv. Extr. Ion Exch.*, 1:97 (1983).

63. J. R. Feuga and T. Kikindai, *Compt. Rend. Acad. Sci.*, Paris, *C264*:8 (1967).

64. K. A. Kraus, H. O. Phillips, T. A. Carlson, J. S. Johnson, *Proc. 2nd Int. Conf. Peaceful Uses Atom. Energ.*, Geneva, No. 15, 1958, p. 1832.

65. L. Szirtes and L. Zsinka, *Radiochem. Radioanal. Lett.*, 7:61 (1971).

66. K. V. Lad and D. R. Bax, *Indian J. Technol.*, 10:224 (1971).

67. M. Qureshi and S. A. Nabi, *J. Inorg. Nucl. Chem.*, 32:2059 (1970).

68. M. Qureshi, R. Kumar, and H. S. Rathore, *Anal. Chem.*, 44:1081 (1972).

69. M. Qureshi, J. P. Rawat, and V. Sharma, *Talanta*, 20:267 (1973).

70. M. Qureshi and R. Kumar, *J. Chem. Soc. A.*, 1488 (1970).

71. L. Zsinka and L. Szirtes, *Radiochem. Radioanal. Lett.*, 2:257 (1969).

72. J. S. Gill and S. N. Tandon, *J. Radioanal. Chem.*, 20:5 (1974).

73. M. Abe, R. Chitrakar, M. Tsuji, and K. Fukumoto, *Solv. Extr. Ion Exch.*, 3:149 (1985).

74. R. Chitrakar and M. Abe, *The Analyst*, 111:339 (1986).

75. M. Qureshi and H. S. Rathore, *J. Chem. Soc. A.*, 2515 (1969).

76. M. Qureshi and J. P. Gupta, *J. Chem. Soc. A.*, 1755 (1969).

77. M. Qureshi and J. P. Gupta, *J. Chromatogr.*, 62:439 (1971).

78. M. Qureshi, K. G. Varshney, and S. K. Kabiruddin, *Can. J. Chem.*, 50:2071 (1972).

79. M. Qureshi and W. Husain, *J. Chem. Soc. A.*, 1204 (1970).

80. K. H. König and E. Meyn, *J. Inorg. Nucl. Chem.*, 29:1153 (1967).

81. G. Alberti, M. A. Massucci, and E. Torracca, *J. Inorg. Nucl. Chem.*, 30:579 (1967).

82. K. H. König and G. Eckstein, *J. Inorg. Nucl. Chem.*, 35:1359 (1973).

83. G. Alberti, U. Costantino, F. Di Gregorio, and E. Torracca, *J. Inorg. Nucl. Chem., 31*:3195 (1969).
84. J. S. Gill and S. N. Tandon, *J. Inorg. Nucl. Chem., 34*:3885 (1972).
85. G. Alberti, Giamnari and G. Grassini-Strazza, *J. Chromatogr., 28*:118 (1967).
86. J. D. Donaldson and M. J. Fuller, *J. Inorg. Nucl. Chem., 33*:4311 (1971).
87. Y. Inoue, *J. Inorg. Nucl. Chem., 26*:2241 (1964).
88. Y. Inoue, *Bull. Chem. Soc. Jpn., 36*:1325 (1963).
89. M. Qureshi and S. A. Nabi, *Talanta, 19*;1033 (1972).
90. J. Piret, J. Henry, G. Balon and C. Beaudet, *Bull. Soc. Chim.,* France, 3590 (1965).
91. M.Qureshi, H. S. Rathore, and R. Kumar, *J. Chem. Soc. A.,* 1986 (1970).
92. M. Abe and N. Furuki, *Solv. Extr. Ion Exch., 4*:547 (1986).
93. M. Qureshi and K. G. Varshney, *J. Inorg. Nucl. Chem., 30*:3081 (1968).
94. D. Betteridge and G. N. Stradling, *J. Inorg. Nucl. Chem., 29*:2652 (1967).
95. T. Akiyama and I. Tomita, *J. Inorg. Nucl. Chem., 35*:2971 (1973).
96. M. Nomura, M. Abe, and T. Ito, *Nippon Kagaku Zasshi,* 529 (1972).
97. M. Qureshi and S. A. Nabi, *Talanta, 20*:609 (1973).
98. T. Ito and M. Abe, *Bull. Chem. Soc. Jpn., 34*:1736 (1961).
99. T. Ito and M. Abe, *Dennki Kagaku, 33*:175 (1965).
100. R. Caletka and C. Konečný, *Radiochem. Radioanal. Lett., 9*:285 (1972).
101. J. van R. Smit, *Nature, 181*:1530 (1958).
102. S. Z. Roginsky, M. I. Janovsky, O. V. Altshuler, A. E. Morokhorets and E. T. Malimina, *Radiokhim., 2*:431, 438 (1960).
103. H. S. Sherry, p. 120, Ref. 1.
104. M. Abe, in *Recent Developments in Ion Exchange* (P. A. Williams and M. J. Hudson, eds.), Elsevier, London, 1987, p. 227.
105. G. A. Parks, *Chem. Rev., 65*:177 (1965).
106. M. Abe, in *Inorganic Ion Exchange Materials* (A. Clearfield, ed.), CRC Press, Boca Raton, FL., 1982, chap. 6.
107. J. A. Dean, *Chemical Separation Methods,* Van Nostrand Reinhold, New York, 1969, p. 142.
108. M. Abe and T. Ito, *Nippon Kagaku Zasshi, 86*:817 (1965).
109. T. Ito and T. Kenjo, *Nippon Kagaku Zasshi, 88*:1120 (1967).
110. T. Ito and T. Yoshida, *Nippon Kagaku Zasshi, 91*:1054 (1970).
111. M. Abe and T. Ito, *Nippon Kagaku Zasshi, 86*:1259 (1965).
112. M. J. Fuller, *Chromatogr. Rev., 14*:45 (1971).
113. G. A. Parks, *Adv. Chem. Ser.,* No. 67, 121, 1967; *Chem. Abstr., 67,* 47538e (1967).
114. T. W. Healy, A. P. Herring, and D. W. Fuerstenau, *J. Colloid Interf. Sci., 21*:435 (1966).
115. R. D. Shannon and C. T. Prewitt, *Acta Chrystallogr., B25,* 925 (1965).
116. B. G. Novikov, E. A. Mateova, and F. A. Belinskaya, *Zh. Neorg. Khim., 20*:1566 (1975).
117. M. Abe and T. Ito, *Bull. Chem. Soc. Jpn., 41*:333 (1969).
118. R. Chitrakar and M. Abe, *Mat. Res. Bull., 23*:1231 (1989).
119. R. Chitrakar and M. Abe, *Solv. Extr. Ion Exch., 7*:721 (1989).

120. M. Tsuji and M. Abe, *Solv. Extr. Ion Exch., 2*:253 (1984).
121. V. V. Vol'khin, G. V. Leont'eva, and S. A. Onolin, *Izv. Akad. Nauk. SSSR Neog. Mater., 9*:1041 (1972).
122. K. Ooi, Y. Miyai, S. Katoh, H. Maeda, and M. Abe, *Langmuir, 5,* 150 (1989).
123. M. Abe and T. Ito, *Kogyo Kagaku Zasshi, 79*:440 (1967).
124. A. Clearfield and G. D. Smith, *Inorg. Chem., 8*:431 (1969).
125. G. Alberti, in *Study Week on Membranes,* (R. Passino, ed.), Pontificiae Acad. Sci. Scripta, Varia, Rome, 1976, p. 629.
126. A. Clearfield, Å. Oskarsson, and C. Oskarsson, *Ion Exch. Membr., 1*:9 (1972).
127. A. Clearfield and A. S. Medina, *J. Phys. Chem., 75*:3750 (1971).
128. L. Kullberg and A. Clearfield, *J. Phys. Chem., 84*:165 (1980).
129. A. Weiss and E. Sextl, in *Ion Exchangers* (K. Dorfner, ed.), Walter de Gruyter, Berlin and New York, 1991, chap. 1, p. 504.
130. H.-J. Werner, K. Beneke, and G. Lagaly, *Z. Anorg. Allg. Chem., 470*: 118 (1980).
131. G. T. Kerr, *J. Phys. Chem., 72*:2594 (1968).
132. H. K. Beyer and I. Belenykaya, in *Catalysis of Zeolites* (B. Imelik et al., eds.), Elsevier, Amsterdam, 1980, p. 203.
133. J. D. Sherman, in *Adsorption and Ion Exchange Separations,* (J. D. Sherman, ed.), A. I. Ch. Symp. Series, *74,* 1978, p. 98.
134. R. M. Barrer and D. Falconer, *Proc. Roy. Soc. (London), A236*:227 (1956).
135. H. S. Sherry, *J. Colloid Interface Sci., 28,* 288 (1968).
136. R. M. Barrer, R. Papadopoulos, and L. V. C. Rees, *J. Inorg. Nucl. Chem., 29*:2047 (1967).
137. M. Abe, K. Yoshigasaki and T. Sugita, *J. Inorg. Nucl. Chem., 42*:1753 (1980).
138. M. Abe, in *Ion Exchange Processes: Advances and Applications* (A. Dyer, M. J. Hudson, and P. A. Williams, eds.) The Royal Society of Chemistry, Science Park, Cambridge, UK, 1993, p. 199.
139. A. Clearfield, *Inorganic Ion Exchange Materials,* CRC Press, Boca Raton, FL., 1982, p. 4.
140. G. Alberti, U. Costantino, S. Alli, and N. Tomassini, *J. Inorg. Nucl. Chem., 40*:1113 (1978).
141. D. W. Breck, *Zeolite Molecular Sieves, Structure, Chemistry and Use,* Wiley, New York, 1974, p. 533.
142. A. Dyer and M. Jamil, in *Ion Exchange Advances—Proceedings of IEX '92* (M. J. Slater, ed.), SCI Elsevier London, 1992, p. 350.
143. G. H. Nancollas and R. Paterson, *J. Inorg. Nucl. Chen., 29*:565 (1967).
144. A. Ruvarac and M. Trtaji, *J. Inorg. Nucl. Chem., 34*:3893 (1972).
145. A. Ruvarac, in *Inorganic Ion Exchange Materials,* (A. Clearfield, ed.), CRC Press, Boca Raton, FL. 1982, chap. 5.
146. B. G. Novikov, F. A. Belinskaya, and E. A. Mateova, *Vestnik Leningrad Univ. Fiz. Khim., No. 4*:29 (1971).
147. M. Abe and K. Sudoh, *J. Inorg. Nucl. Chem., 42*:1051 (1980).
148. B. G. Novikov, F. A. Belinskaya, and E. A. Mateova, *Vestnik Leningrad Univ. Fiz. Khim., No. 4*:35 (1971).

149. A. Clearfield and J. A. Stynes, *J. Inorg. Nucl. Chem., 26*:117 (1964).
150. E. Torracca, G. Alberti, R. Platania, P. Scala, and P. Galli, *Soc. Chem. Ind. (London)*, 315 (1970).
151. G. Alberti, S. Allulli, U. Costantino, M. A. Massucci, and E. Torracca, *Soc. Chem. Ind. (London)*, 318 (1970).
152. E. Torracca, *J. Inorg. Nucl. Chem., 31*:1189 (1969).
153. L. Kulberg and A. Clearfield, *J. Phys. Chem., 85*:1578 (1981).
154. C. B. Amphlett, P. Eaton, L. A. McDonald, and A. J. Miller, *J. Inorg. Nucl. Chem., 26*:297 (1964).
155. A. Ruvarac and V. Veselý, *J. Inorg. Nucl. Chem., 32*:3939 (1970).
156. S. Allulli, M. A. Massucci, U. Costantino, and R. Bertrami, *J. Inorg. Nucl. Chem., 39*:659 (1977).
157. E. Torracca, *J. Inorg. Nucl. Chem., 31*:1189 (1969).
158. G. Alberti, U. Costantino, S. Allulli, and M. A. Massucci, *J. Inorg. Nucl. Chem., 35*:1339 (1973).
159. G. Alberti, M. Casciola, U. Costantino, and M. L. Luciani-Giavagnotti, *J. Chromatogr., 128*:289 (1976).
160. M. Abe, Y. Kanzaki, and R. Chitrakar, *J. Phys. Chem., 91*:2997 (1987).
161. G. Alberti, M. Casciola, U. Costantino, and R. K. Biswas, in *Ion Exchange Processes: Advances and Applications*, (A. Dyer, M. J. Hudson, and P. A. Williams, eds.), 1993, p. 253.
162. R. M. Barrer, L. V. C. Rees, and D. J. Ward, *Proc. Roy. Soc., A273*, 180 (1963).
163. V. A. Federov, A. M. Tolmachev, and G. M. Panchenkov, *Russ. J. Phys. Chem., 38*:679 (1964).
164. R. M. Barrer and W. M. Meier, *Trans. Faraday Soc., 55*:130 (1959).
165. H. S. Sherry and F. H. Walton, *J. Phys. Chem., 71*:1457 (1967).
166. D. W. Breck, W. G. Eversole, R. M. Milton, T. B. Reed, and T. L. Thomas, *J. Am. Chem. Soc., 78*:5963 (1956).
167. S. A. I. Barri and L. V. C. Rees, *J. Chromatogr., 201*:21 (1980).
168. F. Danes and F. Wolf, *Z. Phys. Chem. (Leipzig), 251*:329 (1972).
169. B. H. Wiers, R. J. Grosse, and W. A. Cilley, *Environ. Sci. Technol., 16*:617 (1982).
170. B. W. Mercer and L. L. Ames, Unclassified Hanford Laboratories Report HW-78461 (1963).
171. F. Danes and F. Wolf, *Z. Phys. Chem. (Leipzig), 251*:339 (1972).
172. R. M. Barrer and L. Hinds, *J. Chem. Soc.*, 1879 (1953).
173. L. L. Ames, *Am. Mineral, 51*:903 (1966).
174. R. M. Barrer, J. A. Davies, and L. V. C. Rees, *J. Inorg. Nucl. Chem., 31*:219 (1969).
175. L. L. Ames, in *Proc. 13th Pacific NW Indus. Waste Conf.*, Washington State Univ., 1967, p. 135.
176. B. W. Mercer, L. L. Ames, and C. J. Touhill, *Am. Chem. Soc. Div. Water, Air, Waste Chem., Gen. Papers*, 1968.
177. R. B. Barrer and B. M. Munday, *J. Chem. Soc. A.*, 2904 (1971).
178. L. L. Ames, *Am. Mineral, 49*:1099 (1964).

179. H. S. Sherry, *Clay and Clay Minerals, 27*:231 (1979).
180. P. A. Newell and L. V. C. Rees, *Zeolites, 3*:22 (1983).
181. J. D. Sherman and R. J. Ross, in *Proc. 5th Int. Conf. Zeolites* (L. V. C. Rees, ed.), Heyden, Philadelphia, 1980, p. 823.
182. F. Wolf, H. Fürtig, and K. Knoll, *Chem. Technol., 23*:273 (1971).
183. T. C. Golden and R. G. Jenkins, *J. Chem. Eng. Data, 26*:366 (1981).
184. S. Hocevar and B. Drzaj, in *Proc. 5th Int. Conf. Zeolites* (L. V. C. Rees, ed.), Heyden, Philadelphia, 1980, p. 301.
185. H. S. Sherry, *Ion Exch. Process Ind., Intl. Conf., Soc. Ind.,* 1969, p. 329.
186. Y. Shibue, *Clay and Clay Minerals, 29*:397 (1981).
187. L. L. Ames, *Can. Mineral., 8*:582 (1966).
188. R. M. Barrer, L. V. C. Rees, and M. Shamsuzzoha, *J. Inorg. Nucl. Chem., 28*:629 (1966).
189. H. S. Sherry, p. 89 from Ref. 1.
190. D. W. Breck, *J. Chem. Ed., 41*:678 (1964).
191. E. N. Rosolovskaja, K. V. Topchieva, and S. P. Dorozhko, *Russ. J. Phys. Chem., 51*:861 (1977).
192. H. S. Sherry, *J. Phys. Chem., 72*:4086 (1968).
193. L. L. Ames, *J. Inorg. Nucl. Chem., 27*:885 (1965).
194. A. Maes and A. Cremers, *J. Chem. Soc., Faraday Trans., 171*:265 (1975).
195. L. L. Ames, *Can. Mineral, 8*:325 (1965).
196. L. L. Ames, *Can. Mineral, 8*:574 (1966).
197. A. Maes, J. Verlinden, and A. Cremers, in *The Properties and Applications of Zeolites* (R. P. Townsend, ed.), The Chemical Society, Burlington House, London, 1979, p. 269.
198. I. V. Baranova and A. M. Tolmachev, *Russ. J. Phys. Chem., 51*:416 (1977).
199. R. M. Barrer and B. M. Munday, *J. Chem. Soc. A.,* 2914 (1971).
200. K. A. Kraus and R. J. Raridon, *J. Phys. Chem., 63*:1901 (1959).
201. G. L. Gaihs, Jr., and H. C. Thomas, *J. Chem. Phys., 21*:714 (1953).
202. J. Kielland, *J. Soc. Chem. Ind. (London), 54*:232 T (1935).
203. W. J. Argersinger, Jr., A. W. Davidson, and O. D. Bonner, *Trans. Kansas Acad. Sci., 53*:404 (1950).
204. D. R. Rosseinsky, *Chem. Rev.,* 467 (1965).
205. L. L. Dugger, J. H. Stanton, B. N. Irby, B. L. McConnell, W. W. Cummings, and R. W. Maatman, *J. Phys. Chem., 68*:757 (1964).
206. T. Sasaki, M. Watanabe, Y. Komatsu, and Y. Fujiki, *Inorg. Chem., 24*:2265 (1985).
207. M. Abe, R. Chitrakar, and M. Tsuji, *Kagaku to Kogyo, 42*:1224 (1989).
208. Y. Kanzaki, R. Chitrakar, and M. Abe, *J. Phys. Chem., 94*:2206 (1990).
209. M. Edstrand and N. Ingri, *Acta Chem. Scand., 8*:1021 (1954).
210. Y. Kanzaki, R. Chitrakar, T. Ohsaka, and M. Abe, *Nippon Kagaku Kaisi,* 1299 (1993).
211. Y. Inoue, M. Abe, and M. Tsuji, in *62th Annual Meeting Japan Chemical Soc.,* 1991, p. 77.

Index

Milton Keynes UK
Ingram Content Group UK Ltd.
UKHW021859071024
449327UK00021B/1591